A New Perspective on Nonmonotonic Logics

Dov M. Gabbay · Karl Schlechta

A New Perspective on Nonmonotonic Logics

Springer

Dov M. Gabbay
Department of Computer Science
King's College London
London
UK

and

Department of Computer Science
Bar-Ilan University
Ramat-Gan
Israel

and

Computer Science and Communications,
Faculty of Sciences
Université de Luxembourg
Esch-sur-Alzette
Luxembourg

Karl Schlechta
CNRS, LIF UMR 7279
Aix-Marseille Université
Marseille
France

and

Frammersbach
Germany

ISBN 978-3-319-46815-0 ISBN 978-3-319-46817-4 (eBook)
DOI 10.1007/978-3-319-46817-4

Library of Congress Control Number: 2016952896

Printed on acid-free paper

This Springer imprint is published by Springer Nature
The registered company is Springer International Publishing AG
The registered company address is: Gewerbestrasse 11, 6330 Cham, Switzerland

Contents

Part II Background Material

Part I
Introduction

Chapter 1
Introduction

Throughout, unless said otherwise, we will work in propositional logic.

1.1 Overview

1.1.1 Purpose and Organisation of the Book

The purpose of this book is to present new results and approaches. They are found in Part III.

The new results are on

(1) interpolation for nonmonotonic logics, (Chap. 9),
(2) abstract (function) independence (Chap. 10),
(3) the Talmudic KAL Vachomer rule (Chap. 12),
(4) an equational solution of contrary-to-duty obligations (Chap. 13),
(5) Chapter 11 is the conceptual core of the book. It combines the ideas of several types of nonmonotonic logics, and our analysis of "natural" concepts, into the construction of a special preferential structure.

In more detail:

- Chapter 9 discusses an improvement over the results in Gabbay and Schlechta (2011), in the sense that the present result does not require a language change. In addition, we show a result for interpolation between equivalent formulas.
- The main result of Chap. 10 is the fact that this kind of abstract function independence has no finite characterization.
- The Kal Vachomer rule is based on the hypothesis that the world is relatively orderly. Chapter 12 presents an alternative criterion of "simplicity" of a partial

D.M. Gabbay and K. Schlechta, *A New Perspective on Nonmonotonic Logics*,
DOI 10.1007/978-3-319-46817-4_1

order. This criterion is much simpler than the one discussed in Abraham et al. (2009), yet, surprisingly, gives often (but not always) the same results as the latter one.
- Chapter 13 presents an alternative solution to contrary-to-duty obligations to the one discussed in Gabbay and Schlechta (2009d). The latter works with a special preferential order on the models, the former with an order on the set of truth values.
- Chapter 11 is the main chapter of the book. It does not present new formal results, but combines many concepts into a formal logic. As this logic is a special preferential construction, it has all the nice formal properties of these structures. The special construction, however, combines the formal clarity with the intuitive advantages of Reiter defaults, defeasible inheritance, theory revision and epistemic considerations.

The ideas of this chapter go back, partially, to the second author's Schlechta (1997a), and are the result of comparisons of the various systems, and foundational reflections. In particular, we analyse the hidden assumptions of Reiter defaults and defeasible inheritance systems - which are, in these assumptions, very similar - on the one side, and autoepistemic logic on the other side, and construct a logic, which is based, among other things, on this analysis.

A closer look shows that these assumptions are based on the "naturalness" of our concepts, which are created by our brain, in an interplay with the world. Therefore, it is reasonable to look at the brain's functioning and this interplay - and, we think, more than has often been done by logicians.

An introductory discussion of these assumptions and concepts is found in Sect. 1.2.

1.1.1.1 Intended Audience and Suggested Sequence of Reading

This book is mainly written for experts of at least some of the fields discussed. Still, we do not assume that the reader has active and recent knowledge of all prerequisites. This motivates Part II. Chapters 2–8 give fast introductions to the background material used in Part III. Chapter 6 has an additional function, as it presents in $\mathcal{A}$-ranked structures a much simpler precursor to the construction in Chap. 11. This was already published in Gabbay and Schlechta (2009d). These chapters are probably not sufficient for the reader new to the field of nonmonotonic logic, but, we hope, suitable for the intended audience. Thus, this part is a compromise between no introduction at all, obliging many readers to consult several additional texts, and a fully self-contained textbook, containing new results. The latter would be much longer, at least twice as long, so the compromise we chose seems reasonable. The material in Part II was mostly taken from Gabbay and Schlechta (2009d, 2011). The same considerations apply to Sects. 1.1.3–1.1.11: They tell the expert what to expect in the corresponding chapters, and if this will be enough to bring him up to speed. Chapter 8 stands a bit apart, it discusses aspects of neurology that are important for our ideas. We are not experts in the field, but want to point out the importance of neuroscience for an analysis of human reasoning.

Sections 1.1.12–1.1.16 are, of course, different, as they present an introduction to our new results; they will be more detailed and contain an outlook on the new results. Still, as in Part III in general, they assume certain familiarity with the subject, so the discussion is not too leisurely.

In Sect. 1.2, we introduce the main concepts to be discussed, they are the basis of the formal construction in Chap. 11. A central concept is homogeneousness (or rarity of influence change), respectively, the adequacy of our categories.

We suggest that the reader takes first a very short look at Chap. 2 presenting our notation. Next, the reader could choose any chapter in Part III, depending on his or her interests, they are largely independent (in the sense that they can be read independently, but, of course, their contents is not, and they all treat the same or similar ideas in different forms) and come back to the chapters in Part II and the corresponding sections of Part I, whenever necessary. An exception is Chap. 11, which should best be read only after Sect. 1.2.

For the convenience of the reader, we now give a concise overview. We also point out the connections between the different Chapters.

1.1.2 Concise Overview

- Chapter 2
  This chapter contains basic algebraic and logic definitions and notations.
- Chapter 3, see Gabbay and Schlechta (2009d).
  This chapter presents the basic ideas and problems of defeasible inheritance. Defeasible inheritance diagrams have a simple language, with the basic entities arrows, "formulas", which can either express something like "most A's are B's" or "most A's are not B's", conjunction is implicit, there is no usual negation and no disjunction. Still, many problems, and solutions, present or absent in more complicated systems can already be observed in defeasible inheritance systems.
- Chapter 4
  This is a short introduction to Reiter defaults. Reiter defaults are rules, they have the form $\phi : \psi/\sigma$, with the meaning: if ϕ holds and if ψ is consistent (i.e. $\neg\psi$ does not hold), then assume σ. An important difficulty lies in possible circularity: σ and ψ are on the same "level", e.g. nothing prevents us to write defaults like $\phi : \psi/\neg\psi$, etc. Such circularity may, of course, involve several defaults.
  Often defaults of the type $\phi : \psi/\psi$ are interpreted as "most ϕ's are ψ's", in a (semi)quantitative way. An alternative interpretation is (auto)epistemic: if we know ϕ, and unless we know $\neg\psi$, then assume ψ.
  Section 4.2 is short, as we will solve the circularity problem by a theory revision technique, and the epistemic questions left are treated differently in Chap. 11.
- Chapter 5, see Gabbay and Schlechta (2009d), for Sect. 5.2.1.2 also Schlechta (2004).

Our idea is to construct a special preferential structure in Chap. 11, based on ideas from inheritance, defaults and theory revision. So a thorough understanding of preferential structures is necessary.

- Section 5.2
  Preferential structures are semantical constructions, using a binary relation on model sets, similar to Kripke structures. This relation is supposed to express (degrees of) normality: $m \prec m'$ expresses that m is more normal than m', and $\mu(X)$ will be the set of maximally normal elements of X. If X is the set of models of ϕ, then $\phi \mid\sim \psi$ iff ψ holds in $\mu(X)$. (This is the minimal version, a more subtle version, the limit version, discussed in Sect. 5.2.2.)
- Section 5.3
  It is very useful to consider the laws of preferential logics as laws about size: $\mu(X)$ is supposed to be the set of "normal" elements of X, or "most elements of X". Thus, we are interested in the properties of most or the normal elements of X. "Most" or "a big subset of X" can be interpreted by a (weak) filter. This is discussed in the present section.
- Section 5.3.1
  This section discusses "most" as a weakening of "all" in a first order context, introducing a new quantifier, ∇, which corresponds to a weak filter in a suitable axiomatisation.
- Sections 5.3.2 and 5.3.3
  In these sections, we first elaborate additive laws of our qualitative notion of size, and their relation to laws for preferential structures. We then discuss multiplicative laws for the notion of size, its relation to the logical rule of "rational monotony", and prepares the connection to logical interpolation for nonmonotonic logics, see Chap. 9.
- Section 5.3.4
  Hamming relations form an important part of our approach in Chap. 11. In our context, they count or compare propositional variables or default properties. They will ensure, among other things, graceful degradation. Our use of theory revision can also be seen as based on Hamming relations.
  Their relation to size and interpolation is discussed in this section, too.

- Chapter 6, see Gabbay and Schlechta (2009d). Deontic logic is about obligations. It describes the world as it should be. This is similar to the idea of normality of preferential logic, which describes the normal state of things. As a matter of fact, minimal models were first introduced as a semantics for deontic logic.
  This chapter is not only an introduction to Chap. 13, but also the structure presented in Sect. 6.4 and used in Sect. 6.5 is a simple version of that used in Chap. 11, so it is a good preparation for the latter chapter.

  - Section 6.3
    In this section, we discuss contrary-to-duty obligations. There is a primary obligation, which we cannot, or are not willing to, satisfy, but for this case, there are

secondary obligations. It is natural to represent this idea in a multilayered preferential structure.

  - Section 6.4
    This section introduces and discusses such multilayered structures, and Sect. 6.5 applies them to contrary-to-duty obligations, including representation theorems.

- Chapter 7, see Gabbay and Schlechta (2009d).
  We use in Chap. 11 theory revision techniques to obtain a consistent set of defaults, from which we construct our preferential structure, so we introduce the basic ideas thereof.

  - Section 7.2
    Theory revision is a principled way to integrate new information, say ϕ, into existing knowledge, say T, while preserving consistency. This can be seen as a two-step process: first eliminating enough information from T, and then adding ϕ. The elimination process is called theory contraction, and discussed in Sect. 7.3.
  - Section 7.4
    Finally, and basically for completeness' purposes, we shortly present counterfactual conditionals and theory update. Both have a semantics that is very close to a distance-based semantics for theory revision.

- Chapter 8. In this chapter, we try to introduce the reader to Edelman's theory of organisation of the brain and the resulting phenomena. His ideas seem important for the further development of theories of common sense reasoning, in particular concerning concept formation and abstract constructions.
- Chapter 9, see Gabbay and Schlechta (2011).
  Multiplicative laws of size and the related interpolation property of nonmonotonic logics are important notions of independence, and thus a case of restricted influence, the central topic of this book.
  We first present semantical interpolation, and then, Sect. 9.3, relate interpolation for nonmonotonic logics to a multiplicative law about size: roughly $\mu(X \times Y) = \mu(X) \times \mu(Y)$, where X and Y are language fragments and their resulting model sets. More precisely, we discuss two variants of this rule, its relation to properties of the defining relation of a preferential structure, and to interpolation.
- Chapter 10
  probabilistic independence abstract independence
  This chapter treats probabilistic and abstract independence for sets and general function sets.
  The abstract independence relation $\langle X \mid Y \mid Z\rangle$ was first investigated by W. Spohn, J. Pearl and others, see e.g. Spohn (1980), Pearl (1988). After a very short discussion of the probabilistic case, mainly as motivation, we turn to set and function independence, Sect. 10.6.
  Our main result here is that there is no fixed size finite characterisation of function set independence:
  Let $\langle X \mid Y \mid Z\rangle$ mean for a fixed function set $\mathcal{F} : U \to V$, with $X \cup Y \cup Z \subseteq U$

the following: for all $f, g \in \mathcal{F}$ such that $\forall y \in Y.f(y) = g(y)$ there is $h \in \mathcal{F}$ such that $\forall x \in X.h(x) = f(x)$, $\forall y \in Y.h(y) = f(y) = g(y)$, $\forall z \in Z.h(z) = g(z)$.
There are arbitrarily long-loop conditions that are valid and cannot be reduced to finite conditions.

- Chapter 11. This is the central chapter of the book, where we bring together the main concepts presented in Chap. 1, Sect. 1.2, and further discussed in Sects. 11.2 and 11.3. In particular, we have nice logical properties, as guaranteed by a preferential structures. We have graceful degradation by the particular type of preference relation, we have epistemic properties by treating surprise failure of defaults differently. And, we will base everything (informally) on rarity of influence change.
  - Section 11.5 The construction as well as the resulting properties are discussed here.
  - Section 11.6 Some possible extensions are presented in this section. Among others, we discuss a strategy to work first with normal cases in a restricted language only, and expand when necessary.
- Chapter 12
  This short chapter discusses some problems with an approach to the Talmudic law of Kal Vachomer. This law is based on an assumption of regularity of the world. We do not expect "wild" changes of properties, thus it fits into the general context of this book. In particular, we show that a straightforward induction is impossible, as we might have to "look backward" to lower levels of the construction.
  Moreover, we describe a new and extremely simplistic approach by just counting the number of pairs in the (transitive) relation. In quite a few examples, this gives the same result as above approach, but not always, as examples show.
- Chapter 13
  The strategy described here is an illustration of the fact that one may describe graded information by, either, a relation on the information or the semantics itself, as in preferential structures, or, by a relation on the truth values. Here, we take the latter approach to solve problems with contrary-to-duty obligations.

1.1.3 Nonmonotonic Logics

Nonmonotonic logics (NML) were created to deal with principled reasoning about "normal" situation. Thus, "normal" birds will (be able to) fly, but there are many exceptions, like penguins, roasted chicken, etc. and it is usually difficult to enumerate all exceptions, so they will be treated in bulk as "abnormal" birds.

There are different interpretations of such logics: they can be seen as describing the "normal" case or the majority of cases. They can also be seen as reasoning with information of different quality or reliability. In the above example, the information that penguins do not fly is seen as more reliable than the information that birds fly. Specificity decides here. There is also the (auto)epistemic foundation: We assume that we know about exceptions, so we can work with the normal case, knowledge will tell us when to stop.

The standard example is—as we began to describe already—that "normal" birds will (be able to) fly, that there are exceptions, like penguins, that "normal" penguins will not fly, but that there might be exceptions to the exceptions, that some abnormal penguin might be able to fly—due to a jet pack on its back, some genetic tampering, etc. Then, if we know that some animal is a bird, call it "Tweety" as usual, and if we want to keep it as a pet, we should make sure that its cage has a roof, as it might fly away otherwise. If, however, we know that Tweety is not only a bird, but also a penguin, then its cage need not have a roof.

In classical logic, adding prerequisites will never eliminate a consequence: if $T \vdash \phi$, and $T \subseteq T'$, then $T' \vdash \phi$. In this sense, classical logic is monotone.

The above reasoning is nonmonotonic: From the fact "Tweety is a bird", we conclude that it will (normally) fly, but from the facts that "Tweety is a bird" and "Tweety is a penguin", we will not conclude that it will (normally) fly any more, we will even conclude the contrary, that it will (normally) not fly.

We can also see here a general principle at work: more specific information (Tweety is a penguin) and its consequences (Tweety will not fly) will usually be considered more reliable than the more general information (Tweety is a bird) and its consequences (Tweety will fly). Then, NML can also be considered as a principled treatment of information of different quality or reliability. The classical information is the best one, and the conjecture that the case at hand is a normal one is less reliable.

Note that normality is absolute here in the following sense: normal birds will be normal with respect to all "normal" properties of birds, i.e. they will fly, lay eggs, build nests, etc. In this treatment, there are no birds normal with respect to flying, but not laying eggs, etc.

It is sometimes useful to introduce a generalised quantifier ∇. In a first-order (FOL) setting $\nabla x \phi(x)$ will mean that $\phi(x)$ holds almost everywhere, in a propositional setting $\nabla \phi$ will mean that in almost all models ϕ holds. Of course, this "almost everywhere" or "almost all" has to be made precise, e.g. by a filter over the FOL universe, or the set of all propositional models.

1.1.3.1 Formal Semantics by Preferential Systems

The semantics for preferential logics—a special kind of nonmonotonic logic—are preferential structures, a set of classical models with an arbitrary binary relation. This relation need not be transitive nor does it need to have any other of the usual properties. If $m \prec m'$, then m is considered more normal (or less abnormal) than m'. m is said to be minimal in a set of models M iff there is no $m' \in M$, $m' \prec m$—a word of warning: there might be $m' \prec m$, but $m' \notin M$!

This defines a semantic consequence relation as follows: we say $\phi \mid\sim \psi$ iff ψ holds in all minimal models of ϕ.

As a model m might be minimal in $M(\phi)$—the set of models of ϕ—but not minimal in $M(\psi)$, where $\models \phi \rightarrow \psi$ classically, this consequence relation $\mid\sim$ is nonmonotonic. Non-flying penguins are normal (= minimally abnormal) penguins, but all non-flying birds are abnormal birds.

Minimal models of ϕ need not exist, even if ϕ is consistent - there might be cycles or infinite descending chains. We will write $M(\phi)$ for the set of $\phi-$models and $\mu(\phi)$ or $\mu(M(\phi))$ for the set of minimal models of ϕ. If there is some set X and some $x' \in X$ s.t. $x' \prec x$, we say that x' minimises x in X, likewise that X minimises x.

One can impose various restrictions on $\prec$, they will sometimes change the resulting logic. The most important one is perhaps rankedness: If m and m' are $\prec$-incomparable, then for all m'' $m'' \prec m$ iff $m'' \prec m'$ and also $m \prec m''$ iff $m' \prec m''$. We can interpret the fact that m and m' are $\prec$-incomparable by putting them at the same distance from some imaginary point of maximal normality. Thus, if m is closer to this point than m'' then so will be m', and if m is farther away from this point than m'' then so will be m'.

Preferential structures are presented and discussed in Chap. 5.

1.1.4 Basic Definitions - Chapter 2

This chapter is relatively long as we use a number of more or less involved concepts, which have to be made precise. In addition, we also want to put our work a bit more in perspective, and make it self-contained, for the convenience of the reader. Most of the materials of this chapter (unless marked as "new") was published previously; see Schlechta (2004), Gabbay and Schlechta (2009a, b, d, e).

We begin with basic algebraic and logical definitions, including in particular many laws of nonmonotonic logics, in their syntactic and semantical variants, showing also the connections between both sides; see Definition 2.2.6 and Tables 2.1 and 2.2.

We would like to emphasise the importance of the definability preservation (*dp*) property. In the infinite case, not all model sets X are definable, i.e. there need not necessarily be some formula ϕ or theory T such that $X = M(\phi)$, the models of ϕ, or $X = M(T)$, the models of T. It is by no means evident that a model choice function μ, applied to a definable model set, gives us back again a definable model set (i.e. μ is definability preserving, or in short, dp). If μ does not have this property, some representation results will not hold, which hold if μ is dp, and representation results become much more complicated; see Schlechta (2004) for positive and impossibility results. In our present context, definability is again an important concept. Even if we have semantical interpolation, if language and operators are not strong enough, we cannot define the semantical interpolants, so we have semantical, but not syntactic interpolation.

1.1.5 Inheritance Systems - Chapter 3

1.1.5.1 Basic Discussion

We give here an informal discussion. The reader unfamiliar with inheritance systems should consult in parallel Definitions 3.4.2 and 3.4.3. As there are many variants of the definitions, it seems reasonable to discuss them before a formal introduction, which, otherwise, would seem to pretend to be definite without being so.

(Defeasible or Nonmonotonic) Inheritance Networks or Diagrams

Nonmonotonic inheritance systems describe situations like "normally, birds fly", written $birds \rightarrow fly$. Exceptions are permitted, "normally penguins don't fly", $penguins \nrightarrow fly$.

Definition 1.1.1
Inheritance systems or diagrams are directed acyclic graphs (DAGs) with two types of arrows, positive and negative ones. Roughly, nodes stand for sets of objects, like birds, penguins, etc. or properties like "able to fly". A positive arrow $a \rightarrow b$ stands for "(almost) all $x \in a$ are also in b"—so it admits exceptions. A negative arrow $a \rightarrow b$ stands for "(almost) all $x \in a$ are *not* in b" - so it also admits exceptions. Negation is thus very strong. The problem is to find the valid paths (concatenations of arrows) in a given diagram, considering contradictions and specificity.

We will use Γ, etc. for such graphs, and σ, etc. for paths—the latter to be defined below.

Nodes at the beginning of an arrow can also stand for individuals, so $Tweety \nrightarrow fly$ means something like: "normally, Tweety will not fly". As always in nonmonotonic systems, exceptions are permitted, so the soft rules "birds fly", "penguins don't fly", and (the hard rule) "penguins are birds" can coexist in one diagram, penguins are then abnormal birds (with respect to flying). The direct link $penguins \nrightarrow fly$ will thus be accepted, or considered valid, but not the composite path $penguins \rightarrow birds \rightarrow fly$, by specificity—see below. This is illustrated in Diagram 1.1.1, where a stands for Tweety, c for penguins, b for birds and d for flying animals or objects.

(Remark: The arrows $a \rightarrow c, a \rightarrow b$, and $c \rightarrow b$ can also be composite paths—see below for the details.)

Tweety diagram

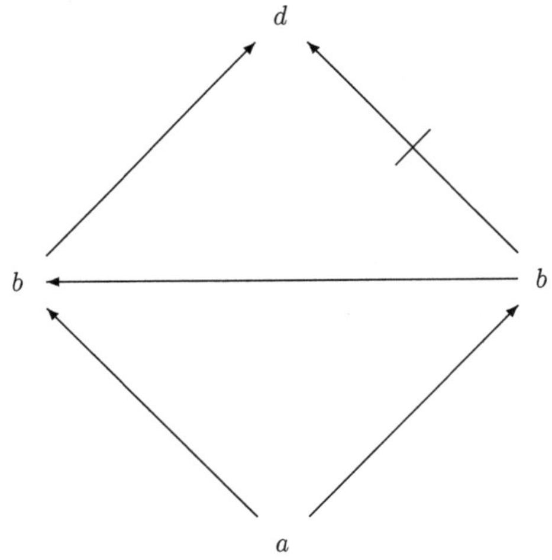

Diagram 1.1.1

(Of course, there is an analogous case for the opposite polarity, i.e. when the arrow from *b* to *d* is negative, and the one from *c* to *d* is positive.)

The main problem is to define in an intuitively acceptable way a notion of valid path, i.e. concatenations of arrows satisfying certain properties.

We will write $\Gamma \models \sigma$, if σ is a valid path in the network Γ, if x is the origin, y the endpoint of σ, and σ is positive, we will write $\Gamma \models xy$, i.e. we will accept the conclusion that x's are y's, and analogously $\Gamma \models x\overline{y}$ for negative paths. Note that we will not accept any other conclusions, only those established by a valid path; therefore many questions about conclusions have a trivial negative answer: there is obviously no path from x to y, e.g., there is no path from b to c in Diagram 1.1.1. Likewise, there are no disjunctions, conjunctions, etc. (directly) in our conclusions, and negation is present only in a strong form: "it is not the case that x's are normally y's" is not a possible conclusion, only "x's are normally not y's" is one. Also, possible contradictions are contained, there is no EFQ.

To simplify matters, we assume that for no two nodes $x, y \in \Gamma$ $x \to y$ and $x \nrightarrow y$ are both in Γ, intuitively, that Γ is free from (hard) contradictions. This restriction is inessential for our purposes. We admit, however, soft contradictions and preclusion, which allows us to solve some soft contradictions—as we already did in the penguins example. We will also assume that all arrows stand for rules with possibly exceptions, again, this restriction is not important for our purposes. Moreover, in the abstract treatment, we will assume that all nodes stand for (nonempty) sets, though this will not be true for all examples discussed.

This might be the place for a remark on absence of cycles. Suppose if we also have a positive arrow from b to c in Diagram 1.1.1, then the concept of preclusion collapses, as there are now equivalent arguments to accept $a \to b \to d$ and $a \to c \nrightarrow d$. Thus, if we do not want to introduce new complications, we cannot rely on preclusion to decide conflicts. It seems that this would change the whole outlook on such diagrams. The interested reader will find more on the subject in Antonelli (1997, 1999, 2005).

Inheritance networks were introduced about 30 years ago (see e.g. Touretzky (1984, 1986), Touretzky et al. (1987)), and exist in a multitude of more or less differing formalisms, see e.g. Schlechta (1997a) for a brief discussion. There still does not seem to exist a satisfying semantics for these networks. The authors' own attempt Schlechta (1990) is an a posteriori semantics, which cannot explain or criticise or decide between the different formalisms. We will give here a conceptual analysis, which provides also at least some building blocks for a semantics, and a translation into (a modified version of) the language of small and big subsets, familiar from preferential structures, see Definition 5.3.2.3.

We will now discuss the two fundamental situations of contradictions, then give a detailed inductive definition of valid paths for a certain formalism so the reader has firm ground under his feet, and then present briefly some alternative formalisms.

As in all of nonmonotonic reasoning, the interesting questions arise in the treatment of contradictions and exceptions. The difference in quality of information is expressed by "preclusion" (or pre-emption). The basic diagram is the Tweety diagram, see Diagram 1.1.1.

Unresolved contradictions give either rise to a branching into different extensions, which may roughly be seen as maximal consistent subsets, or to mutual cancellation in directly sceptical approaches. The basic diagram for the latter is the Nixon Diamond, see Diagram 1.1.2, where $a = Nixon$, $b = Quaker$, $c = Republican$, $d = pacifist$.

In the directly sceptical approach, we will not accept any path from a to d as valid, as there is an unresolvable contradiction between the two candidates.

The Nixon Diamond

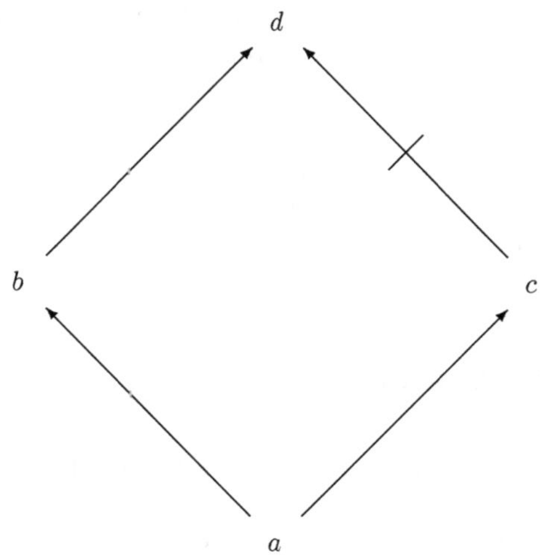

Diagram 1.1.2

The extensions approach can be turned into an indirectly sceptical one, by forming first all extensions, and then taking the intersection of either the sets of valid paths, or of valid conclusions, see [MS91] for a detailed discussion. See also Sect. 3.5 for more discussion on directly versus indirectly sceptical approaches.

1.1.6 Reiter Defaults - Chapter 4

This section introduces Reiter defaults, as presented in Chap. 4. Reiter defaults have the syntactic form $(\alpha : \beta_1, \ldots, \beta_m/\gamma)$, where α, γ, and all β_i are classical (propositional or FOL) formulas. The intuitive meaning is: if we know (can deduce) α, and the β_i are (separately) consistent, then we should assume γ, i.e. add γ to our knowledge. Thus, such a default is a rule, monotone in the first prerequisite, and antitone in the second one.

It is allowed to write defaults like $(true : \phi/\neg\phi)$ or default theories like $\{(true : \phi/\phi), (true : \neg\phi/\neg\phi)\}$, which pose perhaps interesting questions as to their meaning, and lead to complicated semantics, but which do not seem to have any intuitive significance.

In a way, Reiter defaults are similar to inheritance networks, as they have implicit transitivity (by default, of course), but they allow much more complicated theories, too. In particular, one can express circular theories like above, which are excluded in acyclic networks, one has more expressive formulas, etc. Still, at least in the standard cases, the spirit is the same. In particular, we have implicit inheritance in Reiter defaults: The default (bird:fly/fly) inherits downward to blackbirds, but also to penguins, unless we prevent it by $(penguin : \neg fly/\neg fly)$ and some mechanism to make more specific defaults stronger than more general ones.

We have to retain from this discussion, that Reiter defaults are in spirit, similar to inheritance networks, but offer more possibilities in the positive and the negative sense.

For the autoepistemic interpretation of Reiter defaults, we refer directly to Chap. 4.

We will only need the basic ideas of Reiter defaults in this book, and contrast them with those from preferential structures in particular. Thus, our discussion of them will be short, and rather informal.

1.1.7 Preferential Structures - Chapter 5

1.1.7.1 Remarks on Nonmonotonic Logics and Preferential Semantics

Nonmonotonic logics were, historically, studied from two different points of view: the syntactic side, where rules like (AND), (CUM) (see below, Definition 2.2.6) were postulated for their naturalness in reasoning, and from the semantic side, by the introduction of preferential structures (see Definitions 5.2.1 and 5.2.2 below). This work was done on the one hand by Gabbay (1989), Makinson (1994) and others, and on the second hand by Shoham and others, see Yoav Shoham (1987), Bossu and Siegel

(1985). Both approaches were brought together by Kraus, Lehmann, Magidor and others, see Kraus et al. (1990), Lehmann and Magidor (1992), in their completeness results.

A preferential structure $\mathcal{M}$ defines a logic $\mid\!\sim$ by $T \mid\!\sim \phi$ iff ϕ holds in all $\mathcal{M}$-minimal models of T. This is made precise in Definitions 5.2.1 and 5.2.2 below. At the same time, $\mathcal{M}$ also defines a model set function, by assigning to the set of models of T the set of its minimal models. As logics can speak only about definable model sets (here the model set defined by T), $\mathcal{M}$ defines a function from the definable sets of models to arbitrary model sets: $\mu_{\mathcal{M}} : \boldsymbol{D}(\mathcal{L}) \rightarrow \mathcal{P}(M(\mathcal{L}))$. This is the general framework, within which we will work most of the time.

1.1.7.2 Basic Definitions

The following two definitions make preferential structures precise. We first give the algebraic definition, and then the definition of the consequence relation generated by an preferential structure. In the algebraic definition, the set U is an arbitrary set, in the application to logic, this will be the set of classical models of the underlying propositional language.

In both cases, we first present the simpler variant without copies, and then the one with copies. (Note that, e.g. Kraus et al. (1990), Lehmann and Magidor (1992) use labelling functions instead, the version without copies corresponds to injective labelling functions, the one with copies to the general case. These are just different ways of speaking.) We will discuss the difference between the version without and the version with copies below, where we show that the version with copies is strictly more expressive than the version without copies, and that transitivity of the relation adds new properties in the case without copies. In the general case with copies, transitivity can be added without changing properties.

We give here the "minimal version", the much more complicated "limit version" is presented and discussed in Sect. 5.2.2. Recall the intuition that the relation $\prec$ expresses "normality" or "importance" - the $\prec$-smaller, the more normal or important. The smallest elements are those that count.

1.1.7.3 Copies in Preferential Models

Copies in preferential structures (variant (2) in Definition 5.2.1) thus seem to serve to construct universal structures, or code our ignorance, i.e. we know that x is minimised by X, but we do not know by which element of X they are in this view artificial. But they have an intuitive justification, too: They allow minimization by sets of other elements only. We may consider an element m only abnormal in the presence of several other elements together, e.g. considering penguins, nightingales, woodpeckers and ravens all have some exceptional qualities, so we may perhaps not consider a penguin more abnormal than a woodpecker, etc. but seen all these birds together, the penguin stands out as the most abnormal one. But we cannot code minimization by a set, without minimization by its elements, and without the use of copies. Copies will then code the different aspects of abnormality.

1.1.8 Algebraic and Structural Semantics

We make now a major conceptual distinction, between an "algebraic" and a "structural" semantics, which can best be illustrated by an example.

Consider nonmonotonic logics as discussed above. In preferential structures, we only consider the minimal elements, say $\mu(X)$, if X is a set of models. Abstractly, we thus have a choice function μ, defined on the power set of the model set and μ has certain properties, e.g. $\mu(X) \subseteq X$. More important is the following property: $X \subseteq Y \rightarrow \mu(Y) \cap X \subseteq \mu(X)$. (The proof is trivial: suppose there were $x \in \mu(Y) \cap X$, $x \notin \mu(X)$. Then there must be $x' \prec x$, $x' \in X \subseteq Y$, but then x cannot be minimal in Y.)

Thus, all preferential structures generate μ functions with certain properties, and once we have a complete list we can show that any arbitrary model choice function with these properties can be generated by an appropriate preferential structure.

Note that we do not need here the fact that we have a relation between models, just any relation on an arbitrary set suffices. It seems natural to call the complete list of properties of such μ-functions an algebraic semantics, forgetting that the function itself was created by a preferential structure, which is the structural semantics.

This distinction is very helpful, it not only incites us to separate the two semantics conceptually but also to split completeness proof in two parts: One part, where we show correspondence between the logical side and the algebraic semantics, and a second one, where we show the correspondence between the algebraic and the structural semantics. The latter part will usually be more difficult, but any result obtained here is independent from logics itself, and can thus often be re-used in other logical contexts. On the other hand, there are often some subtle problems for the correspondence between the logics and the algebraic semantics (see definability preservation, in particular the discussion in Schlechta (2004)), which we can then more clearly isolate, identify and solve.

Moreover, when we speak about (and not only use) $\mid\sim$, we have to choose the right level of abstraction, e.g. there are many different preferential structures resulting in the same logic, so working on the structural level might not be the right level of abstraction, as we are easily confused by the different possibilities, which should, after all, result in the same outcome. Chosing the level of abstract semantics (abstract size in case of preferential logics, see Sect. 5.3) seems much more suitable.

1.1.8.1 Abstract or Algebraic Semantics

In all cases, we see that the structural semantics define a set operator, and thus an algebraic semantics:

- in nonmonotonic logics (and deontic logic), the function chooses the minimal (morally best) models, a subset, $\mu(X) \subseteq X$
- in (distance based) theory revision, we have a binary operator, say $|$, which chooses the ϕ-models closest to the set of K-models: $M(K) \mid M(\phi)$
- in theory update, the operator chooses the i-th coordinate of all best sequences
- in the logic of counterfactual conditionals, we have again a binary operator $m \mid M(\phi)$, which chooses the ϕ-models closest to m, or, when we consider a whole set X of models as starting points $X \mid M(\phi) = \bigcup\{m \mid M(\phi) : m \in X\}$.
- in modal and intuitionistic logic, seen from some model m, we choose a subset of all the models (thus not a subset of a more restricted model set), those which can be reached from m.

Thus, in each case, the structure "sends" us to another model set, and this expresses the change from the original situation to the "most plausible", "best", "possible", etc. situations. It seems natural to call all such logics "generalised modal logics", as they all use the idea of a model choice function. Such "abstract semantics" are interpretations of the operators of the language (all, flat, top level or not) by functions (or relations in the case of $\mid\sim$), which assign to sets of models sets of models, $\mathcal{O} : \mathcal{P}(\mathcal{M}) \to \mathcal{P}(\mathcal{M})$ - $\mathcal{P}$ the power set operator, and $\mathcal{M}$ the set of basic models -, or binary functions for binary operators, etc. These functions are determined or restricted by the laws for the corresponding operators, e.g. in classical, preferential or modal logic, $\wedge$ is interpreted by $\cap$, etc. in preferential logic ∇ by μ; in modal logic, we interpret $\Box$, etc.

For instance, in preferential structures, it is useful to consider the abstract semantics as a (somehow coherent) system of filters. $\mathcal{O}(X) := \mu(X) \subseteq X$ can be seen as the basis of a principal filter. Thus, $\phi \mid\sim \psi$ iff ψ holds in all minimal models of ϕ, iff there is a "big" subset of $M(\phi)$ where ψ holds, recalling that a filter is an abstraction of size-sets in the filter are big, their complements are small and the other sets have medium size. Thus, the "normal" elements form the smallest big subset. Rules like $X \subseteq Y \to \mu(Y) \cap X \subseteq \mu(X)$ form the coherence between the individual filters, we cannot choose them totally independently. (We use the term "coherence property" to designate (meta) rules of the form $\alpha_1 \mid\sim \beta_1, \ldots, \alpha_n \mid\sim \beta_n \Rightarrow \alpha' \mid\sim \beta'$, i.e. rules about change of the antecedent. An example from preferential logic is: $\alpha \mid\sim \beta, \alpha \mid\sim \gamma \Rightarrow \alpha \wedge \beta \mid\sim \gamma$ -, which is a consequence of some abstract size operations behind preferential logics.) Particularly for preferential structures, the reasoning with small and big subsets can be made very precise and intuitively appealing, and we will come back to this point later. In a first-order context, we can introduce a generalised quantifier, say ∇, with the same meaning, i.e. $\phi \mid\sim \psi$ iff $\nabla(\phi).\psi$, i.e. "almost everywhere", or "in the important cases" where ϕ holds, so will ψ. This is then the syntactic analogue of the semantical filter system.

Operators may be truth-functional or not. $\neg$ is truth-functional. It suffices to know the truth value of ϕ at some point, to know that of $\neg\phi$ at the same point. $\Box$ is not truth-functional: ϕ and ψ may hold, and $\Box\phi$, but not $\Box\psi$, all at the same point (= base model), we have to look at the full picture, not only at some model. Some operators usually have a unique interpretation, like $\wedge$, which is interpreted by $\cap$, $\neg$ by $\boldsymbol{C}$, the set theoretic complement, etc. On the other hand, e.g. ∇ (standing for "most", "the important", etc.) has only restrictions to its interpretation, like $\mu(X) \subseteq X$, etc. Given a set of models without additional structure, we do not know its exact form, we know it only once we have fixed the additional structure (the relation in this case). If the models already "contain" the operator, the function will respect it, i.e. we cannot have ϕ and $\neg\phi$ in the same model, as $\neg$ is interpreted by $\boldsymbol{C}$. Thus, the functions can, at least in some cases, control consistency. As we can apply one function to the result of the other, we can evaluate complicated formulas, using the functions on the set of models. Consequently, if $\mid\!\sim$ or ∇ is evaluated by μ, we can consider $\mu(\mu(X))$, etc. thus, the machinery for the flat case gives immediately an interpretation for nested formulas, too—whether we looked for it, or not. Additional coherence conditions may restrict us, e.g. in the above example, $\mu(\mu(X)) = \mu(X)$.

1.1.8.2 Structural Semantics

Structural semantics generate the abstract or algebraic semantics, i.e. the behaviour of the functions or relations (and of the operators in the language when we work with "rich" basic models). Preferences between models generate corresponding μ-functions, relations in Kripke structures generate the functions corresponding to $\Box$-operators, etc.

Ideally, structural semantics capture the essence of what we want to reason and speak about (beyond classical logic), they come, or should come, first. Next, we try to see the fundamental ingredients and laws of such structures, code them in an algebraic semantics and the language, i.e. extract the functions and operators, and their laws. In a backward movement, we make the roles of the operators (or relations) precise (should they be nested or not?, etc.), and define the basic models and the algebraic operators. This may result in minor modifications of the structural semantics (like introduction of copies), but should still be close to the point of outset. In this view, the construction of a logic is a back-and-forth movement.

1.1.9 Deontic Logic - Chapter 6

Deontic logic treats (among other things) the moral acceptability of situations or acts.

For instance, when driving a car, you should not cause accidents and hurt someone. So in all "good" driving situations, there are no accidents and no victims. Yet, accidents unfortunately happen. And if you have caused an accident, you should stop and help the possibly injured. Thus, in the "morally bad" situations where you have caused an accident, the morally best situations are those where you help the victims, if there are any.

So, we have a primary obligation—not to cause accidents—and a secondary obligation—if you caused an accident, you should help the injured. Of course, this can be extended to more complicated situations, e.g. if you cannot help the injured, because you are injured yourself, you should call an ambulance, etc. A semantics for this is a many layered preferential structure, an $\mathcal{A}$-ranked structure, see Sect. 6.5. (This is a much simpler version than the semantics developed in Chap. 11.)

The parallel to above example for NML is obvious, and as a matter of fact, the first preferential semantics was given for deontic and not for nonmonotonic logics.

There is, however, an important difference to be made. Preferential structures for NML describe what holds in the normally best models, those for deontic logic what holds in "morally" best models.

A word of warning: Here, the morally optimal models describe "good" situations, and not directly actions to take. This is already obvious by the law of weakening, which holds for all such structures: If ϕ holds in all minimal models, and $\vdash \phi \rightarrow \psi$ (classically), then so does ψ. But if one should be kind, then it does not follow that one should be kind or kill one's grandmother. Of course, we can turn this reasoning into advice for action: act the way that the outcome of your actions assures you to be in a morally good situation.

1.1.10 Theory Revision, Update and Counterfactuals: Chapter 7

1.1.10.1 Theory Revision

The problem of theory revision is to "integrate" some new information ϕ into an old body of knowledge K such that the result is consistent, even if K together with ϕ (i.e. the union $K \cup \{\phi\}$) is inconsistent. (We will assume that K and ϕ are consistent separately.). This problem is well known in juridical thinking, where a new law might be inconsistent with the old set of laws, and the task is to "throw away" enough, but not too many, of the old laws, so we can incorporate the new law into the old system in a consistent way.

The best examined approach was first published in Alchourron et al. (1985), and is known for the initials of its authors as the AGM approach. The formal presentation of this approach (and more) is in Chap. 7.

We can take up the example for NML, and modify it slightly. Suppose our background theory K is that birds fly, in the form: Blackbirds fly, ravens fly, penguins fly, robins fly, ... and that the new information is that penguins do not fly. Then, of course, the minimal change to the old theory is to delete the information that penguins fly, and replace it with the new information.

Often, however, the situation is not so simple. K might be that ψ holds, and so does $\psi \rightarrow \rho$. The new information might be that $\neg\rho$ holds. The radical—and usually excessive—modification will be to delete all information from K, and just take the new information. More careful modifications will be to delete either ψ or $\psi \rightarrow \rho$, but not both. But there is a decision problem here: which of the two do we throw out? Logic alone cannot tell us, and we will need more information to take this decision.

Formal semantics In many cases, revising K by ϕ is required to contain ϕ, thus, if $*$ denotes the revision operation, then $K * \phi \vdash \phi$ (classically). Dropping this requirement does not change the underlying mechanism enormously, we will uphold it.

Speaking semantically, $K * \phi$ will then be defined by some subset of $M(\phi)$. If we choose all of ϕ, then any influence of K is forgotten. A good way to capture this influence seems to choose those models of ϕ, which are closest to the K-models, in some way, and with respect to some distance d. This codes minimal change. We thus choose those ϕ-models m such that there is $n \in M(K)$ with $d(n, m)$ minimal among all $d(n', m')$, $n' \in M(K)$, $m' \in M(\phi)$. (We assume again that the minimal distance exists, i.e. that there are no infinite descending chains of distances, without any minimum.) Of course, the choice of the distance is left open, and will influence the outcome. For instance, choosing as d the trivial distance, i.e. $d(x, y) = 1$ iff $x \neq y$, and 0 otherwise, will give us just ϕ - if K is inconsistent with ϕ.

This semantic approach corresponds well to the classical, syntactic AGM revision approach in the following sense: When we fix K, this semantics corresponds exactly to the AGM postulates (which leave K fixed). When we allow K to change, we can also treat iterated revision, i.e. something like $(K * \phi) * \psi$, thus go beyond the AGM approach (but pay the price of arbitrarily long axioms). This semantics leaves the order (or distance) untouched, and is thus fundamentally different from, e.g. Spohn's Ordinal Conditional Functions, see Spohn (1988).

1.1.10.2 Theory Update

Theory Update is the problem of "guessing" the results of in some way optimal developments.

Consider the following situation: There is a house, at time 0, the light is on, and so is the deep freezer. At time 1, the light is off. Problem: Is the deep freezer still on? The probably correct answer depends on circumstances. Suppose in situation A, there is someone in the house, and weather conditions are normal. In situation B, there is no one in the house, and there is a very heavy thunderstorm going on. Then, in situation A, we will conjecture that the people in the house have switched the light off, but left the deep freezer on. In situation B, we might conjecture a general power failure, and that the deep freezer is now off, too.

We can describe the states at time 0 and 1 by a triple: light on/off, freezer on/off, power failure yes/no.

In situation A, we will consider the development (light on, freezer on, no power failure) to (light off, freezer on, no power failure) as the most likely (or normal) one.

In situation B, we will consider the development (light on, freezer on, no power failure) to (light off, freezer off, power failure) as the most likely (or normal) one.

Often, we will assume a general principle of inertia: things stay the way as they are, unless they are forced to change. Thus, when the power failure is repaired, freezer and light will go on again.

1.1.10.3 Counterfactual Conditionals

A counterfactual conditional states an implication, where the antecedent is (at the moment, at least) wrong. "If it were to rain, he would open his umbrella." This is comprehensible, the person has an umbrella, and if it were to start to rain now he would open the umbrella. If, however, the rain would fall in the midst of a hurricane, then opening the umbrella would only lead to its destruction. Thus, if, at the moment the sentence was uttered, there was no hurricane, and no hurricane announced, then the speaker was referring to a situation that was different from the present situation only in so far as it is raining, or, in other words, minimally different from the actual situation, but with rain falling. If, however, there was a hurricane in sight at the moment of uttering the sentence, we might doubt the speakers good sense, and point the problem out to him/her. We see here again a reasoning about minimal change, or normal situations.

Formal semantics Stalnaker and Lewis first gave a minimal distance semantics in the following way:

If we are in the actual situation m, then $\phi > \psi$ (read: if ϕ were the case, then ψ would also hold) holds in m, iff in all ϕ-models which are closest to m, ψ also holds. Thus, there might well be ϕ-models where ψ fails, but these are not among the ϕ-models closest to m. The distance will, of course, express the difference between the situation m and the models considered. Thus, in the first scenario, situations where it rains and there is no extreme wind condition are closer to the original one than those where a hurricane blows.

In the original approach, distances from each possible actual situation are completely independent. It can, however, be shown that we can achieve the same results with one uniform distance over the whole structure, see Schlechta and Makinson (1994).

1.1.11 Neurology - Chapter 8

This section introduces the reader some ideas in neurology, mainly those by G. Edelman, which we present in Chap. 8.

Consider Example 1.2.1. We will not expect such artificial concepts to have any reasonable properties, like inheritance, etc. But our concepts are meaningful, and our logics using those concepts are useful, so our concepts have something to do with the world "out there". It is our brain that corresponds sufficiently well to the world so it is able to create meaningful ideas and concepts. This was our starting point, why we took interest in neurology.

G. Edelman received a Nobel prize for his research in the immune system, and turned later to neurology. He stresses the parallels between the immune system and the brain, in particular the development of both in dialogue with the world around. Thus, his approach is particularly interesting for us, as we think that our concepts also evolved in dialogue with the world.

The brain seems to work very differently from a computer:

(1) Already J. von Neumann, Neumann (1958), saw fundamental differences between the functioning of the brain and a computer. The former is massively parallel, but will not have deep calculations. See Sect. 8.2.
(2) There are generally no fixed locations where memories are stored, or where information is processed. Instead, the same (by their effect) processes can happen in different groups of neurons (this is called degeneracy by Edelman), and conversely the same neurons can be involved in different processes. This results in an enormous plasticity of the brain, for instance, the function of destroyed brain regions may be taken over by other groups of neurons.
(3) There is no overall hierarchy, in particular, there is no "homunculus" that supervises and looks at the results.
(4) The functioning of the brain seems—on a higher level—best be seen as firing patters of groups of neurons.

Higher brains have a very flexible "feedback" architecture with reentrant connections (in Edelman's terminology), which are connections going back from "higher" brain regions to "lower" ones in the sense that sensory input usually passes first through the "lower" regions, and then to the "higher" ones (simplified). We think that this reentrant architecture can help explain, e.g. abstract constructions we are able to do, corrections and classifications of sensory input, etc.

Of course, we can give not more than a very superficial treatment, but hope to make the importance of the functioning of the brain for common sense reasoning and logic plausible. In particular, fundamental properties like reasonable behaviour of our concepts probably cannot be explained without some understanding of our brain.

(From a puristic point of view, of course, we see our brain itself through the glasses of the brain, so a theory of the brain itself can not only be about what "is out there", but the same relativizations apply to, say, physics and neurology.)

1.1.12 Interpolation - Chapter 9

1.1.12.1 The New Results of Chapter 9

We discuss interpolation for nonmonotonic logic, and its connection to multiplication properties of abstract size in Sect. 9.3. One of the main results there, Proposition

9.3.23, was shown already in Gabbay and Schlechta (2011), but we proceed here in a more systematic way.

We begin with preparatory definitions and remarks:

(1) We first introduce three variants of interpolation and abstract size properties in Sect. 9.3.1. (Most of this material was presented already in Gabbay and Schlechta (2011).)
(2) We discuss some general and combinatorial results in Sect. 9.3.2, which will be used later.
(3) We turn to conditions and results for abstract multiplication and generating (preferential) relations, see Sect. 9.3.3, in further preparation of our results.
(4) We discuss several examples of simple relations in Sect. 9.3.4.

We finally turn to the interpolation results for nonmonotonic logics themselves:

(1) We show that $(\mu * 1)$ entails interpolation, Proposition 9.3.23, but also that the condition is not necessary, in Sect. 9.3.5. These results were already shown in Gabbay and Schlechta (2011).
(2) D. Makinson posed the question to one of the authors if there is also a (single) interpolant for both directions when the formulas are equivalent, i.e. if $\phi \mid\sim \psi$ and $\psi \mid\sim \phi$, is there some α such that $\phi \mid\sim \alpha \mid\sim \psi$ and $\psi \mid\sim \alpha \mid\sim \phi$? We show in Sect. 9.3.6, Proposition 9.3.25, that this is the case, but we need in our proof, in addition to $(\mu * 1)$, smoothness of the generating relation, more precisely its consequence

   $$\mu(X) \subseteq Y, \mu(Y) \subseteq X \Rightarrow \mu(X) = \mu(Y)$$

   which is shown in Fact 9.3.4.

   We do not know if the smoothness property is necessary to show the result.
(3) Finally, we show in Sect. 9.3.7, Proposition 9.3.26, that interpolation is also a consequence of condition $(\mu * 4)$. Condition $(\mu * 4)$ is conceptually simpler than condition $(\mu * 1)$, as it does not involve a change of the sets where the sizes are calculated. The price to pay is a more complicated notation.

1.1.12.2 Introduction

We treated very general interpolation problems for monotone and antitone, two-valued and many-valued logics in Gabbay and Schlechta (2011), Chap. 4. splitting the question into two parts, "semantical interpolation" and "syntactic interpolation", showing that the first problem, existence of semantical interpolation, has a simple and general answer, and reducing the second question, existence of syntactic interpolation, to a definability problem. We can also show that the semantical problem has two "universal" solutions, which depend only on one formula and the shared variables.

In Gabbay and Schlechta (2011), Chap. 5, we investigated three variants of semantical interpolation for nonmonotonic logics, in syntactic shorthand of the types $\phi \mid\sim \alpha \vdash \psi$, $\phi \vdash \alpha \mid\sim \psi$, and $\phi \mid\sim \alpha \mid\sim \psi$, where α is the interpolant, and see that two variants are closely related to multiplication laws about abstract size, defining (or originating from) the nonmonotonic logics. The syntactic problem is analogous to that of the monotonic case.

1.1.12.3 Problem and Method

In classical logic, the problem of interpolation is to find for two formulas ϕ and ψ such that $\phi \vdash \psi$ a "simple" formula α such that $\phi \vdash \alpha \vdash \psi$. "Simple" is defined as "expressed in the common language of ϕ and ψ".

Working on the semantical level often has advantages:

- results are robust under logically equivalent reformulations;
- in many cases, the semantical level allows an easy reformulation as an algebraic problem, whose results can then be generalised to other situations;
- we can split a problem into two parts: a semantical problem and the problem to find a syntactic counterpart (a definability problem);
- the semantics of many non-classical logics are built on relatively few basic notions like size, distance, etc. and we can thus make connections to other problems and logics;
- in the case of preferential and similar logics, the very definition of the logic is already semantical (minimal models), so it is very natural to proceed on this level.

This strategy—translate to the semantical level, do the main work there, and then translate back—has proved fruitful also in the present case.

Looking back at the classical interpolation problem, and translating it to the semantical level, it becomes, given $M(\phi) \subseteq M(\psi)$ (the models sets of ϕ and ψ), is there a "simple" model set A such that $M(\phi) \subseteq A \subseteq M(\psi)$? Or, more generally, given model sets $X \subseteq Y$, is there a "simple" A such that $X \subseteq A \subseteq Y$? Of course, we have to define in a natural way what "simple" means in our context. This is discussed in Sect. 1.1.12.5.

Our separation of the semantical from the syntactic question pays immediately:

(1) We see that monotonic (and antitonic) logics *always* have a semantical interpolant. But this interpolant may not be definable syntactically.
(2) More precisely, we see that there is a whole interval of interpolants in above situation.
(3) We also see that our analysis generalises immediately to many-valued logics, with the same result (existence of an interval of interpolants).
(4) Thus, the question remains: under what conditions does a syntactic interpolant exist?
(5) In nonmonotonic logics, our analysis reveals a deep connection between semantical interpolation and questions about (abstract) multiplication of (abstract) size.

1.1.12.4 An Example

We present now a very simple example that will illustrate some of our main ideas about modularity and interpolation.

Example 1.1.1
Take a propositional language with three variables, p, q, r. Models are sequences such as $\langle p, q, r\rangle$ and $\langle p, q, \neg r\rangle$. We will write $\sigma_1\sigma_2$ for the concatenation of two (sub)sequences, e.g. $\langle p, q, r\rangle = \langle p\rangle\langle q, r\rangle$. We will often identify $\langle p\rangle$ with p. $\times$ will stand for the product, e.g. $\{\langle p, q, r\rangle, \langle p, q, \neg r\rangle\} = \{\langle p, q\rangle\} \times \{r, \neg r\}$.

(1) Consider classical logic.

We have for $\phi := p \wedge q$, $\psi := q \vee r$, $\phi \models \psi$.

The set of models of $p \wedge q$, $M(p \wedge q)$, is $\{\langle p, q\rangle\} \times \{r, \neg r\} = \{p\} \times \{q\} \times \{r, \neg r\}$. $M(q \vee r)$ is $\{p, \neg p\} \times \{\langle q, r\rangle, \langle q, \neg r\rangle, \langle \neg q, r\rangle\}$; we can write this for clarity as $(\{p, \neg p\} \times \{q\} \times \{r, \neg r\}) \cup (\{p, \neg p\} \times \{\langle \neg q, r\rangle\})$. The model sets are constructed in a *modular* way. It suffices to know the model sets in the sublanguage of the formula, and then we multiply with a fixed set, the set of all possibilities, $\{r, \neg r\}$ in the first case, $\{p, \neg p\}$ in the second case. This is all trivial, but important.

Consider now interpolation. Let $\alpha := q$, so, of course, $\phi \models \alpha \models \psi$, α is defined in the common language, and we are done. Why? Write $M(\alpha)$ as $\{p, \neg p\} \times \{q\} \times \{r, \neg r\}$. Obviously, $M(\phi) \subseteq M(\alpha)$, as the first component of $M(\phi)$, $\{p\}$, is a subset of the first component of $M(\alpha)$. This holds by the modular definition of $M(\alpha)$; as α does not contain p, the first component is the set of all possibilities. Consider now $M(\alpha) \subseteq M(\psi)$. Here, as ϕ did not contain r, already all relevant possibilities had to be models of ψ; it suffices to consider the first part of the union, $\{p, \neg p\} \times \{q\} \times \{r, \neg r\}$, to see that this is true.

Closer inspection shows that we only used the definition of model sets, and model set inclusion of the consequence relation, in our argument. As a matter of fact *all* monotone or antitone logics, even those with more than two truth values, have this *semantical* interpolation property.

Why is this only a *semantical* interpolation, and not necessarily also a *syntactic* interpolation? In general, the model set of ϕ will have the form $\Sigma \times \Pi J''$, where Σ is a set of sequences, each giving a value to all propositional variables in the sublanguage L' used by ϕ, and $\Pi J''$ is the set of *all* such sequences for the rest of the language, $J'' = L - L'$, if L is the whole language. Semantic interpolation is now to "cut off" Σ and replace the cut part by another product: Let $J' \subseteq L'$ be the common language, J be the part of the language occuring only in ϕ, and for each $\sigma \in \Sigma$ $\sigma \restriction J$, the restriction of σ to J, $\Sigma \restriction J$ be the set of all those restrictions, then the interpolant is $\Pi J \times (\Sigma \restriction J') \times \Pi J''$. In general, it is not at all clear that this set is the model set of any formula or set of formulas of the language. This depends on the expressivity of the language. So semantical interpolation may exist, but the interpolant need not be definable; so syntactic interpolation will fail.

(2) Consider a nonmonotonic logic defined by a preference relation $\prec$ on the model set of this language; $\phi \mid\sim \psi$ iff in all minimal models of ϕ, ψ holds. (A model is minimal iff there is no smaller model in the model set.) Let $\phi := p \wedge q, \psi := q \wedge r$. We define two orders resulting in $\phi \mid\sim \psi$; the first will not have interpolation and the second will have interpolation.

(2.1) The first ordering:
$\langle \neg p, q, \neg r \rangle \prec \langle \neg p, q, r \rangle \prec \langle p, q, r \rangle \prec \langle p, q, \neg r \rangle$ and $\langle p, q, \neg r \rangle \prec m$ for all other models m. We then close this relation under transitivity.
We then have $p \wedge q \mid\sim q \wedge r$, as $\langle p, q, r \rangle \prec \langle p, q, \neg r \rangle$. We have four candidates for interpolation, which are definable by q only: FALSE, TRUE, q, $\neg q$. They all fail, as we show now. $p \wedge q \not\mid\sim$ *FALSE*, *TRUE* $\not\mid\sim q \wedge r$ (in fact, *TRUE* $\mid\sim q \wedge \neg r$, as $\langle \neg p, q, \neg r \rangle$ is the smallest model of all, $q \mid\sim q \wedge \neg r$, and $\neg q \not\mid\sim q \wedge r$, the latter as the $\neg q$-models are not ordered among each other. Consequently, we have no interpolation. The crucial observation here is that the $p \wedge q$-models behave differently from the $\neg p \wedge q$-models, the first prefer r-models, and the second prefer $\neg r$-models. We do *not* have a relation between r and $\neg r$, which can be used as a module to construct the global relation.

(2.2) The second ordering:
In this ordering, we basically prefer r to $\neg r$ in a homogenous way: For any two models m and m', set $m \prec m'$ iff $m(p) = m'(p)$, $m(q) = m'(q)$, but $m(r) = \mathit{TRUE}$, $m'(r) = \mathit{FALSE}$.
Again, we have $p \wedge q \mid\sim q \wedge r$, but also $q \mid\sim q \wedge r$, so we have (semantical) interpolation for $p \wedge q \mid\sim q \wedge r$; the interpolant α is q. The important property is that we can define the order componentwise: Let $\sigma = \sigma'\sigma'', \tau = \tau'\tau''$; then $\sigma \preceq \tau$ iff $\sigma' \preceq \tau'$ and $\sigma'' \preceq \tau''$. Thus, independent of the particular form of σ' and τ', if $\sigma' \preceq \tau'$, then the comparison between σ'' and τ'' will give the result, so we obtain the global comparison by looking at the components. Here, the additional structure, the relation, is also defined in a modular way, this is why we have semantical interpolation. (The problem of syntactic interpolation is the same in the monotone and the nonmonotonic cases.)

We turn to a more abstract view of the nonmonotonic case. The choice function $\mu(X) := \{x \in X : x \text{ is } \prec\text{-minimal}\}$ defines a principal filter on model sets: $A \subseteq X$ is big or in the filter iff $\mu(X) \subseteq A$. The important property is now that for $X = X' \times X''$ $\mu(X) = \mu(X') \times \mu(X'')$, so we can obtain $\mu(X)$ by simple multiplication of $\mu(X')$ and $\mu(X'')$ from the components X' and X''. We can capture this by saying: "*big* $*$ *big* $=$ *big*", and have an (abstract) multiplication law for an abstract size. This is important for the following reason: Many laws for nonmonotonic logics have a systematic and concise description as *additive* laws about abstract size. Moreover, the law of Rational Monotony can be seen as a property of independence, and is also best expressed as a property of abstract multiplication. Thus, we have connected interpolation for nonmonotonic logic to a general abstract picture of these logics.

(3) There is a second, equally important, way to consider this example, and independence: language change.

(3.1) Consider the classical logic side. As stated above, we obtain the model of, say, ϕ by multiplying the models in the sublanguage with the full product of the rest: $M(p \wedge q) = \{\langle p, q\rangle\} \times \{r, \neg r\}$. It suffices to know the value in the sublanguage to obtain the value in the full language; language change has no effect. Again, this sounds utterly trivial, but it is not; it is a profound conceptual property, and the whole idea of tables of truth values is based on it.

(3.2) Consider the nonmonotonic side. In the first ordering, we cannot say "r is better than $\neg r$", as suggested by $\langle p, q, r\rangle \prec \langle p, q, \neg r\rangle$, and this carries over to bigger languages. This is *not* true in the first ordering, as, e.g. $\langle \neg p, q, \neg r\rangle \prec \langle \neg p, q, r\rangle$ shows. It *is* true in the second ordering; we look at $r/\neg r$, and can then construct the order in the bigger language of p, q, r. Thus in the sublanguage consisting only of r, we have $TRUE \mid\!\sim_r r$, but we have the same in the bigger language, $TRUE \mid\!\sim_{p,q,r} r$. This may be a welcome property; it may be true, but, again, it is not trivial, and we have to be aware of it.

In the nonmonotonic case, we have to build the robustness under language change into the relation, if we want it. In the classical case, it was there already. We did not pursue this point in detail in this book, yet it is very important, and should be considered in further research—and borne in mind. For some remarks, see Sect. 5.3.3.2 and Table 5.4.

1.1.12.5 Monotone and Antitone Semantic and Syntactic Interpolation

We consider here the semantical property of monotony or antitony, in the following sense (for the two-valued case, the generalisation to the many-valued case being straightforward):

Let $\vdash$ be some logic such that $\phi \vdash \psi$ implies $M(\phi) \subseteq M(\psi)$ (the monotone case) or $M(\psi) \subseteq M(\phi)$ (the antitone case).

In the many-valued case, the corresponding property is that $\rightarrow$ (or $\vdash$) respects $\leq$, the order on the truth values.

Semantic Interpolation The problem (for simplicity, for the two-valued case) reads now as follows:

If $M(\phi) \subseteq M(\psi)$ (or, symmetrically, $M(\psi) \subseteq M(\phi)$), is there a "simple" model set A, such that $M(\phi) \subseteq A \subseteq M(\psi)$ (or $M(\psi) \subseteq A \subseteq M(\phi)$). Obviously, the problem is the same in both cases. We will see that such A will always exist, so all such logics have semantical interpolation (but not necessarily also syntactic interpolation).

The main conceptual problem is to define "simple model set". We have to look at the syntactic problem for guidance. Suppose ϕ is defined using propositional variables p and q, ψ using q and r. α has to be defined using only q. What are the models of α? By the very definition of validity in classical logic, neither p nor r have any influence on whether m is a model of α or not. Thus, if m is a model of α, we can modify m on p and r, and it will still be a model. Classical models are best seen as functions from the set of propositional variables to $\{TRUE, FALSE\}$, $\{t, f\}$, etc. In this terminology, any m with $m \models \alpha$ is "free" to choose the value for p and r, and we can write the model set A of α as $\{t, f\} \times M_q \times \{t, f\}$, where M_q is the set of values for q the α-models may have (here: $\emptyset$, $\{t\}$, $\{f\}$, $\{t, f\}$).

So, the semantical interpolation problem is to find sets which may be restricted on the common variables, but are simply the Cartesian product of the possible values for the other variables. To summarize: Let two model sets X and Y be given, where X itself is restricted on variables $\{p_1, \ldots, p_m\}$ (i.e. the Cartesian product for the rest), Y is restricted on $\{r_1, \ldots, r_n\}$; then we have to find a model set A which is restricted only on $\{p_1, \ldots, p_m\} \cap \{r_1, \ldots, r_n\}$, and such that $X \subseteq A \subseteq Y$, of course.

Formulated this way, our approach, the problem and its solution, has two trivial generalisations:

- for multi-valued logics we take the Cartesian product of more than just $\{t, f\}$;
- ϕ may be the hypothesis and ψ the consequence, but also vice versa; there is no direction in the problem. Thus, any result for classical logic carries over to the core part of, e.g. preferential logics.

The main result for the situation with $X \subseteq Y$ is that there is always such a semantical interpolant A.

We explain and quote the result for the many-valued case.

Suppose we have $f, g : M \to V$, where, intuitively, M is the set of all models, and V the set of all truth values. Thus, f and g give to each model a truth value, and, intuitively, f and g each code a model set, assigning to m TRUE iff m is in the model set, and FALSE iff not. We further assume that there is an order on the truth value set V. $\forall m \in M(f(m) \leq g(m))$ corresponds now to $M(\phi) \subseteq M(\psi)$, or $\phi \vdash \psi$ in classical logic. Each model m is itself a function from L, the set of propositional variables to V. Let now $J \subseteq L$. We say that f is insensitive to J iff the values of m on J are irrelevant: If $m \upharpoonright (L - J) = m' \upharpoonright (L - J)$, i.e. m and m' agree at least on all $p \in L - J$, then $f(m) = f(m')$. This corresponds to the situation where the variable p does not occur in the formula ϕ; then $M(\phi)$ is insensitive to p, as the value of any m on p does not depend on whether m is a model of ϕ, or not.

We need two more definitions:

Let $J' \subseteq L$ then $f^{+}(m, J') := max\{f(m') : m' \upharpoonright J' = m \upharpoonright J'\}$ and $f^{-}(m, J') := min\{f(m') : m' \upharpoonright J' = m \upharpoonright J'\}$.

We quote now Proposition 9.2.1:

Proposition 1.1.1
Let M be rich, $f, g : M \to V$, $f(m) \le g(m)$ for all $m \in M$. Let $L = J \cup J' \cup J''$; let f be insensitive to J, g be insensitive to J''.

Then $f^+(m, J') \le g^-(m, J')$ for all $m \in M$, and any $h : M \restriction J' \to V$ that is insensitive to $J \cup J''$ is an interpolant iff

$f^+(m, J') \le h(m_J m_{J'} m_{J''}) = h(m \restriction J') \le g^-(m, J')$ for all $m \in M$.

(h can be extended to the full M in a unique way, as it is insensitive to $J \cup J''$.)

See Diagram 9.2.1.

The Interval of Interpolants Our result has an additional reading: it defines an interval of interpolants, with lower bound $f^+(m, J')$ and upper bound $g^-(m, J')$. But these interpolants have a particular form. If they exist, i.e. iff $f \le g$, then $f^+(m, J')$ depends only on f and J' (and m), but *not* on g, and $g^-(m, J')$ depends only on g and J', *not* on f. Thus, they are universal, as we have to look only at one function and the set of common variables.

1.1.12.6 Laws About Size and Interpolation in Nonmonotonic Logics

Various Concepts of Size and Nonmonotonic Logics A natural interpretation of the nonmonotonic rule $\phi \mid\sim \psi$ is that the set of exceptional cases, i.e. those where ϕ holds, but not ψ, is a small subset of all the cases where ϕ holds, and the complement, i.e. the set of cases where ϕ and ψ hold, is a big subset of all ϕ-cases.

This interpretation gives an abstract semantics to nonmonotonic logic, in the sense that definitions and rules are translated to rules about model sets, without any structural justification of those rules, as they are given, e.g. by preferential structures, which provide structural semantics. Yet, they are extremely useful, as they allow us to concentrate on the essentials, forgetting about syntactical reformulations of semantically equivalent formulas; the laws derived from the standard proof-theoretical rules incite to generalise and modify the ideas, and reveal deep connections but also differences.

To put this abstract view a little more into perspective, we mention three alternative systems, also working with abstract size as a semantics for nonmonotonic logics.

- the system of one of the authors for a first-order setting, published in Schlechta (1990) and elaborated in Schlechta (1995b),
- the system of S. Ben-David and R. Ben-Eliyahu, published in Ben-David and Ben-Eliyahu (1994),
- the system of N. Friedman and J. Halpern, published in Friedman and Halpern (1996).

The equivalence of the systems of Ben-David and Ben-Eliyahu (1994) and Friedman and Halpern (1996) was shown in Schlechta (1997b); see also Schlechta (2004).

Historical remarks: Our own view of abstract size was inspired by the classical filter approach, as used in mathematical measure theory. The first time that abstract size was related to nonmonotonic logics was, to the best of our knowledge, in the second author's Schlechta (1990) and Schlechta (1995b), and, independently, in Ben-David and Ben-Eliyahu (1994). The approach to size by partial orders is first discussed, to the best of our knowledge, by N. Friedman and J. Halpern; see Friedman and Halpern (1996). More detailed remarks can also be found in Gabbay and Schlechta (2009b, d, e). A somewhat different approach is taken in Hawthorne and Makinson (2007).

Before we introduce the connection between interpolation and multiplicative laws about size, we give some comments on the laws about size themselves.

Additive and Multiplicative Laws About Size We give here a short introduction to and some examples for additive and multiplicative laws about size. A detailed overview is presented in Tables 5.2, 5.3, and 5.4. (The first two tables have to be read together; they are too big to fit on one page.)

They show connections and how to develop a multitude of logical rules known from nonmonotonic logics by combining a small number of principles about size. We can use them as building blocks to construct the rules from. More precisely, "size" is to be read as "relative size", since it is essential to change the base sets.

In the first two tables, these principles are some basic and very natural postulates, (Opt), (iM), $(eM\mathcal{I})$, $(eM\mathcal{F})$, and a continuum of power of the notion of "small", or, dually, "big", from $(1 * s)$ to $(< \omega * s)$. From these, we can develop the rest except, essentially, rational monotony, and thus an infinity of different rules.

The probably easiest way to see a connection between nonmonotonic logics and abstract size is by considering preferential structures. Preferential structures define a principal filter, generated by the set of minimal elements, as follows: if $\phi \mid\sim \psi$ holds in such a structure, then $\mu(\phi) \subseteq M(\psi)$, where $\mu(\phi)$ is the set of minimal elements of $M(\phi)$. According to our ideas, we define a principal filter $\mathcal{F}$ over $M(\phi)$ by $X \in \mathcal{F}$ iff $\mu(\phi) \subseteq X \subseteq M(\phi)$. Thus, $M(\phi) \cap M(\neg\psi)$ will be a "small" subset of $M(\phi)$. (Recall that filters contain the "big" sets, and ideals the "small" sets.)

We can now go back and forth between rules on size and logical rules, e.g.

(1) The "AND" rule corresponds to the filter property (finite intersections of AND big subsets are still big).
(2) "Right weakening" corresponds to the rule that supersets of big sets are still big.
(3) It is natural, but beyond filter properties themselves, to postulate that if X is a small subset of Y, and $Y \subseteq Y'$, then X is also a small subset of Y'. We call such properties "coherence properties" between filters. This property corresponds to the logical rule (wOR).
(4) In the rule (CM_ω), usually called cautious monotony, we change the base set a little when going from $M(\alpha)$ to $M(\alpha \wedge \beta)$ (the change is small by the prerequisite $\alpha \mid\sim \beta$), and still have $\alpha \wedge \beta \mid\sim \beta'$, if we had $\alpha \mid\sim \beta'$. We see here a conceptually very different use of "small", as we now change the base set, over which the filter is defined, by a small amount.
(5) The rule of Rational Monotony is the last one in the first table, and somewhat isolated there. It is better to see it as a multiplicative law, as described in the third

table. It corresponds to the rule that the product of medium (i.e., neither big nor small) sets still has medium size.

(For details, see Tables 5.2, 5.3, and 5.4.)

Interpolation and Size

The connection between nonmonotonic logic and the abstract concept of size was investigated in Gabbay and Schlechta (2009e) (see also Gabbay and Schlechta (2009d)). There, we looked among other things at abstract addition of size. Here, we will show a connection to abstract multiplication of size. Our semantical approach uses decomposition of set-theoretical products. An important step is to write a set of models Σ as a product of some set Σ' (which is a restriction of Σ) and some full Cartesian product. So, when we speak about size, we will have (slightly simplified) some big subset Σ_1 of one product Π_1 and some big subset Σ_2 of another product Π_2, and will now check whether $\Sigma_1 \times \Sigma_2$ is a big subset of $\Pi_1 \times \Pi_2$, in shorthand, whether "$big * big = big$". Such conditions are called coherence conditions, as they do not concern the notion of size itself, but the way the sizes defined for different base sets are connected. Our main result here is Proposition 9.3.23. They say that if the logic under investigation is defined from a notion of size which satisfies sufficiently many multiplicative conditions, then this logic will have interpolation of type 3 or even 2; see Sect. 1.1.12.6.

For the convenience of the reader, we now repeat some of the main definitions of Chap. 9, Definitions 9.3.1, 9.3.2, and 9.3.3.

Definition 1.1.2
Let $I = J \cup J'$ with $J \cap J' = \emptyset$. For each $i \in I$ $X_i \neq \emptyset$, let $\Pi I := \Pi\{X_i : i \in I\}$, ΠJ, $\Pi J'$ analogously, and let $\Sigma \subseteq \Pi J$, $\Gamma \subseteq \Pi J'$.

(Intuitively, I is the set of propositional variables of the language, and ΠI is the set of models of the language.)

We suppose abstract measures are given on ΠI, ΠJ, $\Pi J'$, with, e.g. $\mu_I(\Delta)$ denoting the smallest big subset of $\Delta \subseteq \Pi I$, intuitively the set of minimal elements of Δ, when μ_I is generated by a preferential relation, similar for μ_J and $\mu_{J'}$. When the context is clear, we might omit the indices for better readability.

We define:

(1) $(\mu * 1)$:

$$\mu(\Sigma \times \Gamma) = \mu(\Sigma) \times \mu(\Gamma)$$

or, more precisely,

$$\mu_I(\Sigma \times \Gamma) = \mu_J(\Sigma) \times \mu_{J'}(\Gamma)$$

(2) $(\mu * 4)$:

$$\mu(\Sigma \times \Gamma) = \mu(\Sigma \times \Pi J') \restriction J \times \mu(\Pi J \times \Gamma) \restriction J'.$$

Definition 1.1.3
Let again $I = J \cup J'$ with $J \cap J' = \emptyset$. For each $i \in I$ $X_i \neq \emptyset$, let $\Pi I := \Pi\{X_i : i \in I\}$, ΠJ, $\Pi J'$ analogously, and let $\Sigma \subseteq \Pi J$, $\Gamma \subseteq \Pi J'$.

(Intuitively, I is the set of propositional variables of the language, and ΠI is the set of models of the language.)

Let $\prec$ be a (transitive) relation on ΠI, $\sigma, \sigma' \in \Pi J$, $\tau, \tau' \in \Pi J'$. We define the following absoluteness condition for $\prec$:

($<$-Abs):

$\sigma\tau \prec \sigma\tau'$ iff $\sigma'\tau \prec \sigma'\tau'$ (for all $\sigma, \tau \neq \emptyset$, etc.)

By abuse of language, we can then write $\tau \prec \tau'$, and $\tau \perp \tau'$ iff $\sigma\tau \perp \sigma\tau'$. When necessary, we may index the relation, $\prec_{J'}$, etc.

Note, that ($<$-Abs) implies that always $\tau = \tau'$, $\tau \prec \tau'$, $\tau' \prec \tau$, or $\tau \perp \tau'$ holds.

Definition 1.1.4
Let again, $I = J \cup J'$ with $J \cap J' = \emptyset$. For each $i \in I$ $X_i \neq \emptyset$, let $\Pi I := \Pi\{X_i : i \in I\}$, ΠJ, $\Pi J'$ analogously, and let $\Sigma \subseteq \Pi J$, $\Gamma \subseteq \Pi J'$.

Let $\sigma, \tau \in \Pi J$, $\sigma', \tau' \in \Pi J'$, and let relations $\prec$ be defined on ΠI, ΠJ, ΠJ. (We will not distinguish them as $\prec_I$, etc. as the context will disambiguate.) As usual, $\preceq$ will denote $\prec$ or $=$.

Analogously, σ_i will be an element of X_i, and we suppose relations $\prec$ are defined on all X_i, and on arbitrary ΠJ for (R1) below.

(1) $(GH1)$: $\sigma \preceq \tau \wedge \sigma' \preceq \tau' \wedge (\sigma \prec \tau \vee \sigma' \prec \tau') \Rightarrow \sigma\sigma' \prec \tau\tau'$.
(2) $(GH2)$: $\sigma\sigma' \prec \tau\tau' \Rightarrow \sigma \prec \tau \vee \sigma' \prec \tau'$.
    $(GH2)$ means that some compensation is possible, e.g. $\tau' \prec \sigma'$ might be the case, but $\sigma'' \prec \tau''$ wins in the end, so $\sigma'\sigma'' \prec \tau'\tau''$.
(3) $(GH3)$: $\sigma'\sigma'' \preceq \tau'\tau'' \Leftrightarrow \sigma' \preceq \tau'$ and $\sigma'' \preceq \tau''$.
    (Thus, $\sigma'\sigma'' \prec \tau'\tau''$ iff $\sigma'\sigma'' \preceq \tau'\tau''$ and ($\sigma' \prec \tau'$ or $\sigma'' \prec \tau''$).)
(4) (R1) $\sigma \prec \sigma'$ iff $\forall i(\sigma_i = \sigma'_i$ or $\sigma_i \prec \sigma'_i)$ and $\exists i(\sigma_i \prec \sigma'_i)$.

(The (GHi) were defined in Gabbay and Schlechta (2011).)

Remark 1.1.2
There is a multitude of possible modifications in this context, e.g. from longer to shorter sequences and inversely, which may all be interesting for special situations, e.g. resulting in restricted, but not full, interpolation.

The main results are that the multiplicative size rules $(\mu * 1)$ and $(\mu * 4)$ entail nonmonotonic interpolation of the form $\phi \mid\sim \alpha \mid\sim \psi$; see Propositions 9.3.23, 9.3.25, and 9.3.26, to be repeated briefly in Sect. 1.1.12.7 below.

We take now a closer look at interpolation for nonmonotonic logic.

The Three Variants of Nonmonotonic Interpolation

Consider preferential logic, and a rule like $\phi \mid\sim \psi$. This means that $\mu(\phi) \subseteq M(\psi)$, where $\mu(\phi)$ is the set of minimal models of ϕ. So we go from $M(\phi)$ to $\mu(\phi)$, and then to $M(\psi)$, and, abstractly, we have $M(\phi) \supseteq \mu(\phi) \subseteq M(\psi)$, so we have neither necessarily $M(\phi) \subseteq M(\psi)$ nor $M(\phi) \supseteq M(\psi)$; the relation between $M(\phi)$ and $M(\psi)$ may be more complicated. Thus, we have neither the monotone nor the antitone case. For this reason, our general results for monotone or antitone logics do not hold any more.

But we also see here that classical logic is used. Suppose that there is ϕ', which describes exactly $\mu(\phi)$; then we can write $\phi \mid\sim \phi' \vdash \psi$.

So we can split preferential logic into a core part—going from ϕ to its minimal models—and a second part, which is just classical logic. (Similar decompositions are also natural for other nonmonotonic logics.) Thus, preferential logic can be seen as a combination of two logics, the nonmonotonic core and classical logic. It is thus natural to consider variants of the interpolation problem, where $\mid\sim$ denotes again preferential logic, and $\vdash$ denotes as usual classical logic:

Given $\phi \mid\sim \psi$, is there some "simple" α such that

(1) $\phi \mid\sim \alpha \vdash \psi$, or
(2) $\phi \vdash \alpha \mid\sim \psi$, or
(3) $\phi \mid\sim \alpha \mid\sim \psi$?

In most cases, we will only consider the semantical version, as the problems of the syntactical version are very similar to those for monotonic logics. We turn to the variants.

(1) The first variant, $\phi \mid\sim \alpha \vdash \psi$, has a complete characterization in Gabbay and Schlechta (2011), Proposition 5.3.2, provided we have a suitable normal form (conjunctions of disjunctions). The condition says that the relevant variables of $\mu(\phi)$ have to be relevant for $M(\phi)$.
(2) The second variant, $\phi \vdash \alpha \mid\sim \psi$, is related to very (and in many cases, too) strong conditions about size. We do not have a complete characterization, only sufficient conditions about size. The size conditions we need are (see Definition 5.3.1) the abovementioned $(\mu * 1)$, and
$(\mu * 2)$: $\mu(X) \subseteq Y \Rightarrow \mu(X \upharpoonright A) \subseteq Y \upharpoonright A$

where X need not be a product any more.
The result is given in Gabbay and Schlechta (2011), Proposition 5.3.3.
Example 5.3.1 shows that $(\mu * 2)$ seems too strong when compared to probability-defined size. We repeat this example here for the reader's convenience.

Example 1.1.2
Take a language of five propositional variables, with $X' := \{a, b, c\}$, $X'' := \{d, e\}$. Consider the model set $\Sigma := \{\pm a \pm b \pm cde,\ -a-b-c-d \pm e\}$, i.e. of eight models of de and two models of $-d$. The models of de are 8/10 of all elements

of Σ, so it is reasonable to call them a big subset of Σ. But its projection on X'' is only 1/3 of Σ''.

So we have a potential *decrease* when going to the coordinates.

This shows that weakening the prerequisite about X as done in $(\mu * 2)$ is not innocent.

We should, however, note that sufficiently modular preferential relations guarantee these very strong properties of the big sets; see Sect. 5.3.4.

(3) We turn to the third variant, $\phi \mid\sim \alpha \mid\sim \psi$. This is probably the most interesting one, as (a) it is more general, as it loosens the connection with classical logic, (b) it seems more natural as a rule, and (c) it is also connected to more natural laws about size. Again, we do not have a complete characterization, only sufficient conditions about size. Here, $(\mu * 1)$ or $(\mu * 4)$ (essentially) suffice, and we have our main results about nonmonotonic semantical interpolation, Propositions 9.3.23, 9.3.25, and 9.3.26 that $(\mu * 1)$ or $(\mu * 4)$ entail interpolation of the type $\phi \mid\sim \alpha \mid\sim \psi$.
There need not always be a semantical interpolation for the third variant; this is shown in Gabbay and Schlechta (2011), Example 5.3.1.

So we see that, roughly, semantical interpolation for nonmonotonic logics works when abstract size is defined in a modular way—and we find independence again. In a way, this is not surprising, as we use an independent definition of validity for interpolation in classical logic, and we use an independent definition of additional structure (relations or size) for interpolation in nonmonotonic logic.

Language Change to Obtain Products To achieve interpolation and other results of independence, we often need to write a set of models as a nontrivial product. Sometimes, this is impossible, but an equivalent reformulation of the language can solve the problem.

As this might be interesting also for the non-specialists, we repeat Example 5.2.5 in Gabbay and Schlechta (2011).

Example 1.1.3
Consider $p = 3$, and let

$abc, a\neg bc, a\neg b\neg c, \neg abc, \neg a\neg b\neg c, \neg ab\neg c$ be the $6 = 2 * 3$ positive cases,

$ab\neg c, \neg a\neg bc$ the negative ones. (It is a coincidence that we can factorise positive and negative cases—probably if one of the factors is the full product, here 2, though it could also be 4, etc.)

We divide the cases by three new variables, grouping them together into positive and negative cases. a' is indifferent; we want this to be the independent factor, the negative ones will be put into $\neg b'\neg c'$. The procedure has to be made precise still. In the following, (n) is negative.

Let a' code the set $abc, a\neg bc, a\neg b\neg c, ab\neg c$ (n).
Let $\neg a'$ code $\neg a\neg bc$ (n), $\neg abc, \neg a\neg b\neg c, \neg ab\neg c$.
Let b' code $abc, a\neg bc, \neg a\neg b\neg c, \neg ab\neg c$.
Let $\neg b'$ code $a\neg b\neg c, ab\neg c$ (n), $\neg a\neg bc$ (n), $\neg abc$.

Let c' code $abc, a\neg b\neg c, \neg abc, \neg a\neg b\neg c$.

Let $\neg c'$ code $a\neg bc$, $ab\neg c$ (n), $\neg a\neg bc$ (n), $\neg ab\neg c$.

Then the six positive instances are

$\{a', \neg a'\} \times \{b'c', b'\neg c', \neg b'c'\}$, the negative ones

$\{a', \neg a'\} \times \{\neg b'\neg c'\}$.

As we have three new variables, we code again all possible cases, so expressivity is the same.

Crucial here is that $6 = 3 * 2$, so we can just rearrange the six models in a different way; see Gabbay and Schlechta (2011), Fact 5.2.9.

A similar result holds for the nonmonotonic case, where the structure must be possible; we can then redefine the language.

All details are to be found in Gabbay and Schlechta (2011), Sect. 5.2.5.

1.1.12.7 New Results

We prepare the main results of Chap. 9 by introducing the main definitions of this Chapter, as repeated in Sect. 1.1.12.6, together with their connections and other useful combinatorial results in Sects. 9.3.2 and 9.3.3, and give a number of simple, but interesting examples in Sect. 9.3.4.

We then turn to our main results in Sects. 9.3.5, 9.3.6 and 9.3.7.

(1) The first result, which was already shown in Gabbay and Schlechta (2011), but that is given and proved here in a slightly larger context, shows that $(\mu * 1)$ entails semantical interpolation,
(2) the second shows that $(\mu * 1)$ together with smoothness of the generating relation entails semantical interpolation for equivalent formulas, (we do not know if the additional requirement of smoothness which we use in our proof, is necessary for the result)
(3) the third shows that $(\mu * 4)$ also entails semantical interpolation.

More precisely:

(1) **Proposition 1.1.3**

$(\mu * 1)$ entails semantical interpolation of the form $\phi \mathrel{|\!\sim} \alpha \mathrel{|\!\sim} \psi$ in two-valued non-monotonic logic generated by minimal model sets.

As the model sets might not be definable, syntactic interpolation does not follow automatically.

So, more precisely, we show the following:

Let $(\mu * 1)$ hold, and

(1) $I = J \cup J' \cup J''$

(2) $\mu_I(X) \subseteq Y$

(3) $\mu_I(Y) \subseteq X$

(4) $X = X \upharpoonright J \cup J' \times \Pi J''$

(5) $Y = \Pi J \times Y \upharpoonright J' \cup J''$.
Then
$Z := \Pi J \times (\mu_I(X) \upharpoonright J') \times \Pi J''$.
is an interpolant, i.e. $\mu_I(X) \subseteq Z$, $\mu_I(Z) \subseteq Y$.

This is Proposition 5.3.5 of Gabbay and Schlechta (2011), slightly rewritten.
A short comment on the translation of the syntactic to the semantic version: Let the product be defined on $I := J \cup J' \cup J''$ (intuitively I is the set of propositional variables of the language). Let ϕ be defined on $J \cup J'$, ψ on $J' \cup J''$. This translates to $X = X \upharpoonright (J \cup J') \times \Pi J''$, $Y = \Pi J \times Y \upharpoonright (J' \cup J'')$. In our translation, we have $\mu_I(X) \subseteq Y$, and look for Z such that $\mu_I(X) \subseteq Z$ and $\mu_I(Z) \subseteq Y$, with $Z = \Pi J \times Z \upharpoonright J' \times \Pi J''$.
For clarity, even if it clutters up notation, we will be precise about where μ is formed. Thus, we write $\mu_I(X)$ when we take the minimal elements in the full product, $\mu_J(X)$ when we consider only the product on J, etc.

(2) The question was posed by D. Makinson: If $\phi \mid\sim \psi$ and $\psi \mid\sim \phi$, is there an interpolant α such that $\phi \mid\sim \alpha \mid\sim \psi$ and $\psi \mid\sim \alpha \mid\sim \phi$?

The answer is yes, but we need for our proof that μ is generated by a smooth relation. We do not know if this additional requirement is necessary.

Proposition 1.1.4
Let $(\mu * 1)$ *hold, and*
(1) $I = J \cup J' \cup J''$
(2) μ_I *be generated by a smooth relation*
(3) $\mu_I(X) \subseteq Y$
(4) $\mu_I(Y) \subseteq X$
(5) $X = X \upharpoonright J \cup J' \times \Pi J''$
(6) $Y = \Pi J \times Y \upharpoonright J' \cup J''$.
Then
(7) $Z := \Pi J \times (\mu_I(X) \cap \mu_I(Y)) \upharpoonright J' \times \Pi J''$.
is an interpolant in both directions, i.e. $\mu_I(X) \subseteq Z$, $\mu_I(Z) \subseteq Y$, *and, conversely,* $\mu_I(Y) \subseteq Z$, $\mu_I(Z) \subseteq X$.

(3) In the final result of Chap. 9, we use condition $(\mu * 4)$, which is conceptually simpler, as it does not involve a change of the sets where μ is defined. The price to pay is a more complicated notation, which is partly hidden in Fact 9.3.9.

Proposition 1.1.5
$(\mu * 4)$ *entails semantical interpolation of the form* $\phi \mid\sim \alpha \mid\sim \psi$ *in two-valued non-monotonic logic generated by minimal model sets.*
More precisely, we show the following:
Let $(\mu * 4)$ *hold, and*
(1) $I = J \cup J' \cup J''$
(2) $\mu_I(X) \subseteq Y$

(3) $\mu_I(Y) \subseteq X$
(4) $X = X \upharpoonright J \cup J' \times \Pi J''$
(5) $Y = \Pi J \times Y \upharpoonright J' \cup J''$.
Then
$Z := \Pi J \times (\mu_I(X) \upharpoonright J') \times \Pi J''$.
is an interpolant, i.e. $\mu_I(X) \subseteq Z$, $\mu_I(Z) \subseteq Y$.

1.1.13 Independence - Chapter 10

1.1.13.1 The Main Result of Chapter 10

Consider $\Pi\{U_i : i \in I\}$, the set of functions f from I to $\bigcup\{U_i : i \in I\}$ with $f(i) \in U_i$, and Σ a subset of this product.

Let $X, Y, Z \subseteq I$ be disjoint, and let $\langle X \mid Y \mid Z \rangle$ express the following closure property of Σ:

If $f, f' \in \Sigma$ are such that $f \upharpoonright Y = f' \upharpoonright Y$, i.e. $\forall i \in Y f(i) = f'(i)$, then there is $g \in \Sigma$ such that $g \upharpoonright X = f \upharpoonright X, g \upharpoonright Y = f \upharpoonright Y = f' \upharpoonright Y, g \upharpoonright Z = f' \upharpoonright Z$, i.e. we can combine f and f' to a new function in Σ, as long as they agree on Y. (What happens outside $X \cup Y \cup Z$ is left open.)

Obviously, this ternary relation $\langle X \mid Y \mid Z \rangle$ satisfies certain properties, e.g. if $\langle X \mid Y \mid Z \rangle$ holds, and $Z' \subseteq Z$, then $\langle X \mid Y \mid Z' \rangle$ holds, too, for any suitable I and Σ.

The problem is to find a full characterization for this ternary relation. Our main result here is that there is no characterization of the relation, of any fixed finite size.

Before we show the result, we describe the framework, so the reader can put the problem into context. But it is not necessary to know this context, in order to understand the result, which is very elementary. Thus, the reader may proceed directly to Sect. 10.4 - or to Sect. 1.1.6 of this introduction.

1.1.13.2 Some Basic Definitions and Results

Independence as an abstract ternary relation $\langle X \mid Y \mid Z \rangle$ for probability and other situations has been examined by W. Spohn, see Spohn (1980), A. P. Dawid, see Dawid (1979), J. Pearl, see, e.g. Pearl (1988), where the situation in graphs is discussed, too.

The aim of this chapter is to extend this abstract approach to the set and arbitrary function situation.

Definition 1.1.5
If f is a function, Y a subset of its domain, we write $f \upharpoonright Y$ for the restriction of f to elements of Y.

If F is a set of functions over Y, then $F \upharpoonright Y := \{f \upharpoonright Y : f \in F\}$.

We start with probabilistic independence.

Probabilistic Independence

Definition 1.1.6

(1) Let $I \neq \emptyset$ be an arbitrary (index) set, for $i \in I$ $U_i \neq \emptyset$ arbitrary sets. Let $\mathcal{U} := \Pi\{U_i : i \in I\}$, and $\Sigma \subseteq \mathcal{U}$, an arbitrary set of sequences.
If all $U_i = \{0, 1\}$, then $\mathcal{U} = \mathcal{P}(I)$, and $\Sigma \subseteq \mathcal{P}(I)$.
Recall, that for $X \subseteq I, \mathcal{U} \restriction X := \Pi\{U_i : i \in X\}$, etc.

(2) Let $P : \mathcal{P}(\mathcal{U}) \to [0, 1]$ be a probability measure. (As we treat the finite case, we may assume that P is defined by its value on singletons.)

(3) By abuse of language, for $X \subseteq I, x \in U \restriction X$, let $P(x) := P(\{u \in U : u \restriction X = x\})$. Analogously, for $X, Y \subseteq I$, $X \cap Y = \emptyset$, $x \in U \restriction X$, $y \in U \restriction Y$, let $P(x, y) := P(\{u \in U : u \restriction X = x \text{ and } u \restriction Y = y\})$. We sometimes set $P(xy) := P(x, y)$, when there is no ambiguity.

(4) Finally, for $X, Y, Z \subseteq I$ pairwise disjoint, $x \in U \restriction X, y \in U \restriction Y, z \in U \restriction Z$, let $P(x \mid y) := \frac{P(x,y)}{P(y)}$, $P(x \mid y, z) := \frac{P(x,y,z)}{P(y,z)}$, etc.
(We have, of course, to pay attention that we do not divide by 0.)

Definition 1.1.7
P as above defines a three-place relation of independence on pairwise disjoint $X, Y, Z \subseteq I$ $\langle X \mid Y \mid Z \rangle_P$ by

$$\langle X \mid Y \mid Z \rangle_P :\leftrightarrow \begin{cases} \forall x \in U \restriction X, \forall y \in U \restriction Y, \forall z \in U \restriction Z (P(y,z) > 0 \to P(x \mid y) = P(x \mid y, z)), & \text{if } Y \neq \emptyset \\ i.e. P(x,y)/P(y) = P(x,y,z)/P(y,z) or & \\ P(x,y,z) * P(y) = P(x,y) * P(y,z) & \\ & \\ \forall x \in U \restriction X, \forall z \in U \restriction Z (P(z) > 0 \to P(x) = P(x \mid z)), & \text{if } Y = \emptyset \\ i.e. P(x) = P(x,z)/P(z) or & \\ P(x,z) = P(x) * P(z) & \end{cases}$$

If $Y = \emptyset$, we shall also write $\langle X \mid\mid Z \rangle_P$ for $\langle X \mid Y \mid Z \rangle_P$. When the context is clear, we may omit the "P".

We also write $\neg\langle X \mid\mid Z \rangle_P$, when $\langle X \mid\mid Z \rangle_P$ is not the case.

Here, and for other types of independence, we will sometimes write $\langle X, X' \mid Y \mid Z \rangle$, etc. for $\langle X \cup X' \mid Y \mid Z \rangle$.

When we write $\langle X, X' \mid Y \mid Z \rangle$, etc. we will tacitly assume that all sets X, X', Y, Z are pairwise disjoint.

We call this notion probabilistic independence.

We list now some well-known conditions for probabilistic independence.

Pearl discusses the rules $(Ba) - (Be)$ of Definition 10.2.5 for the relation defined in Definition 10.2.4, see e.g. Pearl (1988).

Definition 1.1.8
The basic rules $(Ba) - (B\emptyset)$:

(Ba) Symmetry: $\langle X \mid Y \mid Z\rangle \leftrightarrow \langle Z \mid Y \mid X\rangle$
(Bb) Decomposition: $\langle X \mid Y \mid Z \cup W\rangle \rightarrow \langle X \mid Y \mid Z\rangle$
(Bc) Weak Union: $\langle X \mid Y \mid Z \cup W\rangle \rightarrow \langle X \mid Y \cup W \mid Z\rangle$
(Bd) Contraction: $\langle X \mid Y \mid Z\rangle$ and $\langle X \mid Y \cup Z \mid W\rangle \rightarrow \langle X \mid Y \mid Z \cup W\rangle$
(Be) Intersection: $\langle X \mid Y \cup W \mid Z\rangle$ and $\langle X \mid Y \cup Z \mid W\rangle \rightarrow \langle X \mid Y \mid Z \cup W\rangle$
($B\emptyset$) Empty outside: $\langle X \mid Y \mid Z\rangle$ if $X = \emptyset$ or $Z = \emptyset$.

Proposition 1.1.6
If P is a probability measure, and $\langle X \mid Y \mid Z\rangle_P$ defined as above, then (Ba) – (Bd) of Definition 10.2.5 hold for $\langle \ldots \rangle = \langle \ldots \rangle_P$, and if P is strictly positive, (Be) will also hold.

We proceed in this introduction directly to function independence.

Definition 1.1.9

(1) If $\Sigma \subseteq \Pi\{U_i : i \in I\}$, then $\langle X \mid Y \mid Z\rangle$ holds for Σ, $\langle X \mid Y \mid Z\rangle_\Sigma$, iff for $f', f'' \in \Sigma$ such that $f' \upharpoonright Y = f'' \upharpoonright Y$ there is $g \in \Sigma$ such that $g \upharpoonright X = f' \upharpoonright X$, $g \upharpoonright Y = f' \upharpoonright Y = f'' \upharpoonright Y$, $g \upharpoonright Z = f'' \upharpoonright Z$, and for $Y = \emptyset$, $\langle X \mid\mid Z\rangle$ holds for Σ iff for $f', f'' \in \Sigma$ there is $g \in \Sigma$ such that $g \upharpoonright X = f' \upharpoonright X$, $g \upharpoonright Z = f'' \upharpoonright Z$.
(2) When $Y = \emptyset$, we will also write $\langle X \mid\mid Z\rangle$, $\langle X \mid\mid Z\rangle_\Sigma$, etc.
(3) Finally, we set for $y \in \mathcal{U} \upharpoonright Y$ $\Sigma_y := \{\sigma \in \Sigma : \sigma \upharpoonright Y = y\}$.

As we will always use the second reading of set independence, there is no need for new notation—context will tell whether we consider two-valued functions, or the more general case.

Y may be empty, then the condition $f \upharpoonright Y = g \upharpoonright Y$ is void.

Note that nothing is said about $I - (X \cup Y \cup Z)$, so we look at the projection to $X \cup Y \cup Z$.

Our main result in this chapter (absence of a finite characterisation) will concern this general case.

Notation 1.1.1
In more complicated cases, we will often write ABC for $\langle A \mid B \mid C\rangle$, and $\neg ABC$ if $\langle A \mid B \mid C\rangle$ does not hold. Moreover, we will often just write $f(A)$ for $f \upharpoonright A$, etc.

For $\langle A \cup A' \mid B \mid C\rangle$, we will then write $(AA')BC$, etc.

Fact 1.1.7
In our interpretation,
(1) rule (Be) does not hold,
(2) all $\langle X \mid Y \mid \emptyset\rangle$ (and thus also all $\langle \emptyset \mid Y \mid Z\rangle$) hold,
(3) rules *(Ba)*-*(Bd)* hold, even when one or both of the outside elements of the triples is the empty set.

Above rules *(Ba)*-*(Bd)* are not the only ones to hold, and we introduce more complicated ones, and show that they hold in our situation. One of them is:

Definition 1.1.10
(Loop1)
$AB_1B_2, \ldots, AB_{i-1}B_i, AB_iB_{i+1}, AB_{i+1}B_{i+2}, \ldots, AB_{n-1}B_n, AB_nB_1 \Rightarrow AB_1B_n$
so we turn AB_nB_1 around to AB_1B_n.

When we have to be more precise, we will denote this condition $(Loop1_n)$ to fix the length.

Fact 1.1.8
(Simplified)
Let $n \geq 1$, then $(Loop1_n)$ does not follow from the rules $(Ba) - (Bd)$, $(B\emptyset)$, and the shorter versions of (Loop1)

1.1.13.3 There Is No Finite Characterization

We turn to our main result.

Consider the following simple, short, loop for illustration:

$ABC, ACD, ADE, AEF, AFG, AGB \Rightarrow ABG$—so we can turn AGB around to ABG.

Of course, this construction may be arbitrarily long.

The idea is now to make ABG false, and, to make it coherent, to make one of the interior conditions false, too, say ADE. We describe this situation fully, i.e. enumerate all conditions that hold in such a situation. If we make now ADE true again, we know this is not valid, so any (finite) characterization must say "NO" to this. But as it is finite, it cannot describe all the interior triples of the type ADE in a sufficiently long loop, so we just change one of them that it does not "see" to FALSE, and it must give the same answer NO, so this fails.

Basically, we cannot describe parts of the loop, as the $\langle || \rangle$-language is not rich enough to express it, we see only the final outcome. Bundling singletons into bigger sets will not help, either, as we will see.

The problem is to fully describe the situation.

A very helpful fact is the following:

Definition 1.1.11
Let Σ_j be function sets over I into some set $K, j \in J$.

Let $\Sigma := \{f : I \to K^J : f(i) = \{\langle f_j(i), j\rangle : j \in J, f_j \in \Sigma_j\}\}$.

So any $f \in \Sigma$ has the form $f(i) = \langle f_1(i), f_2(i), \ldots, f_n(i)\rangle$, $f_m \in \Sigma_m$ (we may assume J to be finite).

Thus, given $f \in \Sigma, f_m \in \Sigma_m$ is defined.

Fact 1.1.9
For the above Σ $\langle A \mid B \mid C\rangle$ holds iff it holds for all Σ_j.

Thus, we can destroy the $\langle A \mid B \mid C\rangle$ independently, and collect the results.

This paves the way to our main result, absence of a fnite characterization, see Sect. 10.6.

1.1.14 *Formal Construction - Chapter 11*

1.1.14.1 The Main Result of Chapter 11

The main result of this Chapter is a special preferential structure, constructed using ideas from inheritance, Reiter defaults, theory revision, and epistemic considerations. The construction itself is simple, and is based on our conceptual analysis, and the reader is invited to read Sect. 1.2, as well as the beginning sections of Chap. 11, to see where the ideas came from. The single most important idea is probably that influence changes are rare, provided our concepts are "reasonable".

The construction itself is found in Sect. 11.4.

1.1.14.2 The Construction

This chapter codes our conceptual reflections in a special preferential structure. It is a five-step construction:

(1) Defaults are supposed to be attached to sets or properties, e.g. the set birds has the default "fly" attached to it. They are downward inherited, e.g. to blackbirds, penguins, etc. See Definition 11.4.1.
(2) Classical consequence determines specificity. Specificity helps to find in a theory revision approach a consistent set of defaults that are considered valid for a certain set, e.g. the set penguins will have the valid default "not fly". See Definition 11.4.1.
(3) The sets to which defaults are attached determine the basic hierarchy. The higher up a set is, the more exceptional it is, e.g. penguins are more exceptional than birds, etc. This is coded in the relation $\lhd$, see Definition 11.4.4.
(4) Elements of a set, which are not in any subset to which new defaults are attached, may satisfy all defaults valid for this set, or not. In the first case, they "sit" at the place of the set in the basic hierarchy. In the second case, they are *very* exceptional, as we have no standard reason why they should be exceptional. A bird that does not fly, but which is a penguin, has a reason not to fly: it is a penguin. A bird that is no penguin, not dead or in a cage, etc. all know exceptions, but still cannot fly, is a surprise, and very exceptional. Therefore, it will sit higher up in the hierarchy than penguins, etc. This is coded in the relation $\prec$, see Construction 11.4.1.
(5) The subideal cases, i.e. those which do not satisfy all defaults valid for the set are ordered by the set of defaults they satisfy. Those which satisfy all but one, will be the best, those which satisfy none will be the worst. This is coded in the relation $\sqsubset$, see Definition 11.4.3

Much of the formal development in this chapter centres around the exact construction of this hierarchy, see and Sect. 11.4.1, presented now in slight simplification.

1.1.14.3 A Simplified Version

Suppose we have a set of defaults of the type $(\alpha : \beta/\beta)$. We then say that the default β is attached to α. The α's to that defaults are attached generate a decomposition of the universe (set of all models) corresponding to every possible truth value for each α.

For instance, consider Example 1.1.4.

Example 1.1.4
Let $A, A' \subseteq U, A'' \subseteq A'$, and let defaults be attached to A, A', A''.

See Diagram 11.4.1 for illustration.

The decomposition is $U - A - A' - A''$ (000), $A - A'$ (100), $A \cap A' - A''$ (110), $A \cap A''$ (111), $A'' - A$ (011), $A' - A - A''$ (010), where the binary codes are written in parentheses. (The sets corresponding to 001 and 101 are empty, as $A'' \subseteq A'$.)

We order these subsets by the set of A's they are attached to, e.g. $A'' - A$ (011) is more exceptional than $A' - A - A''$ (010), as it is in more exception sets. (As defaults are attached to the A's, they are themselves exception sets.) This order is written $\lhd$ in Chap. 11, Definition 11.4.4, see Diagram 1.1.3 for the order resulting from our example.

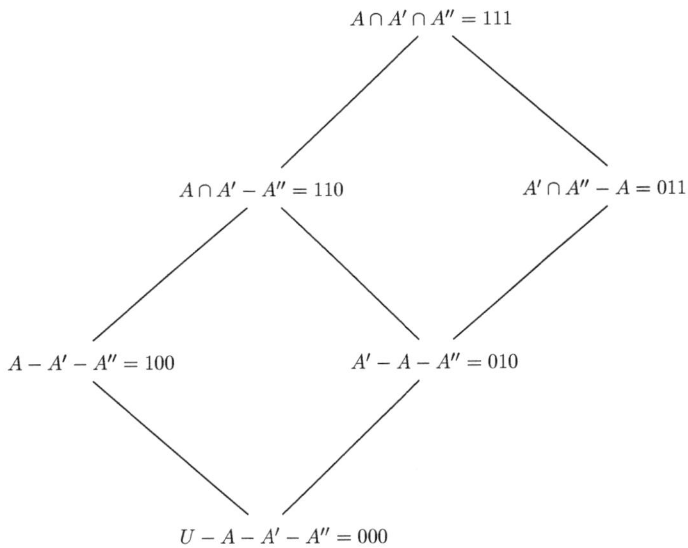

Relation between sets

Diagram 1.1.3

Defaults are inherited downward using specificity. In case of conflict, we take a theory revision approach, see Definition 11.4.1.

Suppose ϕ and ϕ' are defaults attached to A, then a normal element of $A - A'$ should satisfy ϕ and ϕ'. If it does not, then it is an abnormal element, and worse than if it had an "excuse", e.g. if it were a member of $A' \subseteq A$, where the default in question may not hold. The set of normal elements of those subsets will be denoted as $\mu(X)$, the others $\omega(X)$, where X is, e.g. $A - A'$ in our example. The elements of $\omega(X)$ will have an order among themselves: the more defaults they satisfy, the better they are. This is coded by the relation $\sqsubset$, see Definition 11.4.3.

As indicated, elements of $\omega(X)$ are more abnormal than those of $\mu(X)$, so they should sit higher in the hierarchy. This, without the internal $\sqsubset$-structure of the $\omega(X)$, is shown for our Example 1.1.4 in Diagram 1.1.4.

All details are found in Sect. 11.4.1.

Diagram 1.1.4

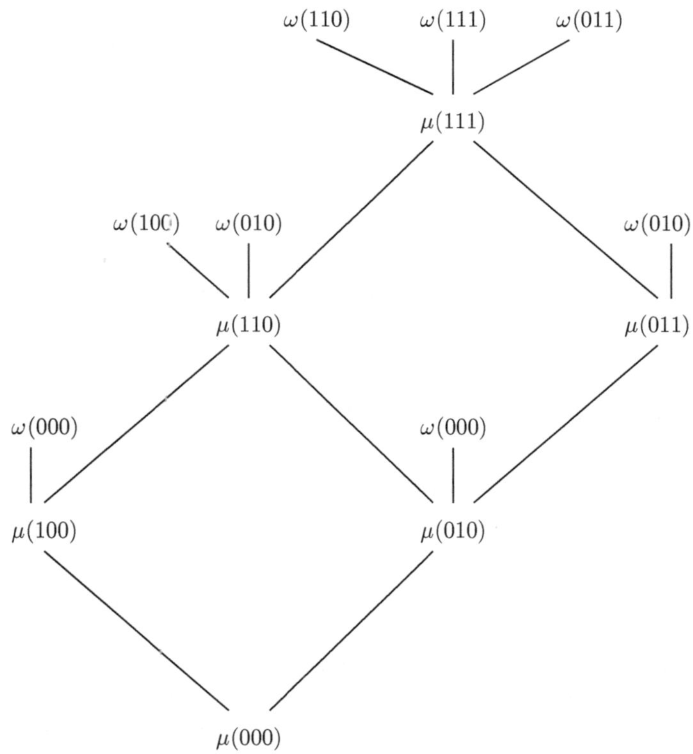

Relation between model sets

1.1.14.4 What We Achieved

(1) A clear and well-examined semantics with nice logical properties in the form of a special preferential structure.

- In contrast to general preferential structures, our construction is not purely algebraic, but guided by language, inheritance and conflict resolution through theory revision.
  General preferential structures can be compared to the bizarre example in Sect. 1.1.11, which has few normal properties.

(2) Throughout, the construction is guided by our basic principle that changes are rare. This is reflected in:

  (2.1) downward inheritance of defaults, see Definition 11.4.1,
  (2.2) the treatment of subideal cases, by the relation $\sqsubset$, see Definition 11.4.3,
  (2.3) resulting general graceful degradation.

(3) A notion of consistency; conflicts are seen and resolved in a principled and intuitive way, thus eliminating cyclic behaviour as is possible with Reiter defaults.
(4) Epistemic considerations are expressed by putting the "unexplained" exceptions higher up.
(5) The construction is modular and flexible, and can be adapted and extended easily, for instance to

  (5.1) negated defaults,
  (5.2) the first-order situation, etc.

(6) We combine the expressive power and natural properties of defaults and inheritance systems with the good logical properties of preferential structures.

1.1.15 Kal Vachomer - Chapter 12

1.1.15.1 The Main Results of Chapter 12

We have two results:

- the first is that there is a problem with the original algorithm in Abraham et al. (2009), which is fundamental, and not easily mended,
- second, we present a very different approach to order comparison, which is very primitive, but, surprisingly, often (but not always) gives the same results as the much more complicated one in Abraham et al. (2009). As we do not see any superiority in results of one approach over the other, simplicity might be a good criterion.

1.1.15.2 Discussion of the Situation

This chapter is basically a footnote to Abraham et al. (2009). Abraham, Gabbay and Schild investigated the Talmudic Kal Vachomer rule. We will call their approach the AGS approach. (The same or similar rules were discussed in Arab and Indian philosophy, Qiyas and Kaimutika Nyaya, respectively, see Abraham et al. (2009).) The Kal Vachomer rule is a rule for principled guessing of an unknown value.

First, we quote an example of this rule from the Bible, as described in Abraham et al. (2009):

Example 1.1.5
A bull can do damage in two ways. It can trample something with its feet or it can use its horns. Also the location of the arena of the damage can either be in a public place or a private place (e.g. a public road or a private garden). The amount of compensation paid depends on these features.

The Bible indicates proportion of the damage to be paid as follows:

(1) Foot action in a public place: 0,
(2) foot action in a private place: 1,
(3) horn action in a public place: 0.5,

but there is no information on horn action in a private place.

The Talmud tries to abduce it from above information by the reasoning: horn action is worse than foot action, and private place is worse than public place; the worse the situation, the bigger the compensation should be.

In a modern setting we might have information about four different computer screens, concerning, e.g. price, size, etc. The information is binary, 0 stands for "not so good", 1 stands for "better", for all qualities, see the table in given Sect. 1.1.15.2, we have One information is lacking, we do not know if Screen 3 has stereophonic speakers, the entry there is "?". The idea is to try both possible values, 0 and 1, and see which one results in the "nicer" partial order between the columns.

Ordering the (vertical) vectors given in the table in Sect. 1.1.15.2 componentwise, (with $1 < 0$, expressing that best is smallest) we have for "?" replaced by 1: $S < D, C, R$; $C < P$; $R < I$ - closed under transitivity. If "?" is replaced by 0, $S < D$ does not hold any more, and D is now an isolated point. We then consider the first order with 1 in the place of "?" as "nicer" than the second order, and assume that the first guess is the better one.

Thus, leaving all details aside for the moment, the Kal Vachomer rule is based on an assumption that the world is rather orderly, and disorder is the exception. We can also consider a more orderly world as simpler, and thus have an example of Occam's Razor, which considers simpler descriptions of the world as more likely to be true. Thus, the Kal Vachomer rule and its discussion is fully in the subject of this book.

The main problem is how to distinguish two partial orders as to their "niceness". What is a nice partial order? A total order is maximally nice, this seems clear, everything grows monotonically. It seems questionable if (partial) equivalences make orders nice. Is $A < B < C$ better or worse than $A = B < C$? The AGS approach says yes, in above example we have just 2 points, $A = B$ and C, the counting approach (see below, Sect. 12.5) leaves the question open, and does not commit.

We now give the formal definition of the problem, and then briefly indicate both approaches.

Definition 1.1.12
A quality matrix is a two-dimensional matrix $a_{i,j}$, $1 \leq i \leq m$, $1 \leq j \leq n$, with $a_{i,j} = 0$ or 1, where columns indicate different qualities, and rows different instances. For instance, for computer screens, columns might be screen size, reaction time, dot size, etc. where 1 is always better than 0 in overall quality. A bigger screen is better, a screen with faster reaction is better, and with smaller dot size is better, too.

Column j is considered at least as good as column j' iff $\forall 1 \leq i \leq m\ a_{i,j} \geq a_{i,j'}$.

In an actual matrix, the order need not be total, as different manufacturers might emphasise different features (in our example).

We are now ready to formulate the reasoning problem:

Suppose one (not more) $a_{i,j}$ is unknown, it is 0 or 1, but we do not know which one, we write the value as "?". Can we guess what it probably is? To obtain an answer, we try the matrix M with 0, say $M(0)$, in place of "?", and the matrix with 1, say $M(1)$, in place of "?", and see if one of them has a nicer order of the columns than the other. The one with the nicer order gives then the better guess.

See Table 1.1.15.2.

Computer screens

	P	C	I	R	D	S
	price over 450	self collection	screen bigger than 24 in.	reaction time below 4ms	dot size less than 0.275	stereo-phonic
Screen 1	0	1	0	1	0	1
Screen 2	0	0	1	1	0	1
Screen 3	0	0	0	0	1	?
Screen 4	1	1	0	0	1	1

1 = yes; 0 = no;? = no data

In the Kal Vachomer context, two quite different ways to measure "nice" of a partial order have been suggested.

The first is the AGS approach, the second is the extremely simple arrow counting approach. As they are quite different, it is not surprising that they do not always give the same results. It is more surprising that they give in all cases discussed in Abraham et al. (2009), and which do not involve equalities, the same result. This might be an argument for the simplistic counting approach.

- The AGS approach
  The AGS approach considers four qualities of a partial order (or, equivalently, a directed graph, where an arrow points to the bigger element).

(1) the number of nodes
(2) the number of connected parts
(3) the maximal number of maximal necessary direction changes in paths, see below
(4) the dimension in a multiset representation.

We refer the reader to the original paper and Sect. 12.3.1.2 for the more involved definitions.

- The arrow counting approach.
  This is extremely simple: we count the numer of arrows (in a graph representation) or the number of ordered pairs $x < y$ of (the transitive closure of) the relation. As a total order seems the best, dividing by the number corresponding to a total order gives a measure of quality of the relation.

It is quite complicated to determine the multiset dimension of a partial order, in particular there is no natural inductive method to determine it, see Sect. 12.4.2, in particular Example 12.4.5. On the other hand, the arrow counting approach is very simple. As we do not see a clear superiority of the much more complicated AGS approach, the arrow counting approach seems to be a serious contender.

1.1.16 Equational CTD - Chapter 13

Chapter 13 deals with modelling contrary-to-duty obligations, see Definition 1.1.17

We explain the problem associated with such modelling and its relation to our book.

Let x be a literal of the form a or $\neg a$, where a is a fact. Assume we can make facts true of false.

Write a set of norms or action clauses of the form

$x \to Oy$
$\neg x \to O\neg y$
$y \to Oz$
$\neg y \to O\neg z$

(with the understanding that $\neg\neg a = a$).

The rule $x \to Oy$ is a demand on execution and indicates a progression of actions. Call this the rule progression. In the case above, the progression is $\pm x, \pm y, \pm z$

Let us also assume that we are in control of executing the facts and so we might do the contrary to what we are asked to do by the rules.

For example, we may do x and then z and then $\neg y$. This gives rise to another progression, call it the fact progression, being the order in which we execute the facts. So in this case, the temporal progression is $\pm x, \pm z, \pm y$.

This fact progression violates the rule progression. There are two violations here:

(1) after x we should have done y
(2) having done $\neg y$ we should have done $\neg z$.

Instead we did z first and then $\neg y$.

These two progressions are independent of each other:

(1) the rules progression, the contrary-to-duties, can be legislated in any way we want,
(2) the facts progression can be temporally executed possibly ignoring the rules.

We also have:

(1) our open future where we do not know which facts will be excuted,
(2) our need to record violations.

We need to model the above situation in logic. We need to have a logic in which we express the rules, express the facts, and have them all pairwise independent from each other. Traditionally modal logic was used and this caused paradoxes because its inability to mix progressions 1 and 2.

There are other solutions like dyadic modal logic or conditionals, we use equations and equational semantics.

Why is equational semantics expected to be successful?

(1) The equations have a progression built into them, when we write

$$y = f(x)$$

$$z = g(y)$$

the direction is clear: first know x, then calculate y, then calculate z. This corresponds to the rules progression. Here the rules are the equations.

(2) We can, on the other hand, substitute for the variables numbers in any order independent of the equations; we can substitute first for z then for x then for y. This corresponds to the facts progression. Obviously these are pairwise independent. We can write any equations we want, and also substitute any values we want. To be crystal clear, if we write in logic the rule $x \to Oy$ and at the same time execute the fact $\neg x$ then the rule and fact are dependent, because we have $\neg x \vdash (x \to Oy)$. While with equations, requiring $y = f(x)$ and setting $x = 7$ are two different mathematical moves.

(3) we can check violations to see if the equations are satisfied.

So if we have a programme of giving semantics for logic via equations, then we can apply equations to model contrary-to-duties, because we have the necessary ingredients built in.

This is what Chap. 13 does, we now carry on a little to give you an idea of it.

We extend in this chapter the equational approach from classical logic and from argumentation Gabbay (2012a), Gabbay (2013) to deontic modal logic and contrary-to-duty obligations Carmo and Jones (2002).

In Sects. 13.2.1 and 13.2.2, we introduce the basic defintions which we repeat here.

Definition 1.1.13 Classical propositional logic has the language of a set of atomic propositions Q (which we assume to be finite for our purposes) and the connectives $\neg$ and $\wedge$. A classical model is an assignment $h : Q \mapsto \{0, 1\}$. h can be extended to all wffs by the following clauses:

- $h(A \wedge B) = 1$ iff $h(A) = h(B) = 1$
- $h(\neg A) = 1 - h(A)$

The set of tautologies are all wffs A such that for all assignments h, $h(A) = 1$.

The other connectives can be defined as usual

$$a \to b = \text{def.} \neg(a \wedge \neg b)$$
$$a \vee b = \neg a \to b = \neg(\neg a \wedge \neg b)$$

Definition 1.1.14

(1) A numerical conjunction is a binary function $\mu(x, y)$ from $[0, 1]^2 \mapsto [0, 1]$ satisfying the following conditions

(1.1) μ is associative and commutative
$\mu(x, \mu(y, z)) = \mu(\mu(x, y), z)$
$\mu(x, y) = \mu(y, x)$
(1.2) $\mu(x, 1) = x$
(1.3) $x < 1 \Rightarrow \mu(x, y) < 1$
(1.4) $\mu(x, y) = 1 \Rightarrow x = y = 1$
(1.5) $\mu(x, 0) = 0$
(1.6) $\mu(x, y) = 0 \Rightarrow x = 0$ or $y = 0$

2. We give two examples of a numerical conjunction

$$\mathbf{n}(x, y) = \min(x, y)$$
$$\mathbf{m}(x, y) = xy$$

Definition 1.1.15 Let I be a set of real numbers $\{0, 1\} \subseteq I \subseteq [0, 1]$. Let μ be a semantics. We say that I supports μ iff the following holds:

(1) For any $x, y \in I$, $\mu(x, y)$ and $\nu(x) = 1 - x$ are also in I.
(2) By a μ expression we mean the following:

(2.1) x is a μ expression, for x atomic
(2.2) If X and Y are μ expressions then so are $\nu(X) = (1 - X)$ and $\mu(X, Y)$

(3) We require that any equation of the form $E_1 = E_2$, where E_1 and E_2 are μ expressions has a solution in I, if it is at all solvable in the real numbers.

Remark 1.1.10 The equational approach allows us to model what is considered traditionally inconsistent theories, if we are prepared to go beyond $\{0, 1\}$ values. Consider the liar paradox $a \leftrightarrow \neg a$. The equation for this is (both for **m** for **n**) $a = 1 - a$ (we are writing 'a' for '$\mathbf{m}(a)$' or '$\mathbf{n}(a)$' f). This solves to $a = \frac{1}{2}$.

The next definition will introduce the methodology involved in the equational point of view.

Definition 1.1.16

(1)

(1.1) A classical equational theory has the form

$$\Delta = \{A_i \leftrightarrow B_i \mid i = 1, 2, \ldots\}$$

where A_i, B_i are wffs.

(1.2) A theory is called a B-theory[1] if it has the form

$$x_i \leftrightarrow A_i$$

where x_i are atomic, and for each atom y there exists at most one i such that $y = x_i$.

(2)

(2.1) A function **f**: wff $\to$ [0, 1] is an μ model of the theory, if we have that **f** is a solution of the system of equations $\mathbf{Eq}(\Delta)$.

$$\mathbf{h}_\mu(A_i) = \mathbf{h}_\mu(B_i), i = 1, 2, \ldots$$

(2.2) Δ is μ consistent if it has an μ model

(3) We say that a theory Δ μ semantically (equationally) implies a theory Γ if every solution of $\mathbf{Eq}(\Delta)$ is also a solution of $\mathbf{Eq}(\Gamma)$.

We write

$$\Delta \vDash_\mu \Gamma.$$

Let $\mathbb{K}$ be a family of functions from the set of wff to [0, 1]. We say that $\Delta \vDash_{(\mu,\mathbb{K})} \Gamma$ if every μ solution **f** of $\mathbf{Eq}(\Delta)$ such that $\mathbf{f} \in \mathbb{K}$ is also an μ solution of $\mathbf{Eq}(\Gamma)$.

(4) We write

$$A \vDash_\mu B$$

iff the theory $\top \leftrightarrow A$ semantically (equationally) implies $\top \leftrightarrow B$. Similarly we write $A \vDash_{(\mu,\mathbb{K})} B$. In other words, if for all suitable solutions $\mathbf{f}, \mathbf{f}(A) = 1$ implies $\mathbf{f}(B) = 1$.

In Sect. 13.3 we use our μ-equational logic to model contrary-to-duty (CTD) sets of obligations. So far such modelling was done in deontic logic and there are difficulties involved. Major among them is the modelling of the Chisholm set Chisholm (1963).

---

[1] B for Brouwer, because we are going to use Brouwer's fixed point theorem to show that theories always have models.

We are going to use our equational semantics and consequence and view the set of contrary-to-duty obligations as a generator for an equational theory. This will give an acceptable paradox free semantics for contrary-to-duty sets.

We shall introduce our semantics in stages. We start with the special case of the generalised Chisholm set and motivate and offer a working semantical solution. Then we show that this solution does not work intuitively well for more general sets where there are loops. Then we indicate a slight mathematical improvement that does work. Then we also discuss a conceptual improvement.

Our strategy is to associate a B-theory $\Delta(\mathbb{C})$ with any contrary-to-duty set $\mathbb{C}$ and examine the associated μ-equations for a suitable μ. This will provide semantics and consequence for the CTD sets and we will discuss how good this representation is.

We repeat here the definition of a CTD system, but then refer the reader directly to Sect. 13.4, as the definitions become rather complicated.

Definition 1.1.17

(1) Consider a language with atoms, the semi-formal $\rightarrow$ and $\neg$ and a semi-formal connective O.
A contrary-to-duty expression has the form $x \rightarrow Oy$ where x and y are literals, i.e. either atoms q or negations of atoms $\neg q$, where we also allow for x not to appear. We might write $\top \rightarrow Oy$ in this case, if it is convenient.

(2) Given a literal x and a set $\mathbb{C}$ of CTD expressions, then the immediate neighbourhood of x in $\mathbb{C}$ is the set $\mathbb{N}_x$ of all expressions from $\mathbb{C}$ of the form

$$z \rightarrow Ox$$

or the form

$$x \rightarrow Oy.$$

(3) A set $\mathbb{F}$ of facts is just a set of literals.
(4) A general CTD system is a pair $(\mathbb{C}, \mathbb{F})$
(5) A Chisholm CTD set $\mathbb{CH}$ has the form

$$\begin{array}{l} x_i \rightarrow Ox_{i+1} \\ \neg x_i \rightarrow O\neg x_{i+1} \\ Ox_1 \end{array}$$

where $1 \leq i \leq m$ and x_i are literals (we understand that $\neg\neg x$ is x).

1.2 Introduction to the Main Concepts

1.2.1 Overview of This Section

This book is to an important extent about explicit and implicit concepts in defeasible or nonmonotonic logics (NML). We describe those concepts here, though we will not pursue all of them in detail.

It is probably best to begin with an example. Reiter defaults (and inheritance systems) permit default transitivity and default inheritance. If people living in northern Europe are mostly Scandinavians, and Scandinavians are mostly tall, then people living in northern Europe are mostly tall. If birds mostly fly, then sparrows (a small minority of birds) will mostly fly, too. These ways of reasoning are useful most of the time, but it is easy to come up with counter examples.

We explore here such default rules and their justifications, semantic principles of nonmonotonic logics like minimal change, as well as different classes of properties. We will turn (part of) this analysis into a formal system in Chap. 11.

From a very abstract point of view, we can say:

- Classical logic is the theory of its operators $\wedge$, $\neg$, $\forall$, etc.
- Preferential (and deontic) logic is the theory of an abstract notion of size.
- Theory revision, theory update and counterfactual conditionals are various forms of a theory of an abstract notion of distance.
- Reiter defaults and defeasible inheritance systems are the theory of homogeneousness or rarity of influence change in the world, and of adequacy of our concepts. This is something new, not a theory of language particles, like "and", "possible" for modal logic, etc. but a theory of the *internal* structure of our concepts, and their adequacy to the world. This might be a reason why the background of the "normal" use of our concepts has been little explored: it is hidden.
- Autoepistemic logic is the theory of adequacy of our knowledge.

Abstract size and distance have been treated before, see e.e. Gabbay and Schlechta (2009d, 2011), see also Chaps. 5 and 7. The reader should keep in mind that the rules and notions of size and distance of preferential logics, theory revision, etc. are *not* default rules, but fixed, and the basis of the logic itself. We concentrate here mostly on homogeneousness/rarity of influence, and adequacy of our concepts and knowledge. Although we tried a certain systematization, and to establish connections, our treatment is still somewhat anecdotal.

1.2.2 Logic and Neuroscience

Basic concepts of logics that try to formalise human common sense reasoning should be intuitively acceptable. Thus, we have to look in two directions: the concepts in logic, and the concepts in human reasoning. We think that the concepts considered here are also basic to human reasoning, and one should look beyond psychology, to the neurological level. The psychological level may be adequate for some questions about how we think, but it cannot answer the question why certain strategies are successful. For this, we have to know about the world, the way we know and reason about it, and about what happens in our brain. This is a programme, and we give only some indications in this book where it might lead us. We can argue for the relevance of the neurological level for NML directly, i.e. from the concepts of nonmonotonic logic, and indirectly, from the concepts of philosophy of science, as the latter is related to the former.

First, the direct way: Our way of reasoning seems to be well adapted to the world, it is successful, as a cat's muscles, senses and reactions are adapted to catching mice. It does not always work, but quite often. Our categories and concepts are adequate, if not, we invent new ones or modify the old ones. They are created in dialogue with the world "out there". (Meta-) Concepts like inheritance, specificity, homogeneousness, default transitivity are "good", they often work, they are good strategies or heuristics, and they seem natural. We think that successful heuristics give a good hint about the correspondence between our thinking and our brain on the one side, the world on the other side. Examples are: the homogeneousness hypothesis, Occam's razor, the "unreasonable effectiveness of mathematics in the natural sciences" (E. Wigner, Wigner (1960)), and mathematical "intuition". The latter is not the full theory, but a useful approximation, a skeleton of known territory. We will see that the discussed properties, like defeasible transitivity, are often heuristics, too. They work normally, but not always.

Second, via philosophy of science: Again, heuristics like homogeneousness of the world (e.g., gravity on the moon works as it does on earth), and more subtle ones like Occam's razor, or the above "effectiveness" are based on an idea of simplicity (of our thoughts, or our abstract constructions), which must be—at least partly—adequate to the world. Finally, we have to show how nonmonotonic logics and philosophy of science are related: In both areas, we have the concepts of homogeneousness, exceptions (which we hope to explain one day in the case of science) and disturbances. Moreover, heuristics are often nonmonotonic rules. In philosophy of science, we have confirmed cases, whose properties are extended by conjecture towards other cases. This procedure is similar to extending a default from the "core" to other cases (see below).

We consider these types of concepts, along with related meta-concepts, first in summary, in later chapters their use in the different variants of NML, and finally their formalisation in Chap. 11. Note that these concepts and meta-concepts are not simply "out there", but are constructions of our brains, in dialogue with the world. We should always keep this very important fact in mind.

1.2.3 Concepts and Properties

Natural classes, concepts, properties are, of course, abstractions. They have (meta-) properties, the way they are used. We now discuss some such classes, concepts and (meta-) properties.

1.2.3.1 Different Classes of Properties

(1) It is very useful to consider—as a contrast to natural concepts—a completely artificial concept, or class of objects:

Example 1.2.1
Enumerate all objects in the universe, and consider the class of objects whose number ends on 3. Now consider the subclass, containing all objects whose number ends on 33, then those ending on 333, etc. We will not expect these concepts to have any reasonable properties. It is likely that all useful properties like homogeneousness will collaps, except those governed by probability.

Thus, the following discussion, as a matter of fact, the whole book, is based on the assumption that the concepts are meaningful.

(2) Classes subject to specificity:
It is reasonable to consider specificity for classes of living things. If we know that a bird is a penguin, properties of penguins will probably overrule properties of birds, in case of conflicts, e.g. we will assume that the bird in question will not fly if it is a penguin. Classes of man made objects (aircraft, tables, etc.) often have similar properties.

(3) Transversal and "discontinuous" properties:
Transversal properties, like "dead". Specificity seems less important here. A dead penguin is not much different from a dead bird in general. They concern many objects in the same way. Discontinuous properties generate many changes. A dead bird will not only not fly any more, but it will not walk any more, either. (The idea of continuity is that there are no "jumps". Going from "alive" to "dead" is a big jump in consequences, and it is so for many objects. See also Schlechta (1995c).)

(4) Spurious properties:
There are spurious properties. Suppose there are some exotic finches, and their white variety does not fly, but walks on the ground. We would (unless informed otherwise!) suppose that being white has nothing to do with the fact that they do not fly, it is just accidental. Living at number 5 in some specific street in Stockholm, might be true of a very small Swede, again, this property (living there) would (probably!) be accidental to being small.

(5) Things and beings "built" to have certain properties:
We know that living things, animals, plants, evolve from an embryo or seed, according to a very elaborate plan. Plans can go wrong, more or less drastically. As living things "want" to live, there are redundancies, and repair mechanisms. Often, there will be graceful degradation of performances. Still, things can go wrong, and the animal might be born with severe handicaps. But we can still see the "intention", also considering parents, etc. Thus, a bird might be "made" to fly, though it has no wings, or lost them in an accident. Human constructions can be similar. Expensive and critical equipment like aircraft also have an elaborate building plan, things can go wrong, and there are redundancies to prevent catastrophes.

We can describe this as a form of inertia, sometimes in graded form, trying to preserve as much as possible.

1.2.3.2 Meta-Properties, Rules and Metarules

(1) Specificity:
This was discussed already above.
(2) homogeneousness and rarity of influence change:
Properties are often supposed to be homogenous with respect to other properties. Swedes are usually blond, and usually tall. But even dark haired Swedes are usually tall, grass on my meadow is similar to grass on my neighbour's meadow. So, "tall" is homogenous with respect to hair colour, grass with respect to ownership. Likewise, homogeneousness is at the basis of science: we assume that gravity works the same way here and on the moon. This is just another way of expressing the same idea, that many properties are independent of each other.
We think that "homogeneousness" is somewhat mysterious, and should be replaced by its origin: Influences and influence changes are rare. We will construct as much as possible from this idea of rarity.
(3) Subideal cases and graceful degradation:
There is a difference between "normal" and "normally". The normal case is the ideal, which may not exist, normally ϕ and normally ψ may hold, even if there is no normal case where ϕ and ψ together hold. (A well-known case is the lottery paradox. Throwing dice, you will normally not obtain 6, normally not 5, ... - but certainly something!) We illustrate this now in a short discussion.
Preferential structures, see Chap. 5, speak about the ideal case, where all properties of the normal case hold. If one property fails, they give no information at all. Reiter defaults, see Chap. 4, (simplified notation: $:\phi/\phi$ means: assume ϕ, unless it is inconsistent to do so) have possible modularity, and thus treatment of the subideal case: $:\phi/\phi$ and $:\psi/\psi$ will give $\phi \wedge \psi$, if consistent, but might still give ϕ, if $\phi \wedge \psi$ is inconsistent, but ϕ alone is not. ($:(\phi \wedge \psi)/(\phi \wedge \psi)$ does not have this modularity, so the way of writing the theory may change the result considerably.) Thus, preferential structures speak only about the ideal

case, whereas Reiter defaults may treat the subideal case, too. The latter have a "graceful degradation" built in, and, in this respect, seem closer to human reasoning and the "real" world. In this context, the reader might look at "continuous logics", as discussed in Schlechta (1997a). On the other hand, : ϕ/ϕ and : $\neg\phi/\neg\phi$ together are a perfectly fine set of Reiter defaults, though this is, of course, just nonsense, and thus a defect of Reiter defaults, and contrary to the intuition (at least for people trained in a minimum of logic).
Thus, we should combine the advantages of both logics: We need a notion of validity (the set of minimal elements provide this for preferential structures) and consistency ($\mu(X) \neq \emptyset$ for preferential structures) of default theories (this is related to a "core" of cases where the full theory holds—this core is $\emptyset$ in above Reiter example), but also an extension from the ideal to the subideal cases, which preferential structures do not provide. The assumption behind Reiter defaults, and behind a suitable extension of preferential structures to treat the subideal case is the homogeneousness hypothesis: dark haired Swedes behave the same way about size as do all Swedes. The same homogeneousness hypothesis is also behind the Talmudic Kal Vachomer logic, see Chap. 12. (There are special cases of treatment of subideal cases in contrary-to-duty obligations, see Chaps. 6 and 13.) The required graceful degradation in the treatment of subideal cases has connections to paraconsistent logics (which try to treat contradictions without the catastrophic EFQ rule of classical logic) and theory revision, see Chap. 7, (another way of treating contradictions) on one side, and the ideas of direct scepticism versus intersection of extensions in defeasible inheritance systems, see Chap. 3.

(4) Adequacy of a logic:
This is a more abstract problem.
We can ask if a logic is adequate for a certain situation, or, better, for the way our brain "sees" a situation. For instance, the autoepistemic aspect of certain nonmonotonic logics assumes that our concepts, knowledge, rules, and abstractions are adequate, and that we know - in an efficient manner - about exceptions. For another example, consider the failure of preferential logics to treat subideal cases. Does this correspond to the world and how we see it? Or would we expect graceful degradation for subideal cases?

(5) Graceful degradation can also be seen as a continuity property.

(5.1) Default transitivity is graceful degradation from full transitivity. Transitivity, i.e. $\alpha \mid\sim \beta$, $\beta \mid\sim \gamma$ imply $\alpha \mid\sim \gamma$, does not hold in general in NML (e.g. penguins are birds, birds fly, but we do not want to conclude that penguins fly), yet Reiter defaults and defeasible inheritance have a defeasible version of it: Unless we have a counterargument, assume that $\alpha \mid\sim \beta$, $\beta \mid\sim \gamma$ imply $\alpha \mid\sim \gamma$. Thus, defeasibility is extendend from the object to the meta-level. We think, this is natural, and it was not (sufficiently) considered, probably, as we enter into second level logic, with its known representation problems in the full predicate logic case. (As long as we treat a finite universe, these problems do not occur.) It is a continuity property, as we go "softly" from classical logic (which has strict transitivity) to NML. (There are also other ways to look at default transitivity, see below.)

(5.2) A similar, but more complicated case of graceful degradation is the problem of the universal quantifier. "Usually $\forall x \phi(x)$", should, if impossible to satisfy, at least give as many instances of $\phi(x)$ as possible. This is in the spirit of (one kind of) defeasibility: Make as many instances true, as possible. We need a cooperation of defeasibility with the logical operators, here with the universal quantifier, which is not guaranteed a priori. Considering $\forall$ as $\bigwedge$, a "big" AND, makes it plausible to treat $\wedge$ in a similar way: if we cannot satisfy $\phi \wedge \psi$, we should at least try to satisfy one of them. This would give a new approach to Reiter style defaults. Likewise, we may want to treat $\vee$ (and $\exists$) in dual fashion: if we can satisfy $\phi \vee \psi$, we should try to satisfy both. We work here "in the spirit" of defeasible reasoning, and consider the logical operators from this perspective.

A more abstract treatment of continuity can be found in Schlechta (1995c).

More generally, any composition of properties, rules, language fragments, etc. should be made in a coherent way, and in the spirit of the logic itself, e.g. in a defeasible way for defeasible logics.

1.2.4 Language and Language Change

The choice of the (logical) language to describe the situation is obviously a profound one, and tells much about our knowledge and idea of the world. Language is an abstraction of the world, and obviously depends on our brain. There are languages that will seem unnatural to us, as illustrated by the artificial concept in Example 1.2.1.

We also conjecture that the human ability of language simplifies reasoning. We can reason in simple languages, and add more predicates when necessary. The use of language allows for simple, small "images" or "scenes" of the world, which will be elaborated when necessary. For some reasoning, knowing that x is a bird, is sufficient, we do not need to know its colour, etc. Not only reasoning of this type is easier, but also other kinds of manipulation, like composing a scene from different objects, are easier. We can add details when necessary, but are not forced to do so. See Sect. 11.6.4.

On the formal side, there are many questions related to language and language change. For instance, it is natural to consider independence of the language elements. This is the idea of theory revision a la Parikh and co-authors, see Chopra and Parikh (2000), Parikh (1999). In preferential structures, we might consider $\mu(X \times Y) = \mu(X) \times \mu(Y)$, alternatively $\mu(X \times Y) = \mu(X) \times Y$ in asymmetrical situations. Such rules are an instance of modularity: we can compute the global result ($\mu(X \times Y)$)

by composing results of simpler situations (e.g. $\mu(X)$ and $\mu(Y)$). They entail logical interpolation, see Sect. 9.3.

1.2.5 Modularity and Independence

We use both words for the same idea: A and B are independent if the situation where they occur is modular with the parts A and B. Independence is a central concept of reasoning, we try to reduce complexity by investigating small parts separately, and putting the results together to obtain the global result. Monotonicity of classical logic can also be seen as a form of independence. Once we know that T entails ϕ, any supplementary information will not change this. This fails, of course, for nonmonotonic logics, and makes other regularities all the more important. There are several approaches to independence, e.g.

- Independence as an abstract ternary relation $\langle X \mid Y \mid Z \rangle$ for probability and other situations has been examined by W. Spohn, see Spohn (1980), A. P. Dawid, see Dawid (1979), J. Pearl, see, e.g. Pearl (1988), where the situation in graphs is discussed, too. The aim of chapter Chap. 10 is to extend this abstract approach to the set and arbitrary function situation.
- In the context of nonmonotonic logic and related areas like theory revision, it was perhaps first investigated formally by R. Parikh and co-authors, see, e.g. Parikh (1999), to obtain "local" conflict solution.
- The present authors investigated its role for interpolation in preferential and other logics in Gabbay and Schlechta (2011), and showed connections to abstract multiplication of size: $\mu(X \times Y) = \mu(X) \times \mu(Y)$, which says that the $\mu-$function preserves independence, see Sect. 9.3.
- In Chap. 11, we use a modular construction of a preferential relation, thus, in most cases, we need only a small part of the relation.

Remark 1.2.1
We can see the EFQ rule of classical logic also as a failure of independence. Argue on the model side. If $A \times B \neq \emptyset$, then we can recover A and B from $A \times B$, $\times$ is injective if $A, B \neq \emptyset$. If, however, $A \times B = \emptyset$, this is not possible, injectivity fails.

1.2.6 Main Concepts and Connections

We summarize here in tabular form the main concepts and their connections, and finally give some examples.

1.2.6.1 Recurrent Concepts

Some recurrent concepts in the above are:

(1) ideal and subideal cases,
(2) graceful degradation and continuity,
(3) specificity,
(4) homogeneousness and rarity of influence (changes),
(5) metarules, coherence properties,
(6) language and abstraction.

1.2.6.2 Main Connections

The meta-concepts mostly describe or are related to heuristics. Recall that they are relative to our concepts, e.g. changes are rare only because our concepts are made to put order into the world. Consider the artificial concept above, Example 1.2.1: changes with respect to this concept will probably be very wild. The reason for the success of our concepts must be in our experiences, and the fundamental adaptation of our brain to the world—thence neuroscience holds the key to our understanding and reasoning.

(1) Rare influence changes generate

(1.1) homogeneousness (there are no disturbances) that generates
- default transitivity

(1.2) "smooth" treatment of subideal cases (we lose only a few properties, there are few disturbances, or they are small), graceful degradation (here in the sense of subideal cases, not in the sense of living beings, etc. see above), continuity and approximation, which all express that changes are small. Coherence conditions are a special case of continuity. (We can see the rules cumulativity and rational monotony, see Chap. 5 also as coherence: change base set X a little, and $\mu(X)$ will change a little, too.)

(1.3) extension of the core of a default (we expect things to "continue as usual")

(1.4) specificity:
Consider the artificial property above, in Example 1.2.1, objects whose number ends on 3. Now consider also those whose number ends on 33, and those whose number ends on 333. Suppose that objects in group $\ldots 3$ mostly have property ϕ, those in group $\ldots 33$ mostly $\neg\phi$. It seems not clear that specificity holds here, i.e. that we conjecture $\neg\phi$ to hold mostly in group $\ldots 333$. So the specificity heuristics also seem to be clearly related to the utility of our concepts.
Our categories are orderly, like animals, plants and lost properties like "flying" do not come back without a reason - influence changes are rare.

(1.5) minimal change of theory revision, update and counterfactual conditionals,

(1.6) weakening of $\forall$ to ∇, "almost all".

(2) Modularity expresses bounderies for influences, it generates for language fragments, e.g.

(2.1) logical interpolation, see Chap. 9,
(2.2) isolation of contradictions, see, e.g. Parikh (1999).

These concepts and connections have to be elaborated, they are the conceptual core of the book, and brought together in Chap. 11.

1.2.6.3 Examples

The following is a list of examples, useful for illustrating our concepts.

(1) *bird* $\mid\!\sim$ *fly*
Birds (can) normally fly. If they do not, there are several standard reasons why not: they are dead, sick, penguins, caught in a cage, etc. There are exotic reasons, too, feathers stuck together due to an oil spill, etc.
Some reasons are reversible, some not, so fly/not fly is not symmetrical.

(2)

(2.1) *bird* $\mid\!\sim$ *fly*
The world of birds is homogenous with respect to flying, colour, size, voice, etc. play a minor role.

(2.1.1) Variant 1:
bird $\mid\!\sim$ *fly*
birds normally fly, blackbirds are birds, so blackbirds normally fly.
Variant 2:
Quakers are normally pacifists. Pacifists normally do not have a gun in their home. So Quakers normally do not have a gun in their home.

(2.2) Graceful degradation:
Swede $\mid\!\sim$ tall
Swedes are normally tall and blond. Even Swedes that are not blond are normally tall.
Swede $\mid\!\sim$ cold-tolerant
Swedes are normally tall, blond and used to cold weather. Even if they are neither tall nor blond, they are normally used to cold weather.
Coherence:
If we consider not only Swedes but also people outside Sweden, but close to Sweden, they will not change much in comparison. (No closed world assumption.)
The less we change the region, the less things will change, a form of continuity.

(2.3) If we know that Swedes in the west of the country are normally tall, we conjecture that also those in the east will have the same property. If we do not see any reason for change, we assume there is none

(2.4) *bird* $\mid\!\sim$ *fly*, *penguin* $\mid\!\sim \neg$*fly*
Birds normally fly, penguins normally do not, so Adelaide penguins probably do not either.

(2.5) Theory revision (see Sect. 7.2), theory update and counterfactual conditionals (see Sect. 7.4) are all based on a notion of closeness of one situation to the next, which is supposed to express little change. The closer A is to B, the more A is like B. We know that something changes, but we assume that the change is small, and changes need an explanation.

(2.6) On most continents, all birds fly. If not all, then most birds fly. This is a graceful degradation, and we should make "all" degrade to "most", when "all" is impossible. (It helps here to see $\forall$ as $\bigwedge$. A suitable treatment has to consider the meaning of $\forall$ in an nonmonotonic context, just as $\forall$ has axioms in FOL, which express its meaning. We cannot expect to get something for nothing.)

1.2.7 Abstract Constructions

Abstract constructions are not part of logic itself, but an important aspect of "higher" reasoning. Therefore, we want to mention them here. Abstract constructions are not only (passive) simplifications but active idealizations. We conjecture that the methods of construction are similar to methods of acting in the real world, e.g. the construction of objects in our brain (see, for instance, the binding problem, Chap. 8). The word "approximation" gives a hint. Going from a dot to an ideal point is similar to refining the tip of an arrow. Trees are often natural abstract constructions.

We conjecture that the reentrant structure of the human brain allows to "consider" abstractions the same way as it considers real objects, and even to mix both in its considerations. There is no special brain structure for abstractions and meta-levels, they all happen in the same brain areas, but the neurological events can be interpreted differently. See Sect. 8.4 for further remarks.

1.3 Previously Published Material

This text builds up on previous research by the authors. To make the text self-contained, it is therefore necessary to repeat some previously published material. We give now the parts concerned and their sources.

Most of the introductory material was already published.

Chapter 2 was published in Gabbay and Schlechta (2009d, 2011), Chap. 2 in both cases. Chapter 3 was published in Gabbay and Schlechta (2009d), Chap. 9 there. Chapter 5 was published in Gabbay and Schlechta (2009d), Chaps. 3, 4, 5 there, and in Gabbay and Schlechta (2011), Chap. 2 there. Section 5.3 in particular was published in Gabbay and Schlechta (2009d), Chap. 3 there, and in Gabbay and Schlechta

(2011), Chap. 5 there. Part of Chap. 9, in particular Proposition 9.3.23, was published in Gabbay and Schlechta (2011), Chap. 4 there. Most of Chap. 7 was published in Gabbay and Schlechta (2009d), Chap. 8 there. Chapter 6 was published in Gabbay and Schlechta (2009d), Sect. 7.3 there. Chapter 13 was published in Gabbay (2014).

Acknowledgments The authors would like to thank David Makinson who, again, has helped to make the text more readable, and Christoph von der Malsburg, who patiently explained neuroscience and its philosophy to the second author.

Part II
Background Material

Chapter 2
Basic Algebraic and Logical Definitions

2.1 Overview of This Chapter

This chapter contains basic notation, definitions and results, sometimes slightly beyond the immediate need of this book, as we want to put our work a bit more in perspective, and make it basically self-contained, for the convenience of the (expert) reader. Most of the material of this chapter (unless marked as "new") was published previously; see Schlechta (2004), Gabbay and Schlechta (2009a, b, d, e, 2011).

We begin with basic algebraic and logical definitions, including in particular many laws of nonmonotonic logics, in their syntactic and semantical variants, showing the connections between both sides; see Definition 2.2.6 and Tables 2.1 and 2.2.

We first recall some standard concepts, such as filters, and introduce our notations for various algebraic and logical notions. This part ends with Fact 2.2.2. We then introduce notation and rules particular to nonmonotonic logics. This might be the place to recall that such logics and their properties were constructed from two sides, the syntactical side, e.g., by D. Gabbay in Gabbay (1989), and the semantical side, e.g., for preferential structures by Y. Shoham in Yoav Shoham (1987). Both approaches were first brought together by S. Kraus, D. Lehmann, M. Magidor in Kraus et al. (1990). The semantical versions of these rules are partly due to one of us (see, e.g., Schlechta (2004)), and described systematically in Gabbay and Schlechta (2009b); see also Gabbay and Schlechta (2009d).

We would like to emphasize the importance of the definability preservation (dp) property. In the infinite case, not all model sets X are definable, i.e. there is some formula ϕ or theory T such that $X = M(\phi)$, the models of ϕ, or $X = M(T)$, the models of T. It is by no means evident that a model choice function μ, applied to a definable model set, gives us back again a definable model set (is definability preserving, or dp). If μ does not have this property, some representation results will not hold, which hold if μ is dp, and representation results become much more complicated; see Schlechta (2004) for positive and for impossibility results. In our present context, definability is again an important concept. Even if we have semantical interpolation, if language and operators are not strong enough, we cannot define the

D.M. Gabbay and K. Schlechta, *A New Perspective on Nonmonotonic Logics*,
DOI 10.1007/978-3-319-46817-4_2

semantical interpolants, so we have semantical, but not syntactic interpolation. The reader is referred to Gabbay and Schlechta (2011), Chap. 5 for the discussion.

2.2 The Definitions

Notation 2.2.1
We use sometimes FOL as an abbreviation for first order logic, and NML for non-monotonic logic. To avoid LaTeX complications in bigger expressions, we replace $\widetilde{xxxxx}$ by $\widetilde{xxxxx}$.

Definition 2.2.1

(1) We use $:=$ and $:\Leftrightarrow$ to define the left-hand side by the right-hand side, as in the following two examples:
$X := \{x\}$ defines X as the singleton with element x.
$X < Y :\Leftrightarrow \forall x \in X \forall y \in Y(x < y)$ extends the relation $<$ from elements to sets.

(2) We use $\mathcal{P}$ to denote the power set operator.
$\Pi\{X_i : i \in I\} := \{g : g : I \to \bigcup\{X_i : i \in I\}, \forall i \in I.g(i) \in X_i\}$ is the general Cartesian product, $X \times X'$ is the binary Cartesian product.
$card(X)$ shall denote the cardinality of X, and V the set-theoretic universe we work in — the class of all sets.
Given a set of pairs $\mathcal{X}$, and a set X, we let $\mathcal{X} \upharpoonright X := \{\langle x, i\rangle \in \mathcal{X} : x \in X\}$. (When the context is clear, we will sometimes simply write X for $\mathcal{X} \upharpoonright X$.)
We will use the same notation $\upharpoonright$ to denote the restriction of functions and in particular of sequences to a subset of the domain.
If Σ is a set of sequences over an index set X, and $X' \subseteq X$, we will abuse notation and write $\Sigma \upharpoonright X'$ for $\{\sigma \upharpoonright X' : \sigma \in \Sigma\}$.
Concatenation of sequences, e.g., of σ and σ', will be denoted by juxtaposition: $\sigma\sigma'$.

(3) $A \subseteq B$ will denote that A is a subset of B or equal to B, and $A \subset B$ that A is a proper subset of B; likewise for $A \supseteq B$ and $A \supset B$.
Given some fixed set U we work in, and $X \subseteq U$, $\boldsymbol{C}(X) := U - X$.

(4) If $\mathcal{Y} \subseteq \mathcal{P}(X)$ for some X, we say that $\mathcal{Y}$ satisfies
$(\cap)$ iff it is closed under finite intersections,
$(\bigcap)$ iff it is closed under arbitrary intersections,
$(\cup)$ iff it is closed under finite unions,
$(\bigcup)$ iff it is closed under arbitrary unions,
$(\boldsymbol{C})$ iff it is closed under complementation,
$(-)$ iff it is closed under set difference.

(5) We will sometimes write $A = B \parallel C$ for: $A = B$, or $A = C$, or $A = B \cup C$.
We make ample and tacit use of the Axiom of Choice.

Definition 2.2.2

$\prec^*$ will denote the transitive closure of the relation $\prec$. If $<$, $\prec$, or a similar relation is given, $a \bot b$ will express that a and b are $<$-incomparable (or $\prec$-incomparable); context will tell. Given any relation $<$, $\leq$ will stand for $<$ or $=$; conversely, given $\leq$, $<$ will stand for $\leq$, but not $=$; similarly for $\prec$, etc.

Definition 2.2.3

Fix a base set X.
A filter (resp. weak filter) on or over X is a set $\mathcal{F} \subseteq \mathcal{P}(X)$ such that $(F1) - (F3)$ (resp. (F1), (F2), $(F3')$) hold:
(F1) $X \in \mathcal{F}$,
(F2) $A \subseteq B \subseteq X$, $A \in \mathcal{F}$ imply $B \in \mathcal{F}$,
(F3) $A, B \in \mathcal{F}$ imply $A \cap B \in \mathcal{F}$,
$(F3')$ $A, B \in \mathcal{F}$ imply $A \cap B \neq \emptyset$.
So a weak filter satisfies $(F3')$ instead of (F3). Note that $\mathcal{P}(X)$ is a filter, but not a weak one, but all proper filters are also weak filters.
A filter is called a principal filter iff there is $X' \subseteq X$ s.t. $\mathcal{F} = \{A : X' \subseteq A \subseteq X\}$.

A (weak) ideal on or over X is a set $\mathcal{I} \subseteq \mathcal{P}(X)$ such that $(I1) - (I3)$ (or (I1), (I2), $(I3')$) hold:
(I1) $\emptyset \in \mathcal{I}$,
(I2) $A \subseteq B \subseteq X$, $B \in \mathcal{I}$ imply $A \in \mathcal{I}$,
(I3) $A, B \in \mathcal{I}$ imply $A \cup B \in \mathcal{I}$,
$(I3')$ $A, B \in \mathcal{I}$ imply $A \cup B \neq X$.
So a weak ideal satisfies $(I3')$ instead of (I3).
A filter is an abstract notion of size; elements of a filter on X are called big subsets of X; their complements are called small, and the rest have medium size. The dual applies to ideals; this is justified by the following trivial fact:

Fact 2.2.1

If $\mathcal{F}$ is a (weak) filter on X, then $\mathcal{I} := \{X - A : A \in \mathcal{F}\}$ is a (weak) ideal on X; if $\mathcal{I}$ is a (weak) ideal on X, then $\mathcal{F} := \{X - A : A \in \mathcal{F}\}$ is a (weak) filter on X.

Definition 2.2.4

(1) V will be the set of truth values when there are more than the classical ones, TRUE and FALSE.
We work here in a classical propositional language $\mathcal{L}$; a theory T will be an arbitrary set of formulas. Formulas will often be named ϕ, ψ, etc., theories T, S, etc.
$v(\mathcal{L})$ or simply L will be the set of propositional variables of $\mathcal{L}$.
$F(\mathcal{L})$ will be the set of formulas of $\mathcal{L}$.
A propositional model m will be a function from the set of propositional variables to the set of truth values V when we have more than two truth values.
$M_{\mathcal{L}}$, or simply M when the context is clear, will be the set of (classical) models for $\mathcal{L}$; $M(T)$ or M_T is the set of models of T; likewise for $M(\phi)$ for a formula ϕ.

(2) $\boldsymbol{D}_{\mathcal{L}} := \{M(T) : T$ a theory in $\mathcal{L}\}$, the set of definable model sets.
Note that, in classical propositional logic, $\emptyset$, $M_{\mathcal{L}} \in \boldsymbol{D}_{\mathcal{L}}$, $\boldsymbol{D}_{\mathcal{L}}$ contains singletons and is closed under arbitrary intersections and finite unions.
An operation $f : \mathcal{Y} \to \mathcal{P}(M_{\mathcal{L}})$ for $\mathcal{Y} \subseteq \mathcal{P}(M_{\mathcal{L}})$ is called definability preserving, (dp) or (μdp) in short, iff for all $X \in \boldsymbol{D}_{\mathcal{L}} \cap \mathcal{Y}$, $f(X) \in \boldsymbol{D}_{\mathcal{L}}$.
We will also use (μdp) for binary functions $f : \mathcal{Y} \times \mathcal{Y} \to \mathcal{P}(M_{\mathcal{L}})$ — as needed for theory revision — with the obvious meaning.
(3) $\vdash$ will be classical derivability, and
$\overline{T} := \{\phi : T \vdash \phi\}$, the closure of T under $\vdash$.
(4) $Con(.)$ will stand for classical consistency, so $Con(\phi)$ will mean that ϕ is classically consistent; likewise for $Con(T)$. $Con(T, T')$ will stand for $Con(T \cup T')$, etc.
(5) Given a consequence relation $\mid\!\sim$, we define
$\overline{\overline{T}} := \{\phi : T \mid\!\sim \phi\}$.
(There is no fear of confusion with $\overline{T}$, as it just is not useful to close twice under classical logic.)
(6) $T \vee T' := \{\phi \vee \phi' : \phi \in T, \phi' \in T'\}$.
(7) If $X \subseteq M_{\mathcal{L}}$, then $Th(X) := \{\phi : X \models \phi\}$; likewise for $Th(m), m \in M_{\mathcal{L}}$. ($\models$ will usually be classical validity.)

In the following, the X_i are arbitrary (nonempty) sets, standing for the set of truth values, and J is intuitively the set of propositional variables. So any element of $\Pi\{X_i : i \in J\}$ stands for a propositional model. Inessential variables in a set Σ are those which do not have any influence on the truth of the formula whose model set is Σ.

Definition 2.2.5
Let $\Sigma \subseteq \Pi := \Pi\{X_i : i \in J\}$. Define:
(1) For $\sigma, \sigma' \in \Sigma$, $J' \subseteq J$, define:
$\sigma \sim_{J'} \sigma' :\Leftrightarrow \forall x \in J' \sigma(x) = \sigma'(x)$.
(2) $I(\Sigma) := \{i \in J : \Sigma = \Sigma \restriction (J - \{i\}) \times X_i\}$ (up to reordering) (they are the irrelevant or inessential i) and
$R(\Sigma) := J - I(\Sigma)$ (the relevant or essential i).

Fact 2.2.2
(1) $\Sigma = \Sigma \restriction R(\Sigma) \times \Pi \restriction I(\Sigma)$ (up to reordering)
(2) $\sigma \restriction R(\Sigma) = \sigma' \restriction R(\Sigma) \wedge \sigma \in \Sigma \Rightarrow \sigma' \in \Sigma$.

Proof
(1) Enumerate $I(\Sigma)$, $I(\Sigma) = \{i : i < \kappa\}$. Define $\Sigma_j := \Sigma \restriction (R(\Sigma) \cup \{i : i < j\})$. We show by induction that $\Sigma_j = \Sigma \restriction R(\Sigma) \times \Pi \restriction (I(\Sigma) \cap j)$ for $j \leq \kappa$ (up to reordering).
$j = 0$ is trivial — there is nothing to show.
$j \to j + 1$: This follows from the induction hypothesis and the definition of $I(\Sigma)$.

j is a limit ordinal: Any sequence of length j can be written as the coherent union of shorter sequences, and these are in both sets, as the result holds for $j' < j$ by induction hypothesis.
(2) Trivial.
□

To put our work more into perspective, we repeat now material from Gabbay and Schlechta (2009b). This gives the main definitions and rules for nonmonotonic logics; see Tables 2.1 and 2.2, "Logical rules, definitions and connections".

Definition 2.2.6
The definitions are given in Table 2.1, "Logical rules, definitions and connections Part I" and 2.2, "Logical rules, definitions and connections Part II", which also show connections between different versions of rules, the semantics, and rules about size. (The tables are split in two, as they would not fit onto one page otherwise.)

Explanation of the tables:

(1) The first table gives the basic properties, the second table those for Cumulativity and Rational Monotony.
(2) The difference between the first two columns is that the first column treats the formula version of the rule, the second the more general theory (i.e. set of formulas) version.
(3) "Corr." stands for "Correspondence".'
(4) The third column, "Corr.", is to be understood as follows:
Let a logic $\mid\!\sim$ satisfy (LLE) and (CCL), and define a function $f : \boldsymbol{D}_{\mathcal{L}} \to \boldsymbol{D}_{\mathcal{L}}$ by $f(M(T)) := M(\overline{\overline{T}})$. Then f is well defined and satisfies (μdp), and $\overline{\overline{T}} = Th(f(M(T)))$.
If $\mid\!\sim$ satisfies a rule on the left-hand side, then — provided the additional properties noted in the middle for $\Rightarrow$ hold, too — f will satisfy the property on the right-hand side.
Conversely:
If $f : \mathcal{Y} \to \mathcal{P}(M_{\mathcal{L}})$ is a function, with $\boldsymbol{D}_{\mathcal{L}} \subseteq \mathcal{Y}$, and we define a logic $\mid\!\sim$ by $\overline{\overline{T}} := Th(f(M(T)))$, then $\mid\!\sim$ satisfies (LLE) and (CCL). If f satisfies (μdp), then $f(M(T)) = M(\overline{\overline{T}})$.
If f satisfies a property on the right-hand side, then — provided the additional properties noted in the middle for $\Leftarrow$ hold, too — $\mid\!\sim$ will satisfy the property on the left-hand side.

(5) We use the following abbreviations for those supplementary conditions in the "Correspondence" columns:
"$T = \phi$" means that if one of the theories (the one named the same way in Definition 2.2.6) is equivalent to a formula, we do not need (μdp). $-(\mu dp)$ stands for "without (μdp)".

Table 2.1 Logical rules, definitions and connections Part I

Logical rule		Corr.	Model set	Corr.	Size Rules
Basics					
(SC) Supraclassicality $\alpha \vdash \beta \Rightarrow \alpha \mid\!\sim \beta$	(SC) $\overline{T} \subseteq \overline{\overline{T}}$	$\Rightarrow$ $\Leftarrow$	$(\mu \subseteq)$ $f(X) \subseteq X$	trivial	(Opt)
(REF) Reflexivity $T \cup \{\alpha\} \mid\!\sim \alpha$					
(LLE) Left logical equivalence $\vdash \alpha \leftrightarrow \alpha', \alpha \mid\!\sim \beta \Rightarrow$ $\alpha' \mid\!\sim \beta$	(LLE) $\overline{T} = \overline{T'} \Rightarrow \overline{\overline{T}} = \overline{\overline{T'}}$		(trivally true)		
(RW) Right weakening $\alpha \mid\!\sim \beta, \vdash \beta \rightarrow \beta' \Rightarrow$ $\alpha \mid\!\sim \beta'$	(RW) $T \mid\!\sim \beta, \vdash \beta \rightarrow \beta' \Rightarrow$ $T \mid\!\sim \beta'$		(upward closure)	trivial	(iM)
(wOR) $\alpha \mid\!\sim \beta, \alpha' \vdash \beta \Rightarrow$ $\alpha \vee \alpha' \mid\!\sim \beta$	(wOR) $\overline{\overline{T}} \cap \overline{T'} \subseteq \overline{\overline{T \vee T'}}$	$\Rightarrow$ $\Leftarrow$	(μwOR) $f(X \cup Y) \subseteq f(X) \cup Y$	$\Leftrightarrow$	$(eM\mathcal{I})$
$(disjOR)$ $\alpha \vdash \neg\alpha', \alpha \mid\!\sim \beta,$ $\alpha' \mid\!\sim \beta \Rightarrow \alpha \vee \alpha' \mid\!\sim \beta$	$(disjOR)$ $\neg Con(T \cup T') \Rightarrow$ $\overline{\overline{T}} \cap \overline{\overline{T'}} \subseteq \overline{\overline{T \vee T'}}$	$\Rightarrow$ $\Leftarrow$	$(\mu disjOR)$ $X \cap Y = \emptyset \Rightarrow$ $f(X \cup Y) \subseteq f(X) \cup f(Y)$	$\Leftrightarrow$	$(I \cup disj)$
(CP) Consistency preservation $\alpha \mid\!\sim \bot \Rightarrow \alpha \vdash \bot$	(CP) $T \mid\!\sim \bot \Rightarrow T \vdash \bot$	$\Rightarrow$ $\Leftarrow$	$(\mu\emptyset)$ $f(X) = \emptyset \Rightarrow X = \emptyset$	trivial	(I_1)
			$(\mu\emptyset fin)$ $X \neq \emptyset \Rightarrow f(X) \neq \emptyset$ for finite X		(I_1)
	(AND_1) $\alpha \not\mid\!\sim \beta \Rightarrow \alpha \not\mid\!\sim \neg\beta$				(I_2)

	(AND_n) $\alpha \mid\!\sim \beta_1, \ldots, \alpha \mid\!\sim \beta_{n-1} \Rightarrow$ $\alpha \not\mid\!\sim (\neg\beta_1 \vee \ldots \vee \neg\beta_{n-1})$				(I_n)
(AND) $\alpha \mid\!\sim \beta, \alpha \mid\!\sim \beta' \Rightarrow$ $\alpha \mid\!\sim \beta \wedge \beta'$	(AND) $T \mid\!\sim \beta, T \mid\!\sim \beta' \Rightarrow$ $T \mid\!\sim \beta \wedge \beta'$		(closure under finite intersection)	trivial	(I_ω)
(CCL) Classical closure	(CCL) $\overline{\overline{T}}$ classically closed		(trivally true)	trivial	$(iM) + (I_\omega)$
(OR) $\alpha \mid\!\sim \beta, \alpha' \mid\!\sim \beta \Rightarrow$ $\alpha \vee \alpha' \mid\!\sim \beta$	(OR) $\overline{\overline{T}} \cap \overline{\overline{T'}} \subseteq \overline{\overline{T \vee T'}}$	$\Rightarrow$ $\Leftarrow$	(μOR) $f(X \cup Y) \subseteq f(X) \cup f(Y)$	$\Leftrightarrow$	$(eM\mathcal{I}) + (I_\omega)$
$\overline{\overline{\alpha \wedge \alpha'}} \subseteq \overline{\overline{\alpha}} \cup \{\alpha'\}$	(PR) $\overline{\overline{T \cup T'}} \subseteq \overline{\overline{\overline{T}} \cup T'}$	$\Rightarrow$ $\Leftarrow (\mu dp) + (\mu \subseteq)$ $\not\Leftarrow -(\mu dp)$ $\Leftarrow (\mu \subseteq)$ $T' = \phi$	(μPR) $X \subseteq Y \Rightarrow$ $f(Y) \cap X \subseteq f(X)$	$\Leftrightarrow$	$(eM\mathcal{I}) + (I_\omega)$
		$\Leftarrow$ $T' = \phi$	$(\mu PR')$ $f(X) \cap Y \subseteq f(X \cap Y)$		
(CUT) $T \mid\!\sim \alpha; T \cup \{\alpha\} \mid\!\sim \beta \Rightarrow$ $T \mid\!\sim \beta$	(CUT) $T \subseteq \overline{\overline{T'}} \subseteq \overline{\overline{T}} \Rightarrow$ $\overline{\overline{T'}} \subseteq \overline{\overline{T}}$	$\Rightarrow$ $\Leftarrow$	(μCUT) $f(X) \subseteq Y \subseteq X \Rightarrow$ $f(X) \subseteq f(Y)$	$\Leftarrow$ $\not\Rightarrow$	$(eM\mathcal{I}) + (I_\omega)$

Table 2.2 Logical rules, definitions and connections Part II

Logical rules, definitions and connections Part II					
Logical rule		Corr.	Model set	Corr.	Size-rule
Cumulativity					
(wCM) $\alpha \mid\!\sim \beta, \alpha' \vdash \alpha, \alpha \wedge \beta \vdash \alpha' \Rightarrow$ $\alpha' \mid\!\sim \beta$				trivial	$(eM\mathcal{F})$
(CM_2) $\alpha \mid\!\sim \beta, \alpha \mid\!\sim \beta' \Rightarrow \alpha \wedge \beta \not\mid\!\sim \neg \beta'$					(I_2)
(CM_n) $\alpha \mid\!\sim \beta_1, \ldots, \alpha \mid\!\sim \beta_n \Rightarrow$ $\alpha \wedge \beta_1 \wedge \ldots \wedge \beta_{n-1} \not\mid\!\sim \neg \beta_n$					(I_n)
(CM) Cautious monotony $\alpha \mid\!\sim \beta, \alpha \mid\!\sim \beta' \Rightarrow$ $\alpha \wedge \beta \mid\!\sim \beta'$	(CM) $T \subseteq \overline{T'} \subseteq \overline{\overline{T}} \Rightarrow$ $\overline{\overline{T}} \subseteq \overline{\overline{T'}}$	$\Rightarrow$ $\Leftarrow$	(μCM) $f(X) \subseteq Y \subseteq X \Rightarrow$ $f(Y) \subseteq f(X)$	$\Leftrightarrow$	$(\mathcal{M}^{+}_{\omega})(4)$
or $(ResM)$ Restricted monotony $T \mid\!\sim \alpha, \beta \Rightarrow T \cup \{\alpha\} \mid\!\sim \beta$		$\Rightarrow$ $\Leftarrow$	$(\mu ResM)$ $f(X) \subseteq A \cap B \Rightarrow$ $f(X \cap A) \subseteq B$		
(CUM) Cumulativity $\alpha \mid\!\sim \beta \Rightarrow$ $(\alpha \mid\!\sim \beta' \Leftrightarrow \alpha \wedge \beta \mid\!\sim \beta')$	(CUM) $T \subseteq \overline{T'} \subseteq \overline{\overline{T}} \Rightarrow$ $\overline{\overline{T}} = \overline{\overline{T'}}$	$\Rightarrow$ $\Leftarrow$	(μCUM) $f(X) \subseteq Y \subseteq X \Rightarrow$ $f(Y) = f(X)$	$\Leftarrow$ $\not\Rightarrow$	$(eM\mathcal{I}) + (I_\omega) + (\mathcal{M}^{+}_{\omega})(4)$
	$(\subseteq\supseteq)$ $T \subseteq \overline{\overline{T'}}, T' \subseteq \overline{\overline{T}} \Rightarrow$ $\overline{\overline{T'}} = \overline{\overline{T}}$	$\Rightarrow$ $\Leftarrow$	$(\mu \subseteq\supseteq)$ $f(X) \subseteq Y, f(Y) \subseteq X \Rightarrow$ $f(X) = f(Y)$	$\Leftarrow$ $\not\Rightarrow$	$(eM\mathcal{I}) + (I_\omega) + (eM\mathcal{F})$

Rationality					
$(RatM)$ Rational monotony	$(RatM)$	$\not\Rightarrow$	$(\mu RatM)$	$\Leftrightarrow$	$(\mathcal{M}^{++})$
$\alpha \mid\!\sim \beta, \alpha \not\mid\!\sim \neg\beta' \Rightarrow$	$Con(T \cup \overline{\overline{T'}}), T \vdash T' \Rightarrow$	$\Leftarrow (\mu dp)$	$X \subseteq Y, X \cap f(Y) \neq \emptyset \Rightarrow$		
$\alpha \wedge \beta' \mid\!\sim \beta$	$\overline{\overline{T}} \supseteq \overline{\overline{\overline{T'}} \cup T}$	$\not\Leftarrow -(\mu dp)$	$f(X) \subseteq f(Y) \cap X$		
		$\Leftarrow T = \phi$			
	$(RatM =)$	$\Rightarrow$	$(\mu =)$		
	$Con(T \cup \overline{\overline{T'}}), T \vdash T' \not\Rightarrow$	$\Leftarrow (\mu dp)$	$X \subseteq Y, X \cap f(Y) \neq \emptyset \Rightarrow$		
	$\overline{\overline{T}} = \overline{\overline{\overline{T'}} \cup T}$	$\not\Leftarrow -(\mu dp)$	$f(X) = f(Y) \cap X$		
		$\Leftarrow T = \phi$			
	$(Log =')$	$\Rightarrow$	$(\mu =')$		
	$Con(\overline{\overline{T'}} \cup T) \Rightarrow$	$\Leftarrow (\mu dp)$	$f(Y) \cap X \neq \emptyset \Rightarrow$		
	$\overline{\overline{T \cup T'}} = \overline{\overline{\overline{T'}} \cup T}$	$\not\Leftarrow -(\mu dp)$	$f(Y \cap X) = f(Y) \cap X$		
		$\Leftarrow T = \phi$			
(DR)	$(Log \parallel)$	$\Rightarrow$	$(\mu \parallel)$		
$\alpha \vee \beta \mid\!\sim \gamma \Rightarrow$	$\overline{\overline{T \vee T'}}$ is one of	$\Leftarrow$	$f(X \cup Y)$ is one of		
$\alpha \mid\!\sim \gamma$ or $\beta \mid\!\sim \gamma$	$\overline{\overline{T}}$, or $\overline{\overline{T'}}$, or $\overline{\overline{T}} \cap \overline{\overline{T'}}$ (by (CCL))		$f(X)$, $f(Y)$ or $f(X) \cup f(Y)$		
	$(Log\cup)$	$\Rightarrow (\mu \subseteq) + (\mu =)$	$(\mu\cup)$		
	$Con(\overline{\overline{T'}} \cup T), \neg Con(\overline{\overline{T'}} \cup \overline{\overline{T}}) \Rightarrow$	$\Leftarrow (\mu dp)$	$f(Y) \cap (X - f(X)) \neq \emptyset \Rightarrow$		
	$\neg Con(\overline{\overline{T \vee T'}} \cup T')$	$\not\Leftarrow -(\mu dp)$	$f(X \cup Y) \cap Y = \emptyset$		
	$(Log\cup')$	$\Rightarrow (\mu \subseteq) + (\mu =)$	$(\mu\cup')$		
	$Con(\overline{\overline{T'}} \cup T), \neg Con(\overline{\overline{T'}} \cup \overline{\overline{T}}) \Rightarrow$	$\Leftarrow (\mu dp)$	$f(Y) \cap (X - f(X)) \neq \emptyset \Rightarrow$		
	$\overline{\overline{T \vee T'}} = \overline{\overline{T}}$	$\not\Leftarrow -(\mu dp)$	$f(X \cup Y) = f(X)$		
			$(\mu \in)$ $a \in X - f(X) \Rightarrow$ $\exists b \in X.a \notin f(\{a, b\})$		

(6) $A = B \parallel C$ will abbreviate $A = B$, or $A = C$, or $A = B \cup C$.

Further comments:

(1) (PR) is also called *infinite conditionalization*; We choose this name for its central role for preferential structures (PR) or (μPR).
(2) The system of rules (AND) (OR) (LLE) (RW) (SC) (CP) (CM) (CUM) is also called system P (for preferential). Adding $(RatM)$ gives the system R (for rationality or rankedness).
Roughly, smooth preferential structures generate logics satisfying system P, while ranked structures generate logics satisfying system R.

(3) A logic satisfying (REF), $(ResM)$, and (CUT) is called a *consequence relation*.
(4) (LLE) and (CCL) will hold automatically, whenever we work with model sets.
(5) (AND) is obviously closely related to filters, and corresponds to closure under finite intersections. (RW) corresponds to upward closure of filters.
More precisely, validity of both depend on the definition, and the direction we consider.
Given f and $(\mu \subseteq)$, $f(X) \subseteq X$ generates a principal filter, $\{X' \subseteq X : f(X) \subseteq X'\}$, with the definition, if $X = M(T)$, then $T \mid\sim \phi$ iff $f(X) \subseteq M(\phi)$. Validity of (AND) and (RW) are then trivial.
Conversely, we can define for $X = M(T)$
$\mathcal{X} := \{X' \subseteq X : \exists \phi (X' = X \cap M(\phi) \text{ and } T \mid\sim \phi)\}$.
(AND) then makes $\mathcal{X}$ closed under finite intersections, and (RW) makes $\mathcal{X}$ upward closed. This is in the infinite case usually not yet a filter, as not all subsets of X need to be definable this way. In this case, we complete $\mathcal{X}$ by adding all X'' such that there is $X' \subseteq X'' \subseteq X$, $X' \in \mathcal{X}$.
Alternatively, we can define
$\mathcal{X} := \{X' \subseteq X : \bigcap\{X \cap M(\phi) : T \mid\sim \phi\} \subseteq X'\}$.

(6) (SC) corresponds to the choice of a subset.
(7) (CP) is somewhat delicate, as it presupposes that the chosen model set is non-empty. This might fail in the presence of ever better choices, without ideal ones; the problem is addressed by the limit versions.
(8) (PR) is an infinitary version of one half of the deduction theorem: Let T stand for ϕ, T' for ψ, and $\phi \wedge \psi \mid\sim \sigma$, so $\phi \mid\sim \psi \rightarrow \sigma$, but $(\psi \rightarrow \sigma) \wedge \psi \vdash \sigma$.
(9) (CUM) (whose more interesting half in our context is (CM)) may best be seen as a normal use of lemmas: We have worked hard and found some lemmas. Now we can take a rest, and come back again with our new lemmas. Adding them to the axioms will neither add new theorems, nor prevent old ones from holding. (This is, of course, a meta-level argument concerning an object-level rule. But also object-level rules should — at least generally — have an intuitive justification, which will then come from a meta-level argument.)

Chapter 3
Defeasible Inheritance

We start with a look at defeasible or nonmonotonic inheritance systems (or networks, diagrams) as they are relatively simple, but already display many desirable properties. They are a kind of "toy system".

3.1 Summary

We discuss, among other things,

- the basic definition of a defeasible inheritance net,
- conflicts and their resolution, as illustrated by the Tweety and the Nixon diagram,
- the concept of preclusion, and on-path and off-path preclusion,
- the directly sceptical and based approach,
- validity of paths and results, and
- the properties important in our context, see Sect. 3.6.

3.2 Conceptual Analysis

Inheritance diagrams are deceptively simple. Their conceptually complicated nature is seen by, e.g. the fundamental difference between direct links and valid paths, and the multitude of existing formalisms, upward versus downward chaining, intersection of extensions versus direct scepticism, on-path versus off-path preclusion (or pre-emption), split validity versus total validity preclusion etc., to name a few, see the discussion in Sect. 3.5. Such a proliferation of formalisms usually hints at deeper problems on the conceptual side, i.e. the underlying ideas are ambiguous, and not sufficiently analysed. Therefore, any clarification and resulting reduction of possible

D.M. Gabbay and K. Schlechta, *A New Perspective on Nonmonotonic Logics*,
DOI 10.1007/978-3-319-46817-4_3

formalisms seems a priori to make progress. Such clarification will involve conceptual decisions, which need not be shared by all, they can only be suggestions. Of course, a proof that such decisions are correct is impossible, and so is its contrary.

For connections with reactive diagrams, big and small subsets and the systems P and R, information flow, and reasoning with prototypes, the reader is referred to Gabbay and Schlechta (2009).

For more on the subject, see, e.g. Thomason and Horty (1989), Thomason et al. (1986), Touretzky et al. (1991).

3.3 Basic Discussion

We give here an informal discussion. The reader unfamiliar with inheritance systems should consult in parallel Definitions 3.4.2 and 3.4.3. As there are many variants of the definitions, it seems reasonable to discuss them before a formal introduction, which, otherwise, would seem to pretend to be definite without being so.

3.3.1 (Defeasible or Nonmonotonic) Inheritance Networks or Diagrams

Nonmonotonic inheritance systems describe situations like "normally, birds fly", written $birds \rightarrow fly$. Exceptions are permitted, "normally penguins don't fly", $penguins \nrightarrow fly$.

Definition 3.3.1
A nonmonotonic inheritance net is a finite Directed Acyclic G raph (DAG), with two types of arrows or links, $\rightarrow$ and $\nrightarrow$, and labelled nodes. We will use Γ, etc. for such graphs, and σ, etc. for paths - the latter to be defined below.

Roughly (and to be made precise and modified below, we try to give here just a first intuition), $X \rightarrow Y$ means that "normal" elements of X are in Y, and $X \nrightarrow Y$ means that "normal" elements of X are not in Y. In a semi-quantitative set interpretation, we will read "most" for "normal", thus "most elements of X are in Y", "most elements of X are not in Y", etc. These are by no means the only interpretations, as we will see – we will use these expressions for the moment just to help the reader's intuition. We should add immediately a word of warning: "most" is here not necessarily, but only by default, transitive, in the following sense. In the Tweety diagram, see Diagram 3.3.1 below, most penguins are birds, most birds fly, but it is not the case that most penguins fly.

According to the set interpretation, we will also use informally expressions like $X \bigcap Y$, X-Y, $\boldsymbol{C}X$ — where $\boldsymbol{C}$ stands for set complement —, etc. But we will also use nodes informally as formulas, like $X \wedge Y$, $X \wedge \neg Y$, $\neg X$, etc. All this will only be used here as an appeal to intuition.

Nodes at the beginning of an arrow can also stand for individuals, so $Tweety \not\rightarrow fly$ means something like: "normally, Tweety will not fly". As always in non-monotonic systems, exceptions are permitted, so the soft rules "birds fly", "penguins don't fly", and (the hard rule) "penguins are birds" can coexist in one diagram, penguins are then abnormal birds (with respect to flying). The direct link $penguins \not\rightarrow fly$ will thus be accepted, or considered valid, but not the composite path $penguins \rightarrow birds \rightarrow fly$, by specificity — see below. This is illustrated by Diagram 3.3.1, where a stands for Tweety, c for penguins, b for birds, d for flying animals or objects.

(Remark: The arrows $a \rightarrow c$, $a \rightarrow b$, and $c \rightarrow b$ can also be composite paths — see below for the details.)

Diagram 3.3.1

Tweety diagram

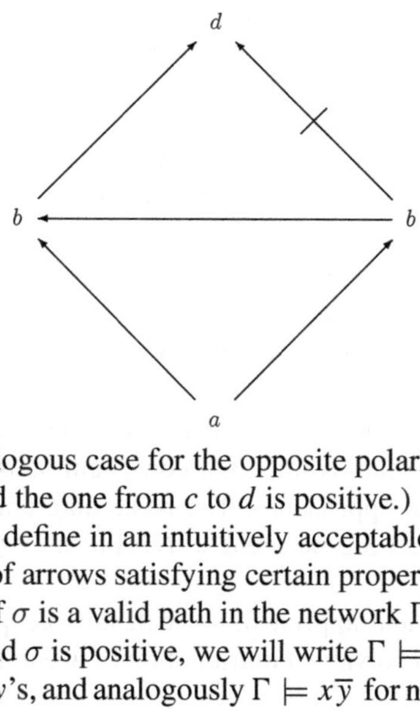

(Of course, there is an analogous case for the opposite polarity, i.e. when the arrow from b to d is negative, and the one from c to d is positive.)

The main problem is to define in an intuitively acceptable way a notion of valid path, , i.e. concatenations of arrows satisfying certain properties.

We will write $\Gamma \models \sigma$, if σ is a valid path in the network Γ, and if x is the origin, and y the endpoint of σ, and σ is positive, we will write $\Gamma \models xy$, i.e. we will accept the conclusion that x's are y's, and analogously $\Gamma \models x\overline{y}$ for negative paths. Note that we will not accept any other conclusions, only those established by a valid path, so many questions about conclusions have a trivial negative answer: there is obviously no path from x to y., e.g. there is no path from b to c in Diagram 3.3.1. Likewise, there are no disjunctions, conjunctions, etc. in our conclusions, and negation is present only in a strong form: "it is not the case that x's are normally y's" is not a possible conclusion, only "x's are normally not y's" is one. Also, possible contradictions are contained, there is no EFQ.

To simplify matters, we assume that for no two nodes $x, y \in \Gamma$ $x \to y$ and $x \not\to y$ are both in Γ, intuitively, that Γ is free from (hard) contradictions. This restriction is inessential for our purposes. We admit, however, soft contradictions, and preclusion, which allows us to solve some soft contradictions — as we already did in the penguins example. We will also assume that all arrows stand for rules with possibly exceptions, again, this restriction is not important for our purposes. Moreover, in the abstract treatment, we will assume that all nodes stand for (nonempty) sets, though this will not be true for all examples discussed.

This might be the place for a remark on absence of cycles. Suppose we also have a positive arrow from b to c in Diagram 3.3.1. Then, the concept of preclusion collapses, as there are now equivalent arguments to accept $a \to b \to d$ and $a \to c \not\to d$. Thus, if we do not want to introduce new complications, we cannot rely on preclusion to decide conflicts. It seems that this would change the whole outlook on such diagrams. The interested reader will find more on the subject in Antonelli (1997, 1999, 2005).

Inheritance networks were introduced in the 1980s (see, e.g. Touretzky et al. 1984, 1986, 1987), and exist in a multitude of more or less differing formalisms, see, e.g. Schlechta (1997) for a brief discussion. There still does not seem to exist a satisfying semantics for these networks. The authors' own attempt (Schlechta 1990) is an a posteriori semantics, which cannot explain or criticise or decide between the different formalisms. We will give here a conceptual analysis, which provides also at least some building blocks for a semantics, and a translation into (a modified version of) the language of small and big subsets, familiar from preferential structures, see Definition 5.3.2.3.

We will now discuss the two fundamental situations of contradictions, then give a detailed inductive definition of valid paths for a certain formalism so the reader has firm ground under his feet, and then present briefly some alternative formalisms.

As in all of nonmonotonic reasoning, the interesting questions arise in the treatment of contradictions and exceptions. The difference in quality of information is expressed by "preclusion" (or pre-emption) . The basic diagram is the Tweety diagram, see Diagram 3.3.1.

Unresolved contradictions give either rise to a branching into different extensions, which may roughly be seen as maximal consistent subsets, or to mutual cancellation in directly sceptical approaches. The basic diagram for the latter is the Nixon Diamond, see Diagram 3.3.2, where $a = Nixon$, $b = Quaker$, $c = Republican$, $d = pacifist$.

In the directly sceptical approach, we will not accept any path from a to d as valid, as there is an unresolvable contradiction between the two candidates.

Diagram 3.3.2

The Nixon Diamond

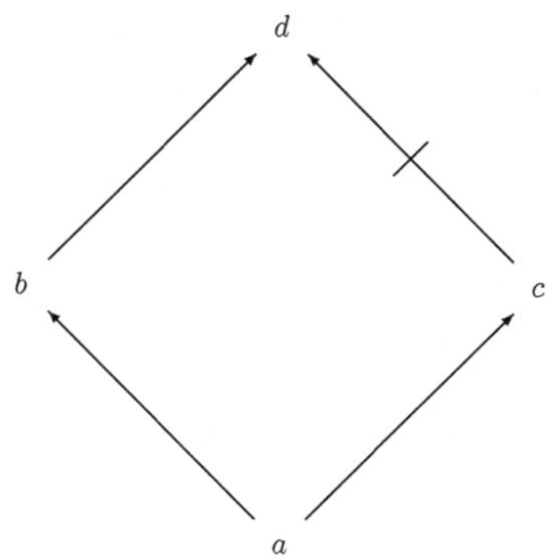

The extensions approach can be turned into an indirectly sceptical one, by forming first all extensions, and then taking the intersection of either the sets of valid paths, or of valid conclusions, see Makinson and Schlechta (1991) for a detailed discussion. See also Sect. 3.5 for more discussion on directly versus indirectly sceptical approaches.

In more detail:

3.3.2 *Preclusion*

In the above example, our intuition tells us that it is not admissible to conclude from the fact that penguins are birds, and that most birds fly that most penguins fly. The horizontal arrow $c \rightarrow b$ together with $c \not\rightarrow d$ bars this conclusion, it expresses specificity. Consequently, we have to define the conditions under which two potential paths neutralise each other, and when one is victorious. The idea is as follows: (1) We want to be sceptical, in the sense that we do not believe every potential path. We will not arbitrarily chose one either. (2) Our scepticism will be restricted, in the sense that we will often make well defined choices for one path in the case of conflict: (a) If a compound potential path is in conflict with a direct link, the direct link wins. (b) Two conflicting paths of the same type neutralise each other, as in the Nixon Diamond, where neither potential path will be valid. (c) More specific information will win over less specific one.

(It is essential in the Tweety diagram that the arrow $c \not\rightarrow d$ is a direct link, so it is in a way stronger than compound paths.) The arrows $a \rightarrow b$, $a \rightarrow c$, $c \rightarrow b$ can also be composite paths: The path from c to b (read $c \subseteq \cdots \subseteq b$, where $\subseteq$ stands here for soft inclusion), however, tells us, that the information coming from c is more specific (and thus considered more reliable), so the negative path from a to d via c will win over the positive one via b. The precise inductive definition will be given below. This concept is evidently independent of the length of the paths, $a \cdots \rightarrow c$

may be much longer than $a \cdots \to b$, so this is not shortest path reasoning (which has some nasty drawbacks, discussed, e.g. in Horty et al. 1987).

A final remark: Obviously, in some cases, it need not be specificity, which decides conflicts. Consider the case where Tweety is a bird, but a dead animal. Obviously, Tweety will not fly, here because the predicate "dead" is very strong and overrules many normal properties. When we generalise this, we might have a hierarchy of causes, where one overrules the other, or the result may be undecided. For instance, a falling object might be attracted in a magnetic field, but a gusty wind might prevent this, sometimes, with unpredictable results. This is then additional information (strength of cause), and this problem is not addressed directly in traditional inheritance networks, we would have to introduce a subclass "dead bird" — and subclasses often have properties of "pseudo-causes", being a penguin probably is not a "cause" for not flying, nor bird for flying, still, things change from class to subclass for a reason.

Before we give a formalism based on these ideas, we refine them, adopt one possibility (but indicate some modifications), and discuss alternatives later.

3.4 Directly Sceptical Split Validity Upward Chaining Off-Path Inheritance

Our approach will be directly sceptical, i.e. unsolvable contradictions result in the absence of valid paths, it is upward chaining, and split-validity for preclusions (discussed below, in particular in Sect. 3.5. We will indicate modifications to make it extension based, as well as for total validity preclusion. This approach is strongly inspired by classical work in the field by Horty, Thomason, Touretzky, and others, and we claim no priority whatever. If it is new at all, it is a very minor modification of existing formalisms.

For the reader's convenience, we give here a very short resume of these ideas: We consider only arrows as information, e.g. $a \to b$ will be considered information b valid at or for a. Valid composed positive paths will not be considered information in our sense. They will be seen as a way to obtain information, so a valid path $\sigma : x \cdots \to a$ makes information b accessible to x, and, secondly, as a means of comparing information strength, so a valid path $\sigma : a \cdots \to a'$ will make information at a stronger than information at a'. Valid negative paths have no function, we will only consider the positive initial part as discussed above, and the negative end arrow as information, but never the whole path.

Choosing direct scepticism is a decision beyond the scope of this section, and we just make it. (The fundamental difference between intersection of extensions and direct scepticism for defeasible inheritance was shown in Schlechta 1995.) See also Sect. 3.5 for more discussion.

We turn now to the announced variants as well as a finer distinction within the directly sceptical approach. Again, see also Sect. 3.5 for more discussion.

Our approach generates another problem, essentially that of the treatment of a mixture of contradictory and concordant information of multiple strengths or truth values. We bundle the decision of this problem with that for direct scepticism into a "plug-in" decision, which will be used in three approaches: the conceptual ideas, the inheritance algorithm, and the choice of the reference class for subset size (and implicitly also for the treatment as a prototype theory). It is thus well encapsulated, and independent from the context.

These decisions (but, perhaps to a lesser degree, (1)) concern a wider subject than only inheritance networks. Thus, it is not surprising that there are different formalisms for solving such networks, deciding one way or the other. But this multitude is not the fault of inheritance theory, it is only a symptom of a deeper question. We first give an overview for a clearer overall picture, and discuss them in detail below, as they involve sometimes quite subtle questions.

(1) Upward chaining against downward or double chaining.

(2.1) Off-path against on-path preclusion.

(2.2) Split validity preclusion against total validity preclusion.

(3) Direct scepticism against intersection of extensions.

(4) Treatment of mixed contradiction and preclusion situations, no preclusion by paths of the same polarity.

(1) When we interpret arrows as causation (in the sense that $X \rightarrow Y$ expresses that condition X usually causes condition Y to result), this can also be seen as a difference in reasoning from cause to effect versus backward reasoning, looking for causes for an effect. (A word of warning: There is a well-known article (Selman and Levesque 1989) from which a superficial reader might conclude that upward chaining is tractable, and downward chaining is not. A more careful reading reveals that, on the negative side, the authors only show that double chaining is not tractable.) We will adopt upward chaining in all our approaches.

(2.1) and (2.2) Both are consequences of our view — to be discussed below in Sect. 4.3 — to see valid paths also as an absolute comparison of truth values, independent of reachability of information. Thus, in Diagram 3.3.1, the comparison between the truth values "penguin" and "bird" is absolute, and does not depend on the point of view "Tweety", as it can in total validity preclusion - if we continue to view preclusion as a comparison of information strength (or truth value). This question of absoluteness transcends obviously inheritance networks. Our decision is, of course, again uniform for all our approaches.

(3) This point, too, is much more general than the problems of inheritance. It is, among other things, a question of whether only the two possible cases (positive and negative) may hold, or whether there might be still other possibilities.

(4) This concerns the treatment of truth values in more complicated situations, where we have a mixture of agreeing and contradictory information. Again, this problem reaches far beyond inheritance networks.

We will group (3) and (4) together in one general, "plug-in" decision, to be found in all approaches we discuss.

Definition 3.4.1
This is an informal definition of a plug-in decision:

We describe now more precisely a situation which we will meet in all contexts discussed, and whose decision goes beyond our problem - thus, we have to adopt one or several alternatives, and translate them into the approaches we will discuss. There will be one global decision, which is (and can be) adapted to the different contexts.

Suppose we have information about ϕ and ψ, where ϕ and ψ are presumed to be independent — in some adequate sense.

Suppose then that we have information sources $A_i : i \in I$ and $B_j : j \in J$, where the A_i speak about ϕ (they say ϕ or $\neg\phi$), and the B_j speak about ψ in the same way. Suppose further that we have a partial, not necessarily transitive (!), ordering $<$ on the information sources A_i and B_j together. $X < Y$ will say that X is better (intuition: more specific) than Y. (The potential lack of transitivity is crucial, as valid paths do not always concatenate to valid paths — just consider the Tweety diagram.)

We also assume that there are contradictions, i.e. some A_i say ϕ, some $\neg\phi$, likewise for the B_j — otherwise, there are no problems in our context.

We can now take several approaches, all taking contradictions and the order $<$ into account.

- (P1) We use the global relation $<$, and throw away all information coming from sources of minor quality, i.e. if there is X such that $X < Y$, then no information coming from Y will be taken into account. Consequently, if Y is the only source of information about ϕ, then we will have no information about ϕ. This seems an overly radical approach, as one source might be better for ϕ, but not necessarily for ψ, too.
  If we adopt this radical approach, we can continue as below, and can even split in analogue ways into (P1.1) and (P1.2), as we do below for (P2.1) and (P2.2).
- (P2) We consider the information about ϕ separately from the information about ψ. Thus, we consider for ϕ only the A_i, for ψ only the B_j. Take now, e.g. ϕ and the A_i. Again, there are (at least) two alternatives.
  - (P2.1) We eliminate again all sources among the A_i for which there is a better $A_{i'}$, irrespective of whether they agree on ϕ or not.
    (a) If the sources left are contradictory, we conclude nothing about ϕ, and accept for ϕ none of the sources. (This is a directly sceptical approach of treating unsolvable contradictions, following our general strategy.)
    (b) If the sources left agree for ϕ,, i.e. all say ϕ, or all say $\neg\phi$, then we conclude ϕ (or $\neg\phi$), and accept for ϕ all the remaining sources.
  - (P2.2) We eliminate again all sources among the A_i for which there is a better $A_{i'}$, but only if A_i and $A_{i'}$ have contradictory information. Thus, more sources may survive than in approach (P2.1).
    We now continue as for (P2.1):
    (a) If the sources left are contradictory, we conclude nothing about ϕ, and accept for ϕ none of the sources.
    (b) If the sources left agree for ϕ,, i.e. all say ϕ, or all say $\neg\phi$, then we conclude ϕ (or $\neg\phi$), and accept for ϕ all the remaining sources.

The difference between (P2.1) and (P2.2) is illustrated by the following simple example. Let $A < A' < A''$, but $A \not< A''$ (recall that $<$ is not necessarily transitive), and $A \models \phi$, $A' \models \neg\phi$, $A'' \models \neg\phi$. Then (P2.1) decides for ϕ (A is the only survivor), (P2.2) does not decide, as A and A'' are contradictory, and both survive in (P2.2).

There are arguments for and against either solution: (P2.1) gives a uniform picture, more independent from ϕ, (P2.2) gives more weight to independent sources, it "adds" information sources, and thus gives potentially more weight to information from several sources. (P2.2) seems more in the tradition of inheritance networks, so we will consider it in the further development.

The reader should note that our approach is quite far from a fixed point approach in two ways: First, fixed point approaches seem more appropriate for extensions-based approaches, as both try to collect a maximal set of uncontradictory information. Second, we eliminate information when there is better, contradicting information, even if the final result agrees with the first. This, too, contradicts in spirit the fixed point approach.

After these preparations, we turn to a formal definition of validity of paths.

3.4.1 The Definition of $\models$ (i.e. of Validity of Paths)

All definitions are relative to a fixed diagram Γ. The notion of degree will be defined relative to all nodes of Γ, as we will work with split validity preclusion, so the paths to consider may have different origins. For simplicity, we consider Γ to be just a set of points and arrows, thus, e.g. $x \to y \in \Gamma$ and $x \in \Gamma$ are defined, when x is a point in Γ, and $x \to y$ an arrow in Γ. Recall that we have two types of arrows, positive and negative ones.

We first define generalised and potential paths, then the notion of degree, and finally validity of paths, written $\Gamma \models \sigma$, if σ is a path, as well as $\Gamma \models xy$, if $\Gamma \models \sigma$ and $\sigma : x \cdots \to y$.

Definition 3.4.2

(1) Generalised paths:

A generalised path is an uninterrupted chain of positive or negative arrows pointing in the same direction, more precisely:

$x \to p \in \Gamma \to x \to p$ is a generalised path,

$x \not\to p \in \Gamma \to x \not\to p$ is a generalised path.

If $x \cdots \to p$ is a generalised path, and $p \to q \in \Gamma$, then $x \cdots \to p \to q$ is a generalised path,

if $x \cdots \to p$ is a generalised path, and $p \not\to q \in \Gamma$, then $x \cdots \to p \not\to q$ is a generalised path.

(2) Concatenation:

If σ and τ are two generalised paths, and the end point of σ is the same as the starting point of τ, then $\sigma \circ \tau$ is the concatenation of σ and τ.

(3) Potential paths (pp):

A generalised path, which contains at most one negative arrow, and this at the end, is a potential path. If the last link is positive, it is a positive potential path, if not, a negative one.

(4) Degree: As already indicated, we shall define paths inductively. As we do not admit cycles in our systems, the arrows define a well-founded relation on the vertices. Instead of using this relation for the induction, we shall first define the auxiliary notion of degree, and do induction on the degree. Given a node x (the origin), we need a (partial) mapping f from the vertices to natural numbers such that $p \to q$ or $p \not\to q \in \Gamma$ implies $f(p) < f(q)$, and define (relative to x):

Let σ be a generalised path from x to y, then $deg_{\Gamma,x}(\sigma) := deg_{\Gamma,x}(y) :=$ the maximal length of any generalised path parallel to σ,, i.e. beginning in x and ending in y.

Definition 3.4.3
Inductive definition of $\Gamma \models \sigma$:

Let σ be a potential path.

- Case I:
  σ is a direct link in Γ. Then $\Gamma \models \sigma$
  (Recall that we have no hard contradictions in Γ.)
- Case II:
  σ is a compound potential path, $deg_{\Gamma,a}(\sigma) = n$, and $\Gamma \models \tau$ is defined for all τ with degree less than n - whatever their origin and endpoint.
- Case II.1:
  Let σ be a positive pp. $x \cdots \to u \to y$, let $\sigma' := x \cdots \to u$, so $\sigma = \sigma' \circ u \to y$
  Then, informally, $\Gamma \models \sigma$ iff

(1) σ is a candidate by upward chaining,
(2) σ is not precluded by more specific contradicting information,
(3) all potential contradictions are themselves precluded by information contradicting them.

Note that (2) and (3) are the translation of (P2.2) in Definition 3.4.1.
Formally, $\Gamma \models \sigma$ iff

(1) $\Gamma \models \sigma'$ and $u \to y \in \Gamma$.
(The initial segment must be a path, as we have an upward chaining approach. This is decided by the induction hypothesis.)
(2) There are no v, τ, τ' such that $v \not\to y \in \Gamma$ and $\Gamma \models \tau := x \cdots \to v$ and $\Gamma \models \tau' := v \cdots \to u$. ($\tau$ may be the empty path, i.e. $x = v$.)
(σ itself is not precluded by split validity preclusion and a contradictory link. Note that $\tau \circ v \not\to y$ need not be valid, it suffices that it is a better candidate (by τ').)
(3) all potentially conflicting paths are precluded by information contradicting them: For all v and τ such that $v \not\to y \in \Gamma$ and $\Gamma \models \tau := x \cdots \to v$ (i.e. for all potentially conflicting paths $\tau \circ v \not\to y$) there is z such that $z \to y \in \Gamma$ and either
$z = x$

(the potentially conflicting pp. is itself precluded by a direct link, which is thus valid)
or
there are $\Gamma \models \rho := x \cdots \to z$ and $\Gamma \models \rho' := z \cdots \to v$ for suitable ρ and ρ'.

- Case II.2: The negative case, i.e. σ a negative pp. $x \cdots \to u \not\to y$, $\sigma' := x \cdots \to u$, $\sigma = \sigma' \circ u \not\to y$ is entirely symmetrical.

Remark 3.4.1
The following remarks all concern preclusion.
(1) Thus, in the case of preclusion, there is a valid path from x to z, and z is more specific than v, so $\tau \circ v \not\to y$ is precluded. Again, $\rho \circ z \to y$ need not be a valid path, but it is a better candidate than $\tau \circ v \not\to y$ is, and as $\tau \circ v \not\to y$ is in simple contradiction, this suffices.
(2) Our definition is stricter than many popular ones, in the following sense: We require—according to our general picture to treat only direct links as information—that the preclusion "hits" the precluded path at the end, i.e. $v \not\to y \in \Gamma$, and ρ' hits $\tau \circ v \not\to y$ at v. In other definitions, it is possible that the preclusion hits at some v',, which is somewhere on the path τ, and not necessarily at its end. For instance, in the Tweety Diagram, see Diagram 3.3.1 (p. xx), if there were a node b' between b and d, we will need the path $c \to b \to b'$ to be valid, (obvious) validity of the arrow $c \to b$ will not suffice.
(3) If we allow ρ to be the empty path, then the case $z = x$ is a subcase of the present one.
(4) Our conceptual analysis has led to a very important simplification of the definition of validity. If we adopt on-path preclusion, we have to remember all paths which led to the information source to be considered: In the Tweety diagram, we have to remember that there is an arrow $a \to b$, it is not sufficient to note that we somehow came from a to b by a valid path, as the path $a \to c \to b \to d$ is precluded, but not the path $a \to b \to d$. If we adopt total validity preclusion, see also Sect. 3.5 for more discussion, we have to remember the valid path $a \to c \to b$ to see that it precludes $a \to c \to d$. If we allow preclusion to "hit" below the last node, we also have to remember the entire path which is precluded. Thus, in all those cases, whole paths (which can be very long) have to be remembered, but NOT in our definition.

We only need to remember (consider the Tweety diagram):
(a) we want to know if $a \to b \to d$ is valid, so we have to remember a, b, d. Note that the (valid) path from a to b can be composed and very long.
(b) we look at possible preclusions, so we have to remember $a \to c \not\to d$, again the (valid) path from a to c can be very long.
(c) we have to remember that the path from c to b is valid (this was decided by induction before).

So in all cases (the last one is even simpler), we need only remember the starting node, a (or c), the last node of the valid paths, b (or c), and the information $b \to d$

or $c \not\rightarrow d$—, i.e. the size of what has to be recalled is ≤ 3. (Of course, there may be many possible preclusions, but in all cases we have to look at a very limited situation, and not arbitrarily long paths.)

We take a fast look forward to Sect. 4.3, where we describe diagrams as information and its transfer, and nodes also as truth values. In these terms— and the reader is asked to excuse the digression—we may note above point (a) as $a \Rightarrow_b d$—expressing that, seen from a, d holds with truth value b, (b) as $a \Rightarrow_c \neg d$, (c) as $c \Rightarrow_c b$—and this is all we need to know. □

We indicate here some modifications of the definition without discussion, which is to be found below.

(1) For on-path preclusion only: Modify condition (2) in Case II.1 to: (2′) There is no v on the path σ (i.e. $\sigma : x \cdots \rightarrow v \cdots \rightarrow u$) such that $v \not\rightarrow y \in \Gamma$.

(2) For total validity preclusion: Modify condition (2) in Case II.1 to: (2′) There are no v, τ, τ' such that $v \not\rightarrow y \in \Gamma$ and $\tau := x \cdots \rightarrow v$ and $\tau' := v \cdots \rightarrow u$ such that $\Gamma \models \tau \circ \tau'$.

(3) For extension based approaches: Modify condition (3) in Case II.1 as follows: (3′) If there are conflicting paths, which are not precluded themselves by contradictory information, then we branch recursively (i.e. for all such situations) into two extensions, one, where the positive non-precluded paths are valid, one, where the negative non-precluded paths are valid.

Definition 3.4.4
Finally, define $\Gamma \models xy$ iff there is $\sigma : x \rightarrow y$ s.th. $\Gamma \models \sigma$, likewise for $x\overline{y}$ and $\sigma : x \cdots \not\rightarrow y$.

Diagram 3.4.1 shows the most complicated situation for the positive case.

Diagram 3.4.1

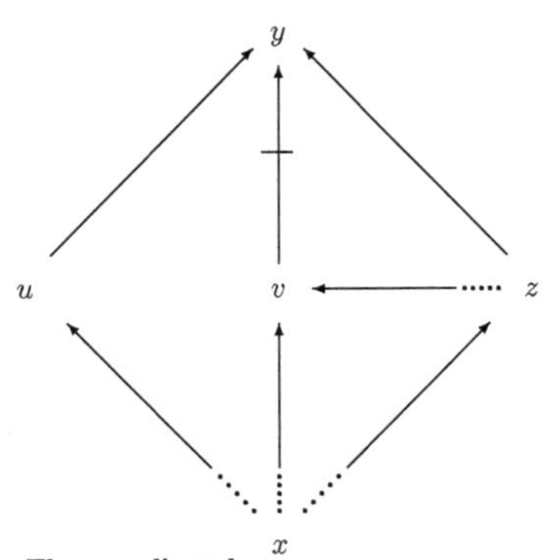

The complicated case

We have to show now that the above approach corresponds to the preceeding discussion.

Fact 3.4.2
The above definition and the informal one outlined in Definition 3.4.1 correspond, when we consider valid positive paths as access to information and comparison of information strength as indicated at the beginning of Sect. 3.4.

3.5 Review of Other Approaches and Problems

We now discuss shortly in more detail some of the differences between various major definitions of inheritance formalisms.

Diagram 6.8, in Schlechta (1997) (which is probably due to folklore of the field) shows requiring downward chaining would be wrong. We repeat it here, see Diagram 3.5.1.

Diagram 3.5.1

The problem of downward chaining

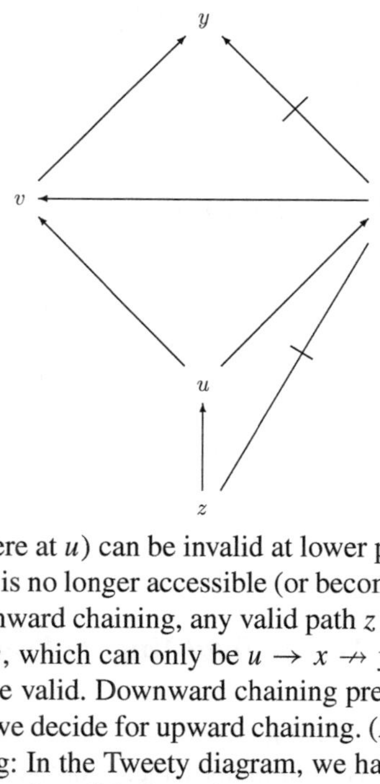

Preclusions valid above (here at u) can be invalid at lower points (here at z), as part of the relevant information is no longer accessible (or becomes accessible). We have $u \to x \not\to y$ valid, by downward chaining, any valid path $z \to u \ldots .y$ has to have a valid final segment $u \ldots .y$, which can only be $u \to x \not\to y$, but intuition says that $z \to u \to v \to y$ should be valid. Downward chaining prevents such changes, and thus seems inadequate, so we decide for upward chaining. (Already preclusion itself underlines upward chaining: In the Tweety diagram, we have to know that the path from bottom up to penguins is valid. So at least some initial subpaths have to be known—we need upward chaining.) (The rejection of downward chaining seems at first sight to be contrary to the intuitions carried by the word "inheritance".)

3.5.1 Extension-Based Versus Directly Sceptical Definitions

As this distinction has already received detailed discussion in the literature, we shall be very brief here. An extension of a net is essentially a maximally consistent and in some appropriate sense reasonable subset of all its potential paths. This can of course be presented either as a liberal conception (focussing on individual extensions) or as a sceptical one (focussing on their intersection—or, the intersection of their conclusion sets). The seminal presentation is that of Touretzky (1986), as refined by Sandewall (1986). The directly sceptical approach seeks to obtain a notion of sceptically accepted path and conclusion, but without detouring through extensions. Its classic presentation is that of Horty et al. (1987). Even while still searching for fully adequate definitions of either kind, we may use the former approach as a useful "control" on the latter. For if we can find an intuitively possible and reasonable extension supporting a conclusion $x\overline{y}$, whilst a proposed definition for a directly sceptical notion of legitimate inference yields xy as a conclusion, then the counter exemplary extension seems to call into question the adequacy of the directly sceptical construction, more readily than inversely. It has been shown in Schlechta (1995) that the intersection of extensions is fundamentally different from the directly sceptical approach.

From now on, all definitions considered shall be (at least) upward chaining.

3.5.2 On-Path Versus Off-Path Preclusion

This is a rather technical distinction, discussed in Touretzky et al. (1987). Briefly, a path σ: $x \to \cdots \to y \to \cdots \to z$ and a direct link $y \not\to u$ is an off-path preclusion of τ: $x \to \cdots \to z \to \cdots \to u$, but an on-path preclusion only iff all nodes of τ between x and z lie on the path σ.

For instance, in the Tweety diagram, the arrow $c \not\to d$ is an on-path preclusion of the path $a \to c \to b \to d$, but the paths $a \to c$ and $c \to b$, together with $c \not\to d$, is an (split validity) off-path preclusion of the path $a \to b \to d$.

3.5.3 Split Validity Versus Total Validity Preclusion

Consider again a preclusion $\sigma : u \to \cdots \to x \to \cdots \to v$, and $x \not\to y$ of $\tau : u \to \cdots \to v \to \cdots \to y$. Most definitions demand for the preclusion to be effective—, i.e. to prevent τ from being accepted—that the total path σ is valid. Some (Geffner and Verma 1989; Krishnaprasad and Kifer 1989; Krishnaprasad et al. 1989a, b) content themselves with the combinatorially simpler separate (split) validity of the lower and upper parts of σ: $\sigma' : u \to \cdots \to x$ and $\sigma'' : x \to \cdots \to v$. In Diagram 3.5.2, taken from Schlechta (1997), the path $x \to w \to v$ is valid, so is $u \to x$, but not the whole preclusion path $u \to x \to w \to v$.

Diagram 3.5.2

Split vs. total validity preclusion

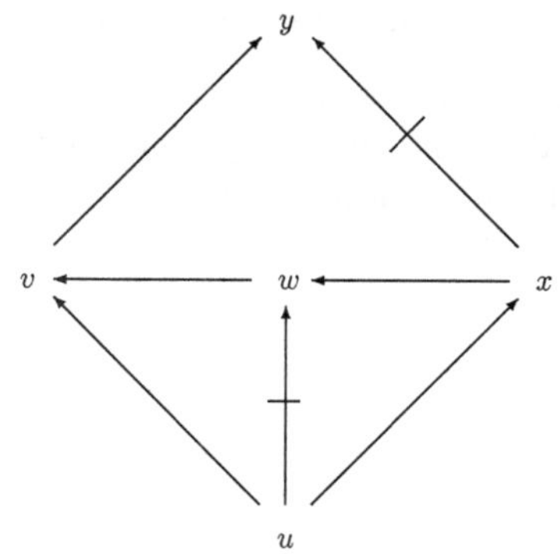

Thus, split validity preclusion will give here the definite result $u\overline{y}$. With total validity preclusion, the diagram has essentially the form of a Nixon Diamond.

3.6 Discussion of Their Properties in Our Context

Defeasible inheritance networks show:

(1) Homogeneousness and rare influence changes:

(1.1) subclasses behave the same way as the original classes, unless explicitly said that they don't,
(1.2) subideal subsets still preserve their unquestioned properties, so we have graceful degradation,
(1.3) we have transitivity by default,
(1.4) specificity decides some conflicts.

(2) They have the following shortcomings:

(2.1) the language is very poor, there is no "or", and the negation is very strong: we cannot negate a property without asserting the opposite,
(2.2) it is not clear whether one should adopt a downward or upward reasoning formalism,
(2.3) if $A \rightarrow B$ and $C \not\rightarrow B$, we might expect something like $A \not\rightarrow C$, but this is not necessarily the case—they seem not complete under reasonable interpretations,
(2.4) it is not clear if $A \rightarrow B$ stands for something like "most A's are B's" or "we have information that A's are B's",
(2.5) (related) it is not clear whether the intersection of extensions or the directly sceptical one is the better approach, (If A is a singleton, then the latter makes

only sense if we take the "information" approach. Nixon cannot mostly be a pacifist and a hawk.)

(2.6) the question of validity of defaults is avoided, as we may always postulate that there are unmentioned subclasses ("dark matter"), for which they hold.

(3) We will see in the coming sections that preferential structures almost solve the issues presented by inheritance systems, and the formers' problems are, conversely, solved by inheritance structures.

Chapter 4
Reiter Defaults and Autoepistemic Logic

4.1 Reiter Defaults

This is a very short introduction to Reiter defaults, see Reiter (1980). We modify notation slightly, and use some simplifications and facts due to Sergot (2010). All material is due to Reiter (this will not be mentioned) or to Sergot, this will be mentioned.

A Reiter default has the syntactic form $(\alpha : \beta_1, \ldots, \beta_m / \gamma)$, where α, γ, and the β_i are classical (propositional or first order) formulas, with the intuitive meaning: if α holds, and we may consistently assume all β_i (separately!), then γ should hold, too.

$(: \beta / \gamma)$ is an abbreviation for $(true : \beta / \gamma)$, etc.

For instance, if we know that Tweety is a bird, and we may assume that it is not a penguin, then Tweety should (be able to) fly.

Reiter starts the formal development with closed formulas, i.e. propositional formulas, or FOL formulas without free variables, and treats open formulas via Skolemization.

We just give the first few steps of his work, to make our text self-contained.

δ, etc. will range over defaults, and D will denote sets of defaults, W sets of classical formulas (which are assumed to be true), and Δ a pair (D, W).

For the moment, we assume that all formulas are closed.

To avoid confusion about the sets relative to which the constructions are carried out, we note them in upper indices, whenever this seemed helpful.

Definition 4.1.1

(Due to Sergot) Fix a set S of classical formulas, and D a set of defaults. Then D^S $:= \{\delta = (\alpha : \beta_1, \ldots, \beta_m / \gamma) \in D$: $\neg\beta_1, \ldots, \neg\beta_m \notin S\}$ is the set of defaults $\delta \in D$, applicable in S, i.e. which are not prevented from firing by S.

Thus, $\delta \in D^S$ is equivalent in S to a simple rule α / γ.

D.M. Gabbay and K. Schlechta, *A New Perspective on Nonmonotonic Logics*,
DOI 10.1007/978-3-319-46817-4_4

Fact 4.1.1
(Sergot) Such rules are not equivalent to classical (material) implication, as they allow neither reasoning by cases, nor contraposition.
(1) Let $R := \{\alpha/\gamma, \beta/\gamma\}$. Then γ will not be a consequence of $\alpha \vee \beta$ under R (in the obvious sense).
(2) Let $R := \{\alpha/\gamma\}$. Then $\neg\alpha$ will not be a consequence of $\neg\gamma$ under R.

Fact 4.1.2
Let (D, W) be given, and S a fixed set of classical formulas.
Let Σ^S be the set of sets of formulas X such that
(D1) $W \subseteq X$
(D2) $\overline{X} = X$
(D3) If $\delta = (\alpha : \beta_1, \ldots, \beta_m/\gamma) \in D^S$, $\alpha \in X$, then $\gamma \in X$.
Let $X_i \in \Sigma^S$, then $\bigcap X_i \in \Sigma^S$.

Proof
Trivial. □

Thus, the following definition is justified:

Definition 4.1.2
Given (D, W) and S, Σ^S as above, define $\Gamma(S)$:= the $\subseteq$-smallest $X \in \Sigma^S$, where Σ^S is as defined above—alternatively, $\Gamma(S) := \bigcap \Sigma^S$.

Note, and this is important, all X are formed relative to the same D^S, and not to D^X. This is illustrated below by Example 4.1.1.

Definition 4.1.3
Let (D, W) and S, Γ as above. S is called an extension for (D, W), iff $\Gamma(S) = S$.

Example 4.1.1
(Sergot) Let $W := \{p\}$, $D := \{p : q/q\}$. If we allow D^X to vary, then $E := \overline{\{p, \neg q\}}$ will be an extension: $W \subseteq E$, $E = \overline{E}$, and, as $D^E = \emptyset$, E is closed under D^E. The only other possible candidate, $E' := \overline{\{p\}}$ is, under the modified condition, not an extension, as $D^{E'} = \{p/q\}$, and E' is not closed under $D^{E'}$ (but under D^E, of course).

Definition 4.1.4
Fix (D, W) and E a set of classical formulas.
$E_0^E := W$
$E_{i+1}^E := \overline{E_i^E} \cup \{\gamma: (\alpha : \beta_1, \ldots, \beta_m/\gamma) \in D^E, \alpha \in E_i^E\}$
$(0 \leq i < \omega)$
$E_\omega^E := \bigcup\{E_i^E : i < \omega\}$.

Fact 4.1.3
Let $E_{\omega+1}^E$ be defined from E_ω^E as E_{i+1}^E is defined from E_i^E. Then $E_{\omega+1}^E = E_\omega^E$, irrespective of the size of the language and the number of defaults, they may both be uncountable. Thus, we can stop the development at stage ω.

Proof
This is trivial by compactness of classical logic. No new α appears in E^E_ω.

Fact 4.1.4
Fix (D, W) and E a set of classical formulas, let E^E_i, $i \leq \omega$ as above. Then
$(D1')$ $W \subseteq E^E_\omega$
$(D2')$ $E^E_\omega = \overline{E^E_\omega}$
$(D3')$ If $(\alpha : \beta_1, \ldots, \beta_m/\gamma) \in D^E$, $\alpha \in E^E_\omega$, then $\gamma \in E^E_\omega$.

Proof
$(D1')$ is trivial, $(D2')$ holds by compactness of classical logic, and $(D3')$ by the inductive construction (γ in the next step after α).

Fact 4.1.5
Fix (D, W) and E a set of classical formulas. Define the E^E_i as above, in Definition 4.1.4, $i \leq \omega$.
Then $\Gamma(E) = E^E_\omega$.

Proof
By Fact 4.1.4, $E^E_\omega \in \Sigma^E$ as defined in Fact 4.1.2. By minimality of $\Gamma(E)$, $\Gamma(E) \subseteq E^E_\omega$.
It remains to show $E^E_i \subseteq X$ for all i and all $X \in \Sigma^E$.
$i = 0$: Trivial by definition of E^E_0 and condition (D1) for Σ^E. $i \to i + 1$: Assume $E^E_i \subseteq X$ for all $X \in \Sigma^E$. Let $\gamma \in E^E_{i+1}$. If $\gamma \in \overline{E^E_i}$, then by induction hypothesis and $X = \overline{X}$, $\gamma \in X$. If γ is introduced by some $\delta \in D^E$, $\delta = (\alpha : \beta_1, \ldots, \beta_m/\gamma)$, then $\alpha \in E^E_i$, so $\gamma \in X$ by (D3).

Corollary 4.1.6
Fix (D, W) and E a set of classical formulas, let E^E_i as above. Then E is an extension iff $E = E^E_\omega$.

Proof
If E is an extension, then $E = \Gamma(E)$. By Fact 4.1.5, $\Gamma(E) = E^E_\omega$. If $E = E^E_\omega$, then again by Fact 4.1.5, $E = \Gamma(E)$.

Example 4.1.2
$W := \{B \to (\neg A \wedge \neg C)\}$, $D = \{(: A/A), (: B/B), (: C/C)\}$ has two extensions, $E_1 := \overline{W \cup \{A, C\}}$, $E_2 := \overline{W \cup \{B\}}$.

Formulas with free variables are treated as follows: We introduce Skolem constants as usual, but also for defaults, like $(: \exists x\phi(x)/\exists x\phi(x))$, and let the variables range over all such constants. Consequently, a default like $(: \phi(x)/\phi(x))$ will make ϕ true for as many elements as consistently possible.

4.2 Autoepistemic Logic

Autoepistemic logic is based on the assumption that we know exceptions. Birds fly, and unless we know that a certain bird does not fly, then it will fly. Thus, we reason explicitely with our knowledge.

Autoepistemic logic was introduced by Moore (1985), who also gave a fixed point semantics (with various types of fixed points), and, later, a possible worlds semantics, Moore (1984). The latter is a set of classical models, where every model is accessible from every other model, and $L\phi$ (read: we know that ϕ) is true (in the whole structure and in every model) iff ϕ holds in every model. Thus, tautologies are always known, contradictions are never known. If $\phi \vee \phi'$ is true in every model, but in some models only ϕ, in some only ϕ', then $\phi \vee \phi'$ is known, but neither ϕ nor ϕ' are known.

Gottlob (1993), has translated default logic into autoepistemic logic, Denecker et al. (2011), have interpreted Reiter's default logic as an autoepistemic logic. Often, a Reiter default is read 'most x are y', but this is not always justified, as the perfectly 'legal' default theory $\{: \phi/\phi,\ : \neg\phi/\neg\phi\}$ shows. They argue that it is better to see Reiter defaults as rules involving auto-theoremhood: the default $: \neg\psi/\phi$ means that we should conclude ϕ, unless we can conclude ψ. (Of course, above example does not make much sense in this interpretation either.) This interpretation is also justified by the definition of extensions for Reiter defaults.

4.3 Discussion of the Properties in Our Context

We do not intend to discuss all problems with Reiter defaults, nor all attempted improvements. We want to show where our main points present themselves in this context.

First, there need not be a core, they have no consistency criterion.

Example 4.3.1

(1) Consider $W := \{p\}$, $D := \{(: \neg p/\neg p)\}$. Of course, the default will never fire.
(2) Or, consider $W := \{\forall x\phi(x)\}$, $D := \{(: \neg\phi(x)/\neg\phi(x))\}$.
If a default is to express something like 'normally ψ', then it fails.
If a default is to express something like 'for efficiency reasons, assume ψ' then it fails, too. Considering to apply those defaults is a waste of resources.
(3) Let $W := \emptyset$, $D := \{(: p/p), (\neg p/\neg p)\}$ is a formally sound default theory, but we do not see any meaning in it, it should be inconsistent.

Other well-known problems include the fact that they are not closed under 'or'. If $\phi : /\psi$, $\phi' : /\psi$, but we know only $\phi \vee \phi'$, then we cannot conclude ψ. Preferential systems do not have these problems. On the other hand, Reiter defaults show, e.g. graceful degradation, as the failure of one default might still preserve other defaults. Subsets are treated reasonably, too.

In our approach, see Chap. 11, we will combine the advantages of rule-based systems like inheritance networks or Reiter defaults with those of model choice systems, like preferential structures. Note that our approach has

(1) a 'core', where the valid defaults have to hold, i.e. we guarantee that the set of defaults is consistent,
(2) we enumerate the known violations (penguins, etc.),
(3) we leave room for 'surprise exceptions', but treat them differently (they sit higher up, are more exceptional), so we differentiate between the 'immediately known' exceptions, and those we 'discover' later, and thus distinguish different epistemic levels,
(4) and finally treat 'uneventful' subsets in the intuitive way: nothing changes, except by losing classical models.

Chapter 5
Preferential Structures and Related Concepts

5.1 Summary

This chapter has two sections, one on preferential structures themselves, and one on the corresponding laws about (abstract) size. The first is the structural semantics, the second the corresponding algebraic semantics, see Sect. 1.1.8 for a discussion.

We discuss among other things

- for preferential structures:
  - the minimal variant with and without copies,
  - the logic defined by such structures,
  - the smoothness and rankedness properties of the relation,
  - a summary of representation results,
  - the limit variant, with and without copies,
  - the logic defined by the limit variant, and its properties.
- for laws about abstract size:
  - "most" as a weak filter, with corresponding generalised quantifier ∇ in a first-order setting,
  - additive laws about size, and their corresponding logical properties,
  - multiplicative laws about size, and their corresponding logical properties, for subsets and subspaces, and
  - Hamming relations and distances, and their relation to laws about size, and to theory revision.

D.M. Gabbay and K. Schlechta, *A New Perspective on Nonmonotonic Logics*,
DOI 10.1007/978-3-319-46817-4_5

5.2 Preferential Structures

Preferential structures are among the best examined semantics for non-monotonic logics. For historical remarks and various results, see, e.g., Schlechta (2004) and Gabbay and Schlechta (2009). We omit proofs, and refer the reader to former publications, e.g., Gabbay and Schlechta (2011).

For more on the subject, see, e.g., Bossu and Siegel (1985), Schlechta (1992), Schlechta (1996), Lehmann and Magidor (1992).

5.2.1 The Minimal Variant

Definition 5.2.1

Fix $U \neq \emptyset$, and consider arbitrary X. Note that this X has not necessarily anything to do with U, or the $\mathcal{U}$ below. Thus, the functions $\mu_{\mathcal{M}}$ below are in principle functions from V to V, where V is the set-theoretical universe we work in.

Note that we work here often with copies of elements (or models). In other areas of logic, most authors work with valuation functions. Both definitions—copies and valuation functions—are equivalent; a copy $\langle x, i\rangle$ can be seen as a state $\langle x, i\rangle$ with valuation x. In the beginning of research on preferential structures, the notion of copies was widely used, whereas, e.g., Kraus et al. (1990) used that of valuation functions. There is perhaps a weak justification for the former terminology. In modal logic, even if two states have the same valid classical formulas, they might still be distinguishable by their valid modal formulas. But this depends on the modality being in the object language. In most work on preferential structures, the consequence relation is outside the object language, so different states with same valuation are in a stronger sense copies of each other.

(1) *Preferential models* or *structures.*

(1.1) The version without copies:
A pair $\mathcal{M} := \langle U, \prec\rangle$ with U an arbitrary set and $\prec$ an arbitrary binary relation on U is called a *preferential model* or *structure*.

(1.2) The version with copies:
A pair $\mathcal{M} := \langle \mathcal{U}, \prec\rangle$ with $\mathcal{U}$ an arbitrary set of pairs and $\prec$ an arbitrary binary relation on $\mathcal{U}$ is called a *preferential model* or *structure*.
If $\langle x, i\rangle \in \mathcal{U}$, then x is intended to be an element of U, and i the index of the copy.
We sometimes also need copies of the relation $\prec$. We will then replace $\prec$ by one or several arrows α attacking non-minimal elements, e.g., $x \prec y$ will be written $\alpha : x \to y$, $\langle x, i\rangle \prec \langle y, i\rangle$ will be written $\alpha : \langle x, i\rangle \to \langle y, i\rangle$, and finally we might have $\langle \alpha, k\rangle : x \to y$ and $\langle \alpha, k\rangle : \langle x, i\rangle \to \langle y, i\rangle$.

(2) *Minimal elements*, the functions $\mu_{\mathcal{M}}$

(2.1) The version without copies:
Let $\mathcal{M} := \langle U, \prec\rangle$, and define

$\mu_{\mathcal{M}}(X) := \{x \in X : x \in U \wedge \neg\exists x' \in X \cap U . x' \prec x\}$.
$\mu_{\mathcal{M}}(X)$ is called the set of *minimal elements* of X (in $\mathcal{M}$).
Thus, $\mu_{\mathcal{M}}(X)$ is the set of elements such that there is no smaller one in X.

(2.2) The version with copies:
Let $\mathcal{M} := \langle \mathcal{U}, \prec \rangle$ be as above. Define
$\mu_{\mathcal{M}}(X) := \{x \in X : \exists \langle x, i \rangle \in \mathcal{U} . \neg\exists \langle x', i' \rangle \in \mathcal{U}(x' \in X \wedge \langle x', i' \rangle' \prec \langle x, i \rangle)\}$.
Thus, $\mu_{\mathcal{M}}(X)$ is the projection on the first coordinate of the set of elements such that there is no smaller one in X.
Again, by abuse of language, we say that $\mu_{\mathcal{M}}(X)$ is the set of *minimal elements* of X in the structure. If the context is clear, we will write just μ.
We sometimes say that $\langle x, i \rangle$ "*kills*" or "*minimizes*" $\langle y, j \rangle$ if $\langle x, i \rangle \prec \langle y, j \rangle$. By abuse of language, we also say a set X *kills* or *minimizes* a set Y if for all $\langle y, j \rangle \in \mathcal{U}$, $y \in Y$ there is $\langle x, i \rangle \in \mathcal{U}$, $x \in X$ s.t. $\langle x, i \rangle \prec \langle y, j \rangle$.
$\mathcal{M}$ is also called *injective* or 1-copy iff there is always at most one copy $\langle x, i \rangle$ for each x. Note that the existence of copies corresponds to a non-injective labelling function—as is often used in nonclassical logic, e.g., modal logic.

We say that $\mathcal{M}$ is *transitive*, *irreflexive*, etc., iff $\prec$ is.
Note that $\mu(X)$ might well be empty, even if X is not.

Definition 5.2.2
We define the consequence relation of a preferential structure for a given propositional language $\mathcal{L}$.

(1) (1.1) If m is a classical model of a language $\mathcal{L}$, we say by abuse of language
$\langle m, i \rangle \models \phi$ iff $m \models \phi$,
and if X is any set of such pairs, that
$X \models \phi$ iff for all $\langle m, i \rangle \in X$ $m \models \phi$.
(1.2) If $\mathcal{M}$ is a preferential structure, and X is a set of $\mathcal{L}$-models for a classical propositional language $\mathcal{L}$, or is a set of pairs $\langle m, i \rangle$ where the m are such models, we call $\mathcal{M}$ a *classical preferential structure* or *model*.

(2) *Validity* in a preferential structure, or the *semantical consequence relation* defined by such a structure:
Let $\mathcal{M}$ be as above.
We define
$T \models_{\mathcal{M}} \phi$ iff $\mu_{\mathcal{M}}(M(T)) \models \phi$, i.e., $\mu_{\mathcal{M}}(M(T)) \subseteq M(\phi)$.

(3) $\mathcal{M}$ will be called *definability preserving* iff for all $X \in \boldsymbol{D}_{\mathcal{L}}$ $\mu_{\mathcal{M}}(X) \in \boldsymbol{D}_{\mathcal{L}}$.

As $\mu_{\mathcal{M}}$ is defined on $\boldsymbol{D}_{\mathcal{L}}$, but need by no means always result in some new definable set, this is (and reveals itself as a quite strong) additional property.

Definition 5.2.3
Let $\mathcal{Y} \subseteq \mathcal{P}(U)$. (In applications to logic, $\mathcal{Y}$ will be $\boldsymbol{D}_{\mathcal{L}}$.)

A preferential structure $\mathcal{M}$ is called $\mathcal{Y}$-smooth iff for every $X \in \mathcal{Y}$ every element $x \in X$ is either minimal in X or above an element which is minimal in X. More precisely:

(1) The version without copies:
If $x \in X \in \mathcal{Y}$, then either $x \in \mu(X)$ or there is $x' \in \mu(X).x' \prec x$.

(2) The version with copies:
If $x \in X \in \mathcal{Y}$, and $\langle x, i\rangle \in \mathcal{U}$, then either there is no $\langle x', i'\rangle \in \mathcal{U}, x' \in X, \langle x', i'\rangle \prec \langle x, i\rangle$ or there is a $\langle x', i'\rangle \in \mathcal{U}, \langle x', i'\rangle \prec \langle x, i\rangle, x' \in X$, s.t. there is no $\langle x'', i''\rangle \in \mathcal{U}$, $x'' \in X$, with $\langle x'', i''\rangle \prec \langle x', i'\rangle$.
(Writing down all details here again might make it easier to read applications of the definition later on.)

When considering the models of a language $\mathcal{L}$, $\mathcal{M}$ will be called *smooth* iff it is $\boldsymbol{D}_{\mathcal{L}}$-smooth; $\boldsymbol{D}_{\mathcal{L}}$ is the default.

Obviously, the richer the set $\mathcal{Y}$ is, the stronger the condition $\mathcal{Y}$-smoothness will be.

A remark for the intuition: Smoothness is perhaps best motivated through Gabbay's concept of reactive diagrams; see, e.g., Gabbay (2016) and Gabbay (2008), and also Gabbay and Schlechta (2009), Gabbay and Schlechta (2009). In this concept, smaller, or "better", elements attack bigger, or "less good", elements. But when a attacks b, and b attacks c, then one might consider the attack of b against c weakened by the attack of a against b. In a smooth structure, for every attack against some element x, there is also an uncontested attack against x, as it originates in an element y, which is not attacked itself.

Fact 5.2.1
This is Fact 2.3.1 of Gabbay and Schlechta (2011), which also contains the proof.

Let $\prec$ be an irreflexive, binary relation on X; then the following two conditions are equivalent:

(1) There is an Ω and an irreflexive, total, binary relation $\prec'$ on Ω and a function $f : X \to \Omega$ s.t. $x \prec y \Leftrightarrow f(x) \prec' f(y)$ for all $x, y \in X$.

(2) Let $x, y, z \in X$ and $x \bot y$ with respect to $\prec$ (i.e., neither $x \prec y$ nor $y \prec x$); then $z \prec x \Rightarrow z \prec y$ and $x \prec z \Rightarrow y \prec z$.

Definition 5.2.4
We call an irreflexive, binary relation $\prec$ on X which satisfies (1) (equivalently (2)) of Fact 5.2.1 ranked. By abuse of language, we also call a preferential structure $\langle X, \prec\rangle$ ranked iff $\prec$ is.

We quote from Schlechta (2004) the following summary for preferential structures:

Table 5.1, "Preferential representation", summarises the more difficult half of a full representation result for preferential structures. It shows equivalence between certain abstract conditions for model choice functions and certain preferential structures. They are shown in the respective representation theorems.

"Singletons" means that the domain must contain all singletons, "1 copy" or "≥ 1 copy" means that the structure may contain only one copy for each point, or several, and "$(\mu\emptyset)$" for the preferential structure means that the μ-function of the structure has to satisfy this property.

We call a characterization "normal" iff it is a universally quantified Boolean combination (of any fixed, but perhaps infinite, length) of rules of the usual form. We do not go into details here.

In the second column from the left, "$\Rightarrow$" means, for instance for the smooth case, that for any $\mathcal{Y}$ closed under finite unions, and any choice function f which satisfies the conditions in the left-hand column, there is a (here $\mathcal{Y}$-smooth) preferential structure $\mathcal{X}$ which represents it, i.e., for all $Y \in \mathcal{Y}$, $f(Y) = \mu_{\mathcal{X}}(Y)$, where $\mu_{\mathcal{X}}$ is the model choice function of the structure $\mathcal{X}$. The inverse arrow $\Leftarrow$ means that the model choice function for any smooth $\mathcal{X}$ defined on such a $\mathcal{Y}$ will satisfy the conditions on the left.

For more details on preferential logics and size, the reader is referred to Gabbay and Schlechta (2009) and Gabbay and Schlechta (2016).

5.2.1.1 More on the Minimal Variant of Preferential Structures (Bubble Structures)

We discuss now a variant of preferential structures which will prove to be useful. We call them bubble structures; see Diagram 5.2.1 for illustration.

The basic idea is as follows: We have one global structure, and parts of it, "bubbles", behave in a uniform way. Thus, with respect to the "outside world", elements inside an individual bubble are indistinguishable.

Formally, if x is outside a given bubble, and b, b' are inside the bubble, then $x \prec b$ iff $x \prec b'$, and $b \prec x$ iff $b' \prec x$. Thus, seen from the outside, a bubble behaves like a layer in a ranked structure. But we do not require the inside of the bubble to consist of only incomparable elements, like a layer in a ranked structure. Thus, ranked structures are special cases of bubble structures, with the layers being the bubbles.

We may identify the bubbles with single points, and, as long as we do not look into the bubbles, we have just the usual preferential structure.

We can see the whole structure as an abstraction, where details are encapsulated in the bubbles. We can also work with different languages, a "global" language for the big structure, and one or more sublanguages which are used only inside the bubbles. When all bubbles have the same internal structure, we can also push this structure into the truth values. The basic idea is, of course, another expression of modularity, where we try to isolate different aspects of reasoning as much as possible to simplify the task—divide et impera!

We will not give a full formal representation result here, but point out the important ingredients. The cases to consider (different sublanguages, Cumulativity or not inside/outside the bubbles, etc.) may be quite varied, and we leave it to future research to elaborate on the details. We turn now to a list of things to consider.

(1) Copies of the same element should probably be in the same bubble.
(2) We will probably work with one copy of each bubble, so we add a 1-copy condition to the global structure; see Schlechta (2004) for a discussion.
(3) Singletons without copies can always be considered to be bubbles.

Table 5.1 Preferential representation

Preferential representation				
μ-function		Pref. Structure		Logic
$(\mu \subseteq)$	$\Leftrightarrow$	Reactive	$\Leftrightarrow$	$(LLE) + (CCL) + (SC)$
$(\mu \subseteq) + (\mu CUM)$	$\Rightarrow (\cap)$	Reactive + essentially smooth		
$(\mu \subseteq) + (\mu \subseteq\supseteq)$	$\Rightarrow$	Reactive + essentially smooth	$\Leftrightarrow$	$(LLE) + (CCL) + (SC) + (\subseteq\supseteq)$
$(\mu \subseteq) + (\mu CUM) + (\mu \subseteq\supseteq)$	$\Leftarrow$	Reactive + essentially smooth		
$(\mu \subseteq) + (\mu PR)$	$\Leftarrow$	General	$\Rightarrow (\mu dp)$	$(LLE) + (RW) + (SC) + (PR)$
	$\Rightarrow$		$\Leftarrow$	
			$\not\Rightarrow$ without (μdp)	
			$\not\Leftrightarrow$ without (μdp)	Any "normal" characterization of any size
$(\mu \subseteq) + (\mu PR)$	$\Leftarrow$	Transitive	$\Rightarrow (\mu dp)$	$(LLE) + (RW) + (SC) + (PR)$
	$\Rightarrow$		$\Leftarrow$	
			$\not\Rightarrow$ without (μdp)	
			$\Leftrightarrow$ without (μdp)	using "small" exception sets
$(\mu \subseteq) + (\mu PR) + (\mu CUM)$	$\Leftarrow$	Smooth	$\Rightarrow (\mu dp)$	$(LLE) + (RW) + (SC) + (PR) + (CUM)$
	$\Rightarrow (\cup)$		$\Leftarrow (\cup)$	
	$\not\Rightarrow$ without $(\cup)$		$\not\Rightarrow$ without (μdp)	

$(\mu \subseteq) + (\mu PR) + (\mu CUM)$	$\Leftarrow$	Smooth + Transitive	$\Rightarrow (\mu dp)$	$(LLE) + (RW) +$ $(SC) + (PR) +$ (CUM)
	$\Rightarrow (\cup)$		$\Leftarrow (\cup)$	
			$\not\Rightarrow$ without (μdp)	
			$\not\Leftrightarrow$ without (μdp)	Using "small" exception sets
$(\mu \subseteq) + (\mu =) + (\mu PR) +$ $(\mu =') + (\mu \parallel) + (\mu \cup) +$ $(\mu \cup') + (\mu \in) + (\mu RatM)$	$\Leftarrow$	Ranked, $\geq$1 copy		
$(\mu \subseteq) + (\mu =) + (\mu PR) +$ $(\mu \cup) + (\mu \in)$	$\not\Rightarrow$	Ranked		
$(\mu \subseteq) + (\mu =) + (\mu \emptyset)$	$\Leftrightarrow$, $(\cup)$	Ranked, 1 copy + $(\mu \emptyset)$		
$(\mu \subseteq) + (\mu =) + (\mu \emptyset)$	$\Leftrightarrow$, $(\cup)$	Ranked, smooth, 1 copy + $(\mu \emptyset)$		
$(\mu \subseteq) + (\mu =) + (\mu \emptyset fin) +$ $(\mu \in)$	$\Leftrightarrow$, $(\cup)$, singletons	Ranked, smooth, $\geq$ 1 copy + $(\mu \emptyset fin)$		
$(\mu \subseteq) + (\mu PR) + (\mu \parallel) +$ $(\mu \cup) + (\mu \in)$	$\Leftrightarrow$, $(\cup)$, singletons	Ranked $\geq$ 1 copy	$\not\Rightarrow$ without (μdp)	$(RatM)$, $(RatM =)$, $(Log\cup)$, $(Log\cup')$
			$\not\Leftrightarrow$ without (μdp)	any "normal" characterization of any size

(4) Given a global structure, the decomposition into bubbles is usually not unique: Take a linear order; then all elements of any interval $[a, b]$ behave to the outside world in the same way, so any such interval can be considered a bubble. But, of course, we want bubbles to be disjoint, or, at least, forming systems of bubbles, superbubbles, etc.

(5) Similarly to ranked structures, any element of a bubble replaces any other for the order relation:
Suppose $A \cap B = \emptyset$, B is a bubble; then $\mu(A \cup B) = \mu(A) \cup \mu(B)$, or $= \mu(A)$, or $= \mu(B)$, provided the relation is transitive.
For a counter example to the non-transitive case, consider $b \prec a \prec b'$, but not $b \prec b'$, with $B = \{b, b'\}$, $A = \{a\}$; then $\mu(B) = \{b, b'\}$, $\mu(A \cup B) = \{b\}$.

(6) The full rankedness conditions like $A \subseteq B$, $\mu(B) \cap A \neq \emptyset \Rightarrow \mu(A) = \mu(B) \cap A$ are generally too strong, as the insides of the bubbles need not consist of incomparable elements.

Diagram 5.2.1

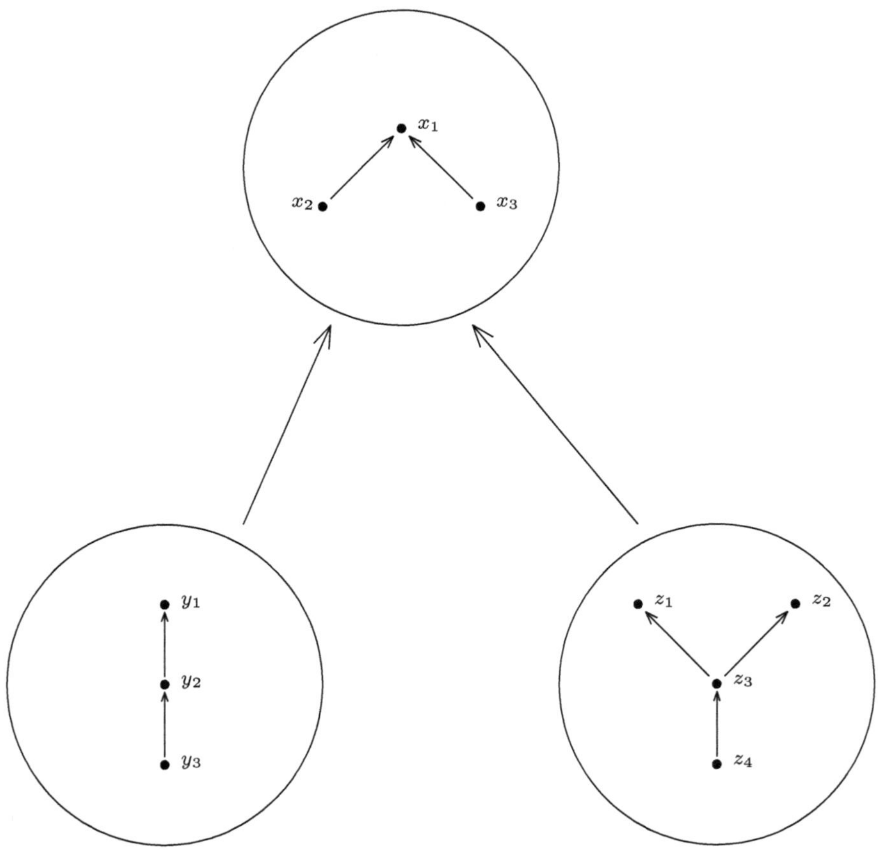

5.2.1.2 One-Copy Version

The essential property of preferential structures with at most one copy each is that we never need two or more elements to kill one other element. This is expressed by the following property, which we give in a finitary and an infinitary version:

Definition 5.2.5
(1-fin) Let $X = A \cup B_1 \cup B_2$ and $A \cap \mu(X) = \emptyset$. Then $A \subseteq (A \cup B_1 - \mu(A \cup B_1)) \cup (A \cup B_2 - \mu(A \cup B_2))$.

(1-infin) Let $X = A \cup \bigcup\{B_i : i \in I\}$ and $A \cap \mu(X) = \emptyset$. Then $A \subseteq \bigcup\{A \cup B_i - \mu(A \cup B_i)\}$.

It is obvious that both hold in 1-copy structures, it is equally obvious that the second guarantees the 1-copy property (consider $X = \{\{x\} : x \in X\}$, if $x \notin \mu(X)$, we find at least one $x' \in X$ s.t. $x \notin \mu(\{x, x'\})$, and this gives the construction for representation, too. It is almost as obvious that the finitary version does not suffice:

Example 5.2.1
Take an infinitary language $\{p, q_i : i \in \omega\}$, and let every p-model be killed by any infinite set of $\neg p$-models, and nothing else. Now, if A is minimized by $B_1 \cup B_2$, $B_1 \cup B_2$ contains an infinite number of $\neg p$-models, so either B_1 or B_2 does, so (1-fin) holds, but, obviously, the structure is not equivalent to any structure with one copy at most. □

We turn to transitivity in the 1-copy case. Consider $a \prec b \prec c$, but $a \not\prec c$. So $\mu(\{a, c\}) = \{a, c\}$. By $\mu(\{a\}) = \{a\}$, $\mu(\{b\}) = \{b\}$, $\mu(\{c\}) = \{c\}$, we see that all three elements are present, so each has to be there as one copy. By $\mu(\{a, b\}) = \{a\}$ and $\mu(\{b, c\}) = \{b\}$, we see that $a \prec b \prec c$ has to hold. But then transitivity requires $a \prec c$, thus $\mu(\{a, c\}) = \{a\}$ has to hold, so the present structure is not equivalent to any transitive structure with the 1-copy property. Thus, to have transitivity, we need a supplementary condition:

Definition 5.2.6
(T) $\mu(A \cup B) \subseteq A$, $\mu(B \cup C) \subseteq B \to \mu(A \cup C) \subseteq A$.

Taking $a \prec b \prec c$ and $A := \{a\}$, $B := \{b\}$, $C := \{c\}$, we see that (T) imposes transitivity on 1-copy structures.

Note, however, that (T) does not necessarily hold in transitive structures with more than one copy—see above example.

5.2.2 *The Limit Variant*

5.2.2.1 Motivation for the Limit Variant

Distance-based semantics give perhaps the clearest motivation for the limit variant. For instance, the Stalnaker–Lewis semantics for counterfactual conditionals defines $\phi > \psi$ to hold in a (classical) model m iff in those models of ϕ which are closest to m, ψ holds. For this to make sense, we need, of course, a distance d on the model set. We call this approach the minimal variant. Usually, one makes a limit assumption: The set of ϕ-models closest to m is not empty if ϕ is consistent, i.e., the ϕ-models are not arranged around m in a way that they come closer and closer, without a minimal distance. This is, of course, a very strong assumption, and probably difficult to justify philosophically. It seems to have its only justification in the fact that it avoids degenerate cases, where, in the above example, for consistent ϕ, $m \models \phi > FALSE$ holds. As such, this assumption is unsatisfactory.

The limit version avoids such assumptions. It will still work in the above situation, i.e., when there are not always optimal (closest) elements, it defines what happens when we get "better and better", i.e., approach the limit (the "best" case).

We will have to define what a suitable "neighbourhood" of the best cases is; in our context, this will roughly be a set of elements which minimizes all other elements and is downward closed, i.e., contains all elements better than some x already in the set. We call such sets MISE for minimizing initial segment. We will see (Example 5.2.2) that this definition will not always do what we want it to do, and we will have to impose additional properties.

Essentially, we want MISE sets to reflect the properties of the sets of minimal elements, if they exist. Thus, the set of minimal elements should be a special case of a MISE. But we also want MISE sets to be closed under finite intersection, to have the logical (AND) property; see again Example 5.2.2. If our definition is such that its properties are sufficiently close to those of the ideal (the minimal elements), then we will also have the desired algebraic and logical properties, but avoid pathologies originating from the empty set (when there are no best elements)—and this is what we wanted. Of course, our definition still has to correspond to the intuition of what an approximation to the ideal case should be.

We give now the basic definitions for the limit version of preferential and ranked preferential structures.

5.2.2.2 The Limit Variant

Definition 5.2.7

(1) General preferential structures.

(1.1) The version without copies:

Let $\mathcal{M} := \langle U, \prec \rangle$. Define

$Y \subseteq X \subseteq U$ is a minimizing initial segment, or MISE, of X iff:

(a) $\forall x \in X \exists x \in Y.y \preceq x$, where $y \preceq x$ stands for $x \prec y$ or $x = y$ (i.e., Y is minimizing), and

(b) $\forall y \in Y, \forall x \in X(x \prec y \Rightarrow x \in Y)$ (i.e., Y is downward closed or an initial part).

(1.2) The version with copies:

Let $\mathcal{M} := \langle \mathcal{U}, \prec \rangle$ be as above. Define for $Y \subseteq X \subseteq \mathcal{U}$

Y is a minimizing initial segment or MISE of X iff:

(a) $\forall \langle x, i \rangle \in X \exists \langle y, j \rangle \in Y. \langle y, j \rangle \preceq \langle x, i \rangle$

and

(b) $\forall \langle y, j \rangle \in Y, \forall \langle x, i \rangle \in X \; (\langle x, i \rangle \prec \langle y, j \rangle \Rightarrow \langle x, i \rangle \in Y)$.

(1.3) For $X \subseteq \mathcal{U}$, let $\Lambda(X)$ be the set of a MISE of X.

(1.4) We say that a set $\mathcal{X}$ of a MISE is cofinal in another set of a MISE $\mathcal{X}'$ (for the same base set X) iff for all $Y' \in \mathcal{X}'$, there is $Y \in \mathcal{X}$, $Y \subseteq Y'$.

(1.5) A MISE X is called definable iff $\{x : \exists i. \langle x, i \rangle \in X\} \in \boldsymbol{D}_{\mathcal{L}}$.

(1.6) $T \models_{\mathcal{M}} \phi$ iff there is $Y \in \Lambda(\mathcal{U} \upharpoonright M(T))$ such that $Y \models \phi$.

($\mathcal{U} \upharpoonright M(T) := \{\langle x, i \rangle \in \mathcal{U} : x \in M(T)\}$; if there are no copies, we simplify in the obvious way.)

(2) Ranked preferential structures.

In the case of ranked structures, we may assume without loss of generality that the MISE sets have a particularly simple form:

For $X \subseteq U$, $A \subseteq X$ is a MISE iff $X \neq \emptyset$ and $\forall a \in A \forall x \in X(x \prec a \vee x \bot a \Rightarrow x \in A)$. (A is downward and horizontally closed.)

(3) Theory revision.

Recall that we have a distance d on the model set, and are interested in $y \in Y$ which are close to X.

Thus, given X, Y, we define analogously:

$B \subseteq Y$ is a MISE iff

(1) $B \neq \emptyset$ and

(2) there is a d' such that $B := \{y \in Y : \exists x \in X.d(x, y) \leq d'\}$ (we could also have chosen $d(x, y) < d'$, but this is not important).

And we define $\phi \in T * T'$ iff there is a $B \in \Lambda(M(T), M(T'))$ $B \models \phi$.

There are basic problems with the limit in general preferential structures, as we shall see now:

Example 5.2.2

Let $a \prec b$, $a \prec c$, $b \prec d$, $c \prec d$ (but $\prec$ not transitive!); then $\{a, b\}$ and $\{a, c\}$ are such S and S', but there is no $S'' \subseteq S \cap S'$ which is an initial segment. If, for instance, in a and b ψ holds, in a and c, ψ', then "in the limit" ψ and ψ' will hold, but not $\psi \wedge \psi'$. This does not seem right. We should not be obliged to give up ψ to obtain ψ'. □

We will therefore require it to be closed under finite intersections, or at least that if S, S' are such segments, then there must be $S'' \subseteq S \cap S'$ which is also such a segment.

We make this official. Let $\Lambda(X)$ be the set of initial segments of X; then we require

$(\Lambda\cap)$. If $A, B \in \Lambda(X)$ then there is a $C \subseteq A \cap B$, $C \in \Lambda(X)$.

To familiarise the reader with the limit version, we show two easy but important results.

Fact 5.2.2

(Taken from Schlechta (2004), Fact 3.4.3, Proposition 3.10.16, see also Fact 2.3.2 of Gabbay and Schlechta (2011) which also contains the proof.)

Let the relation $\prec$ be transitive. The following hold in the limit variant of general preferential structures:

(1) If $A \in \Lambda(Y)$ and $A \subseteq X \subseteq Y$, then $A \in \Lambda(X)$.

(2) If $A \in \Lambda(Y)$, $A \subseteq X \subseteq Y$, and $B \in \Lambda(X)$, then $A \cap B \in \Lambda(Y)$.

(2a) We summarise to make finitary semantical Cumulativity evident: Let $A \in \Lambda(Y)$, $A \subseteq X \subseteq Y$. Then, if $B \in \Lambda(Y)$, $A \cap B \in \Lambda(X)$. Conversely, if $B \in \Lambda(X)$, then $A \cap B \in \Lambda(Y)$.

(3) If $A \in \Lambda(Y)$, $B \in \Lambda(X)$, then there is $Z \subseteq A \cup B$ $Z \in \Lambda(Y \cup X)$.

The following hold in the limit variant of ranked structures without copies, where the domain is closed under finite unions and contains all finite sets.

(4) $A, B \in \Lambda(X) \Rightarrow A \subseteq B$ or $B \subseteq A$,

(5) $A \in \Lambda(X)$, $Y \subseteq X$, $Y \cap A \neq \emptyset \Rightarrow Y \cap A \in \Lambda(Y)$,

(6) $\Lambda' \subseteq \Lambda(X)$, $\bigcap \Lambda' \neq \emptyset \Rightarrow \bigcap \Lambda' \in \Lambda(X)$,

(7) $X \subseteq Y$, $A \in \Lambda(X) \Rightarrow \exists B \in \Lambda(Y).B \cap X = A$.

We have as immediate logical consequence:

Fact 5.2.3

(Fact 3.4.4 of Schlechta (2004).)

If $\prec$ is transitive, then in the limit variant hold:

(1) (AND),

(2) (OR).

We also have the following:

Fact 5.2.4

(Fact 3.4.5 in Schlechta (2004).)

Finite Cumulativity holds in transitive limit structures:

If $\phi \mathrel{|\!\sim} \psi$, then $\overline{\overline{\phi}} = \overline{\overline{\phi \wedge \psi}}$.

See Schlechta (2004) for a direct proof, or above Fact 5.2.2, (2a). □

We repeat now (without proof) our main logical trivialization results on the limit variant of general preferential structures, Propositions 3.4.7 and 3.10.19 from Schlechta (2004):

Proposition 5.2.5
(1) Let the relation be transitive. If we consider only formulas on the left of $\mid\!\sim$*, the resulting logic of the limit version can also be generated by the minimal version of a (perhaps different) preferential structure. Moreover, this structure can be chosen smooth.*
(2) Let a logic $\phi \mid\!\sim \psi$ *be given by the limit variant of a ranked structure without copies. Then there is a ranked structure which gives exactly the same logic, but interpreted in the minimal variant.* □

(The negative results for the general, not definability preserving minimal case apply also to the general limit case; see Sect. 5.2.3 in Schlechta (2004) for details.)

Consider the following analogon to (μPR) $(A \subseteq B \Rightarrow \mu(B) \cap A \subseteq \mu(A))$:

Fact 5.2.6
This is Fact 2.3.6 of Gabbay and Schlechta (2011), which also contains the proof.

Let $\prec$ be transitive, $\Lambda(X)$ the MISE systems over X.

Let $A \subseteq B$, $A' \in \Lambda(A) \Rightarrow \exists B' \in \Lambda(B).B' \cap A \subseteq A'$.

We have immediately the following:

Corollary 5.2.7
This is Corollary 2.3.7 of Gabbay and Schlechta (2011), which also contains the proof.

$\overline{\overline{\phi \wedge \phi'}} \subseteq \overline{\overline{\overline{\phi}} \cup \{\phi'\}}$.

5.3 Laws About Size

For more on the subject, see, e.g., Ben–David and Ben–Eliyahu (1994), Friedman and Halpern (1996), Makinson (1994).

5.3.1 Defaults as Generalised Quantifiers

We first recall an interpretation of defaults as a generalised quantifier, ∇, informally read as "almost all", and made precise by a weak filter.

"Weak filters", were defined in Definition 2.2.3, we briefly recall it:

Fix a base set X. A weak filter on or over X is a set $\mathcal{F} \subseteq \mathcal{P}(X)$ such that the following conditions hold:

(F1) $X \in \mathcal{F}$

(F2) $A \subseteq B \subseteq X$, $A \in \mathcal{F}$ imply $B \in \mathcal{F}$

$(F3')$ $A, B \in \mathcal{F}$ imply $A \cap B \neq \emptyset$.

We use weak filters on the semantical side, and add the following axioms on the syntactical side to a FOL axiomatisation:

1. $\nabla x \phi(x) \wedge \forall x(\phi(x) \rightarrow \psi(x)) \rightarrow \nabla x \psi(x)$,
2. $\nabla x \phi(x) \rightarrow \neg \nabla x \neg \phi(x)$,
3. $\forall x \phi(x) \rightarrow \nabla x \phi(x)$ and $\nabla x \phi(x) \rightarrow \exists x \phi(x)$.

A model is now a pair, consisting of a classical FOL model M, and a weak filter over its universe. Both sides are connected by the following definition, where $\mathcal{N}(M)$ is the weak filter on the universe of the classical model M :

$\langle M, \mathcal{N}(M) \rangle \models \nabla x \phi(x)$ iff there is an $A \in \mathcal{N}(M)$ such that $\forall a \in A$ $(\langle M, \mathcal{N}(M) \rangle \models \phi[a])$.

Soundness and completeness are shown in Schlechta (1995); see also Schlechta (2004).

The extension to defaults with prerequisites by restricted quantifiers is straightforward.

5.3.2 *Additive Laws About Size*

This section recalls the abstract semantics of preferential logics. For many purposes, this semantics seems to be the right level of abstraction to treat preferential systems. We think that, ultimately, rare influence changes (and adequate concepts) are the right level of abstraction for common sense reasoning (leaving aside details of the various approaches), but we are still far from a rigorous development. Thus, this section can be seen as a standard to achieve in future development of our main idea.

The section presents *additive* laws for abstract size, whereas our emphasis in this book is on *multiplicative* laws. We present them mainly to put our work in perspective, to connect this book to more familiar laws about size, and to make the text more self-contained, i.e., as a service to the reader.

We now give the main additive rules for manipulation of abstract size from Gabbay and Schlechta (2016); see Tables 5.2 and 5.3, "Rules on size".

These laws allow us to give very concise abstract descriptions of laws about non-monotonic logics, and, by their modularity, to construct new logics by changing some parameters. Moreover, they reveal deep connections between laws like AND and Cumulativity, which are not so obvious in their usual formulations.

The notation is explained with some redundancy, so the reader will not have to go back and forth to Chap. 2.

5.3.2.1 Notation

(1) $\mathcal{P}(X)$ is the power set of X; $\subseteq$ is the subset relation, $\subset$ the strict part of $\subseteq$, i.e., $A \subset B$ iff $A \subseteq B$ and $A \neq B$. The operators $\wedge$, $\neg$, $\vee$, $\rightarrow$ and $\vdash$ have their usual, classical interpretation.

(2) $\mathcal{I}(X) \subseteq \mathcal{P}(X)$ and $\mathcal{F}(X) \subseteq \mathcal{P}(X)$ are dual abstract notions of size; $\mathcal{I}(X)$ is the set of "small" subsets of X, $\mathcal{F}(X)$ the set of "big" subsets of X. They are dual in the sense that $A \in \mathcal{I}(X) \Leftrightarrow X - A \in \mathcal{F}(X)$. "$\mathcal{I}$" evokes "ideal", "$\mathcal{F}$" evokes "filter" though the full strength of both is reached only in $(< \omega * s)$. "s" evokes "small", and "$(x * s)$" stands for "x small sets together are still not everything".

(3) If $A \subseteq X$ is neither in $\mathcal{I}(X)$, nor in $\mathcal{F}(X)$, we say it has medium size, and we define $\mathcal{M}(X) := \mathcal{P}(X) - (\mathcal{I}(X) \cup \mathcal{F}(X))$. $\mathcal{M}^+(X) := \mathcal{P}(X) - \mathcal{I}(X)$ is the set of subsets which are not small.

(4) $\nabla x \phi$ is a generalised first-order quantifier; it is read "almost all x have property ϕ". $\nabla x(\phi : \psi)$ is the relativized version, read "almost all x with property ϕ have also property ψ". To keep the table "Rules on size" simple, we write mostly only the non-relativized versions. Formally, we have $\nabla x \phi :\Leftrightarrow \{x : \phi(x)\} \in \mathcal{F}(U)$, where U is the universe, and $\nabla x(\phi : \psi) :\Leftrightarrow \{x : (\phi \wedge \psi)(x)\} \in \mathcal{F}(\{x : \phi(x)\})$. Soundness and completeness results on ∇ can be found in Schlechta (1995).

(5) Analogously, for propositional logic, we define
$\alpha \mid\!\sim \beta :\Leftrightarrow M(\alpha \wedge \beta) \in \mathcal{F}(M(\alpha))$,
where $M(\phi)$ is the set of models of ϕ.

(6) In preferential structures, $\mu(X) \subseteq X$ is the set of minimal elements of X. This generates a principal filter by $\mathcal{F}(X) := \{A \subseteq X : \mu(X) \subseteq A\}$. Corresponding properties about μ are not listed systematically.

(7) The usual rules (AND), etc., are named here (AND_ω), as they are in a natural ascending line of similar rules, based on strengthening of the filter/ideal properties.

(8) For any set of formulas T, and any consequence relation $\mid\!\sim$, we will use $\overline{T} := \{\phi : T \vdash \phi\}$, the set of classical consequences of T, and $\overline{\overline{T}} := \{\phi : T \mid\!\sim \phi\}$, the set of consequences of T under the relation $\mid\!\sim$.

(9) We say that a set X of models is definable by a formula (or a theory) iff there is a formula ϕ (or a theory T) such that $X = M(\phi)$ (or $X = M(T)$), the set of models of ϕ (or T).

(10) Most rules are explained in the table "Logical rules", and "RW" stands for Right Weakening.

5.3.2.2 The Groups of Rules

The rules concern properties of $\mathcal{I}(X)$ or $\mathcal{F}(X)$, or dependencies between such properties for different X and Y. All X, Y, etc., will be subsets of some universe, say V. Intuitively, V is the set of all models of some fixed propositional language. It is not necessary to consider all subsets of V; the intention is to consider subsets of V, which are definable by a formula or a theory. So we assume all X, Y, etc., taken from some $\mathcal{Y} \subseteq \mathcal{P}(V)$, which we call the domain. In the former case, $\mathcal{Y}$ is closed under set difference, in the latter case not necessarily so. (We will mention it when we need some particular closure property.)

The rules are divided into five groups:

(1) (Opt), which says that "All" is optimal, i.e., when there are no exceptions, then a soft rule $\mid\!\sim$ holds.
(2) Three monotony rules:

  (2.1) (iM) is inner monotony; a subset of a small set is small;
  (2.2) $(eM\mathcal{I})$ is external monotony for ideals: enlarging the base set keeps small sets small;
  (2.3) $(eM\mathcal{F})$ is external monotony for filters: a big subset stays big when the base set shrinks.

  These three rules are very natural if "size" is anything coherent over a change of base sets. In particular, they can be seen as weakening.
(3) $(\approx)$ keeps proportions; it is present here mainly to point the possibility out.
(4) a group of rules $x * s$, which say how many small sets will not add to the base set. The notation "$(< \omega * s)$" is an allusion to the full filter property, that filters are closed under *finite* intersections.
(5) Rational Monotony, which can best be understood as robustness of $\mathcal{M}^+$; see $(\mathcal{M}^{++})$, (3).

We will assume all base sets to be nonempty in order to avoid pathologies and in particular clashes between (Opt) and $(1 * s)$.

Note that the full strength of the usual definitions of a filter and an ideal is reached only in line $(< \omega * s)$.

Regularities

(1) The group of rules $(x * s)$ use ascending strength of $\mathcal{I}/\mathcal{F}$.
(2) The column $(\mathcal{M}^+)$ contains interesting algebraic properties. In particular, they show strengthening from $(3 * s)$ up to Rationality. They are not necessarily equivalent to the corresponding (I_x) rules, not even in the presence of the basic rules. The examples show that care has to be taken when considering the different variants.
(3) Adding the somewhat superfluous (CM_2), we have increasing Cautious Monotony from (wCM) to full (CM_ω).
(4) We have increasing "or" from (wOR) to full (OR_ω).
(5) The line $(2 * s)$ is only there because there seems to be no $(\mathcal{M}_2^+)$; otherwise we could begin $(n * s)$ at $n = 2$.

Summary

We can obtain all rules except $(RatM)$ and $(\approx)$ from (Opt), the monotony rules—(iM), $(eM\mathcal{I})$, $(eM\mathcal{F})$—and $(x * s)$ with increasing x.

5.3.2.3 Table

The following table is split into two, as it is too big for printing on one page.
(See Tables 5.2, "Rules on size—Part I" and 5.3, "Rules on size—Part II".

5.3.3 *Multiplicative Laws About Size*

We are mainly interested in non-monotonic logic. In this domain, independence is strongly connected to multiplication of abstract size, and an important part of the present text treats this connection and its repercussions.

We have at least two scenarios for multiplication; one is described in Diagram 5.3.1, the second in Diagram 5.3.2. In the first scenario, we have nested sets, in the second, we have set products. In the first scenario, we consider subsets which behave as the big set does; in the second scenario we consider subspaces, and decompose the behaviour of the big space into behaviour of the subspaces. In both cases, this results naturally in multiplication of abstract sizes. When we look at the corresponding relation properties, they are quite different (rankedness vs. some kind of modularity). But this is perhaps to be expected, as the two scenarios are quite different.

Other scenarios which might be interesting to consider in our framework are as follows:

- When we have more than two truth values, say three, and two is considered a big subset, and we have n propositional variables, and m of them are considered many, then 2^m might give a "big" subset of the total of 3^n situations.
- Similarly, when we fix one variable, consider two cases of the possible three, and multiply this with a "big" set of models.
- We may also consider the utility or cost of a situation, and work with a "big" utility, and "many" situations, etc.
- Note that, in the case of distances, subspaces add distances, and do not multiply them: $d(xy, x'y') = d(x, x') + d(y, y')$.

5.3.3.1 Multiplication of Size for Subsets

Here we have nested sets, $A \subseteq X \subseteq Y$, A is a certain proportion of X, and X of Y, resulting in a multiplication of relative size or proportions. This is a classical subject of non-monotonic logic; see the last section, taken from Gabbay and Schlechta (2016); it is partly repeated here to stress the common points with the other scenario.

Table 5.2 Rules on size—Part I

Rules on size—Part I				
	"Ideal"	"Filter"	$\mathcal{M}^+$	∇
Optimal proportion				
(Opt)	$\emptyset \in \mathcal{I}(X)$	$X \in \mathcal{F}(X)$		$\forall x\alpha \to \nabla x\alpha$
Monotony (Improving proportions). (iM): internal monotony, $(eM\mathcal{I})$: External monotony for ideals, $(eM\mathcal{F})$: external monotony for filters				
(iM)	$A \subseteq B \in \mathcal{I}(X)$ $\Rightarrow$ $A \in \mathcal{I}(X)$	$A \in \mathcal{F}(X)$, $A \subseteq B \subseteq X$ $\Rightarrow B \in \mathcal{F}(X)$		$\nabla x\alpha \wedge \forall x(\alpha \to \alpha')$ $\to \nabla x\alpha'$
$(eM\mathcal{I})$	$X \subseteq Y \Rightarrow$ $\mathcal{I}(X) \subseteq \mathcal{I}(Y)$			$\nabla x(\alpha : \beta) \wedge$ $\forall x(\alpha' \to \beta) \to$ $\nabla x(\alpha \vee \alpha' : \beta)$
$(eM\mathcal{F})$		$X \subseteq Y \Rightarrow$ $\mathcal{F}(Y) \cap \mathcal{P}(X) \subseteq$ $\mathcal{F}(X)$		$\nabla x(\alpha : \beta) \wedge$ $\forall x(\beta \wedge \alpha \to \alpha') \to$ $\nabla x(\alpha \wedge \alpha' : \beta)$
Keeping proportions				
$(\approx)$	$(\mathcal{I} \cup disj)$ $A \in \mathcal{I}(X)$, $B \in \mathcal{I}(Y)$, $X \cap Y = \emptyset \Rightarrow$ $A \cup B \in \mathcal{I}(X \cup Y)$	$(\mathcal{F} \cup disj)$ $A \in \mathcal{F}(X)$, $B \in \mathcal{F}(Y)$, $X \cap Y = \emptyset \Rightarrow$ $A \cup B \in \mathcal{F}(X \cup Y)$	$(\mathcal{M}^+ \cup disj)$ $A \in \mathcal{M}^+(X)$, $B \in \mathcal{M}^+(Y)$, $X \cap Y = \emptyset \Rightarrow$ $A \cup B \in \mathcal{M}^+(X \cup Y)$	$\nabla x(\alpha : \beta) \wedge$ $\nabla x(\alpha' : \beta) \wedge$ $\neg \exists x(\alpha \wedge \alpha') \to$ $\nabla x(\alpha \vee \alpha' : \beta)$
Robustness of proportions: $n * small \neq All$				
$(1 * s)$	$(\mathcal{I}_1)$ $X \notin \mathcal{I}(X)$	$(\mathcal{F}_1)$ $\emptyset \notin \mathcal{F}(X)$		(∇_1) $\nabla x\alpha \to \exists x\alpha$
$(2 * s)$	$(\mathcal{I}_2)$ $A, B \in \mathcal{I}(X) \Rightarrow$ $A \cup B \neq X$	$(\mathcal{F}_2)$ $A, B \in \mathcal{F}(X) \Rightarrow$ $A \cap B \neq \emptyset$		(∇_2) $\nabla x\alpha \wedge \nabla x\beta$ $\to \exists x(\alpha \wedge \beta)$

$(n * s)$ $(n \geq 3)$	$(\mathcal{I}_n)$ $A_1, .., A_n \in \mathcal{I}(X)$ $\Rightarrow$ $A_1 \cup . \cup A_n \neq X$	$(\mathcal{F}_n)$ $A_1, .., A_n \in \mathcal{I}(X)$ $\Rightarrow$ $A_1 \cap . \cap A_n \neq \emptyset$	$(\mathcal{M}_n^+)$ $X_1 \in \mathcal{F}(X_2), ..,$ $X_{n-1} \in \mathcal{F}(X_n) \Rightarrow$ $X_1 \in \mathcal{M}^+(X_n)$	(∇_n) $\nabla x \alpha_1 \wedge . \wedge \nabla x \alpha_n$ $\rightarrow$ $\exists x(\alpha_1 \wedge . \wedge \alpha_n)$
$(< \omega * s)$	$(\mathcal{I}_\omega)$ $A, B \in \mathcal{I}(X) \Rightarrow$ $A \cup B \in \mathcal{I}(X)$	$(\mathcal{F}_\omega)$ $A, B \in \mathcal{F}(X) \Rightarrow$ $A \cap B \in \mathcal{F}(X)$	$(\mathcal{M}_\omega^+)$ (1) $A \in \mathcal{F}(X), X \in \mathcal{M}^+(Y)$ $\Rightarrow A \in \mathcal{M}^+(Y)$ (2) $A \in \mathcal{M}^+(X), X \in \mathcal{F}(Y)$ $\Rightarrow A \in \mathcal{M}^+(Y)$ (3) $A \in \mathcal{F}(X), X \in \mathcal{F}(Y)$ $\Rightarrow A \in \mathcal{F}(Y)$ (4) $A, B \in \mathcal{I}(X) \Rightarrow$ $A - B \in \mathcal{I}(X\text{-B})$	(∇_ω) $\nabla x \alpha \wedge \nabla x \beta \rightarrow$ $\nabla x(\alpha \wedge \beta)$
Robustness of $\mathcal{M}^+$				
$(\mathcal{M}^{++})$			$(\mathcal{M}^{++})$ (1) $A \in \mathcal{I}(X), B \notin \mathcal{F}(X)$ $\Rightarrow A - B \in \mathcal{I}(X - B)$ (2) $A \in \mathcal{F}(X), B \notin \mathcal{F}(X)$ $\Rightarrow A - B \in \mathcal{F}(X - B)$ (3) $A \in \mathcal{M}^+(X),$ $X \in \mathcal{M}^+(Y)$ $\Rightarrow A \in \mathcal{M}^+(Y)$	

Table 5.3 Rules on size—Part II

Rules on size—Part II	Various rules	AND	OR	Caut./Rat. Mon.
Optimal proportion				
(Opt)	(SC) $\alpha \vdash \beta \Rightarrow \alpha \mid\!\sim \beta$			
Monotony (Improving proportions)				
(iM)	(RW) $\alpha \mid\!\sim \beta, \beta \vdash \beta' \Rightarrow$ $\alpha \mid\!\sim \beta'$			
$(eM\mathcal{I})$	(PR') $\alpha \mid\!\sim \beta, \alpha \vdash \alpha',$ $\alpha' \wedge \neg\alpha \vdash \beta \Rightarrow$ $\alpha' \mid\!\sim \beta$ (μPR) $X \subseteq Y \Rightarrow$ $\mu(Y) \cap X \subseteq \mu(X)$		(wOR) $\alpha \mid\!\sim \beta, \alpha' \vdash \beta \Rightarrow$ $\alpha \vee \alpha' \mid\!\sim \beta$ (μwOR) $\mu(X \cup Y) \subseteq \mu(X) \cup Y$	
$(eM\mathcal{F})$				(wCM) $\alpha \mid\!\sim \beta, \alpha' \vdash \alpha,$ $\alpha \wedge \beta \vdash \alpha' \Rightarrow$ $\alpha' \mid\!\sim \beta$
Keeping proportions				
$(\approx)$	(NR) $\alpha \mid\!\sim \beta \Rightarrow$ $\alpha \wedge \gamma \mid\!\sim \beta$ or $\alpha \wedge \neg\gamma \mid\!\sim \beta$		$(disjOR)$ $\alpha \mid\!\sim \beta, \alpha' \mid\!\sim \beta'$ $\alpha \vdash \neg\alpha', \Rightarrow$ $\alpha \vee \alpha' \mid\!\sim \beta \vee \beta'$ $(\mu disjOR)$ $X \cap Y = \emptyset \Rightarrow$ $\mu(X \cup Y) \subseteq \mu(X) \cup \mu(Y)$	
Robustness of proportions: $n * small \neq All$				
$(1 * s)$	(CP) $\alpha \mid\!\sim \bot \Rightarrow \alpha \vdash \bot$	(AND_1) $\alpha \mid\!\sim \beta \Rightarrow \alpha \nvdash \neg\beta$		

$(2 * s)$		(AND_2) $\alpha \mid\sim \beta, \alpha \mid\sim \beta' \Rightarrow$ $\alpha \not\vdash \neg\beta \vee \neg\beta'$	(OR_2) $\alpha \mid\sim \beta \Rightarrow \alpha \not\mid\sim \neg\beta$	(CM_2) $\alpha \mid\sim \beta \Rightarrow \alpha \not\mid\sim \neg\beta$
$(n * s)$ $(n \geq 3)$		(AND_n) $\alpha \mid\sim \beta_1, .., \alpha \mid\sim \beta_n$ $\Rightarrow$ $\alpha \not\vdash \neg\beta_1 \vee . \vee \neg\beta_n$	(OR_n) $\alpha_1 \mid\sim \beta, .., \alpha_{n-1} \mid\sim \beta$ $\Rightarrow$ $\alpha_1 \vee . \vee \alpha_{n-1} \not\mid\sim \neg\beta$	(CM_n) $\alpha \mid\sim \beta_1, .., \alpha \mid\sim \beta_{n-1}$ $\Rightarrow$ $\alpha \wedge \beta_1 \wedge . \wedge \beta_{n-2} \not\mid\sim \neg\beta_{n-1}$
$(< \omega * s)$		(AND_ω) $\alpha \mid\sim \beta, \alpha \mid\sim \beta' \Rightarrow$ $\alpha \mid\sim \beta \wedge \beta'$	(OR_ω) $\alpha \mid\sim \beta, \alpha' \mid\sim \beta \Rightarrow$ $\alpha \vee \alpha' \mid\sim \beta$ (μOR) $\mu(X \cup Y) \subseteq \mu(X) \cup \mu(Y)$	(CM_ω) $\alpha \mid\sim \beta, \alpha \mid\sim \beta' \Rightarrow$ $\alpha \wedge \beta \mid\sim \beta'$ (μCM) $\mu(X) \subseteq Y \subseteq X \Rightarrow$ $\mu(Y) \subseteq \mu(X)$
Robustness of $\mathcal{M}^+$				
$(\mathcal{M}^{++})$				$(RatM)$ $\alpha \mid\sim \beta, \alpha \not\mid\sim \neg\beta' \Rightarrow$ $\alpha \wedge \beta' \mid\sim \beta$ $(\mu RatM)$ $X \subseteq Y,$ $X \cap \mu(Y) \neq \emptyset \Rightarrow$ $\mu(X) \subseteq \mu(Y) \cap X$

Diagram 5.3.1

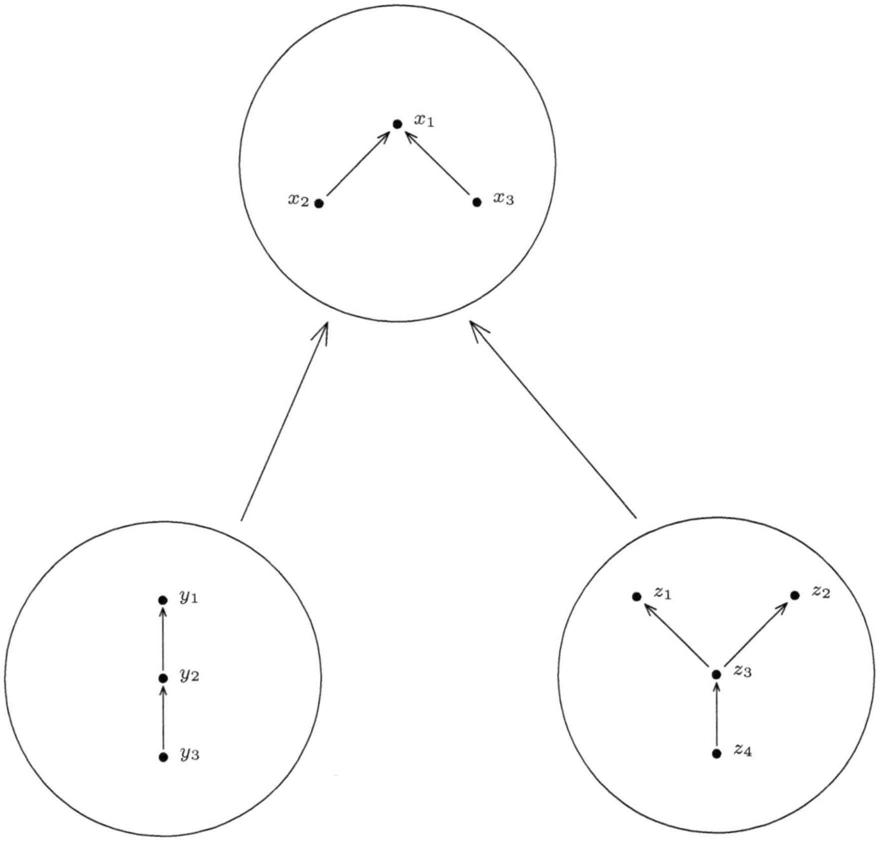

Properties

Diagram 5.3.1 is to be read as follows: The whole set Y is split into X and $Y - X$, X is split into A and $X - A$. X is a small/medium/big part of Y, and A is a small/medium/big part of X. The question is, is A a small/medium/big part of Y?

Note that the relation of A to X is conceptually different from that of X to Y, as we change the base set by going from X to Y, but not when going from A to X. Thus, in particular, when we read the diagram as expressing multiplication, commutativity is not necessarily true.

We looked at this scenario in Gabbay and Schlechta (2016), but from an additive point of view, using various basic properties like (*iM*), (*eM*$\mathcal{I}$), (*eM*$\mathcal{F}$); see Sect. 5.3.2. Here, we use just multiplication—except sometimes for motivation.

We examine different rules:

If $Y = X$ or $X = A$, there is nothing to show, so 1 is the neutral element of multiplication.

If $X \in \mathcal{I}(Y)$ or $A \in \mathcal{I}(X)$, then we should have $A \in \mathcal{I}(Y)$. (Use for motivation (*iM*) or (*eM*$\mathcal{I}$), respectively.)

So it remains to look at the following cases, with the "natural" answers given already:

(1) $X \in \mathcal{F}(Y), A \in \mathcal{F}(X) \Rightarrow A \in \mathcal{F}(Y)$,
(2) $X \in \mathcal{M}^+(Y), A \in \mathcal{F}(X) \Rightarrow A \in \mathcal{M}^+(Y)$,
(3) $X \in \mathcal{F}(Y), A \in \mathcal{M}^+(X) \Rightarrow A \in \mathcal{M}^+(Y)$,
(4) $X \in \mathcal{M}^+(Y), A \in \mathcal{M}^+(X) \Rightarrow A \in \mathcal{M}^+(Y)$.

But (1) is case (3) of $(\mathcal{M}^+_\omega)$ in Gabbay and Schlechta (2016); see Table "Rules on size" in Sect. 5.3.2.

(2) is case (1) of $(\mathcal{M}^+_\omega)$ there,
(3) is case (2) of $(\mathcal{M}^+_\omega)$ there, and finally,
(4) is $(\mathcal{M}^{++})$ there.

So the first three correspond to various expressions of (AND_ω), (OR_ω), (CM_ω), the last one to $(RatM)$.

But we can read them also the other way round, e.g.,

(1) corresponds to $\alpha \mid\!\sim \beta, \alpha \wedge \beta \mid\!\sim \gamma \Rightarrow \alpha \mid\!\sim \gamma$,
(2) corresponds to $\alpha \not\mid\!\sim \neg\beta, \alpha \wedge \beta \mid\!\sim \gamma \Rightarrow \alpha \not\mid\!\sim \neg(\beta \wedge \gamma)$,
(3) corresponds to $\alpha \mid\!\sim \beta, \alpha \wedge \beta \not\mid\!\sim \neg\gamma \Rightarrow \alpha \not\mid\!\sim \neg(\beta \wedge \gamma)$.

All these rules might be seen as too idealistic, so as we did in Gabbay and Schlechta (2016), we can consider milder versions: We might for instance consider a rule which says that *big* $* \cdots *$ *big*, n times, is not small. Consider for instance the case $n = 2$. So we would conclude that A is not small in Y. In terms of logic, we then have $\alpha \mid\!\sim \beta, \alpha \wedge \beta \mid\!\sim \gamma \Rightarrow \alpha \not\mid\!\sim (\neg\beta \vee \neg\gamma)$. We can obtain the same logical property from $3 *$ *small* $\neq$ *all*.

5.3.3.2 Multiplication of Size for Subspaces

Our main interest here is multiplication for subspaces, which we discuss now. The reason for this interest is that it is an abstract view on model sets defined for sublanguages and their composition. Consequently, it offers an abstract view on properties, assuring interpolation for non-monotonic logics. Here we will, however, only look at these laws in general, which are interesting also beyond interpolation. In particular, they also describe what may happen when we go from one language L to a bigger language L', and to the consequence relations in both languages. A priori, they may be totally different, but it often seems reasonable to postulate certain coherence properties between those consequence relations. For instance, if $\alpha \mid\!\sim \beta$ in L, then we will often implicitly assume that also $\alpha \mid\!\sim \beta$ in L'—but this is not trivial. This corresponds to the following coherence property for sizes: If $\Sigma \subseteq \Pi L$ is a big subset, then $\Sigma \times \Pi(L' - L) \subseteq \Pi L'$ is also a big subset. This is the kind of property we investigate here.

Properties
Diagram 5.3.2

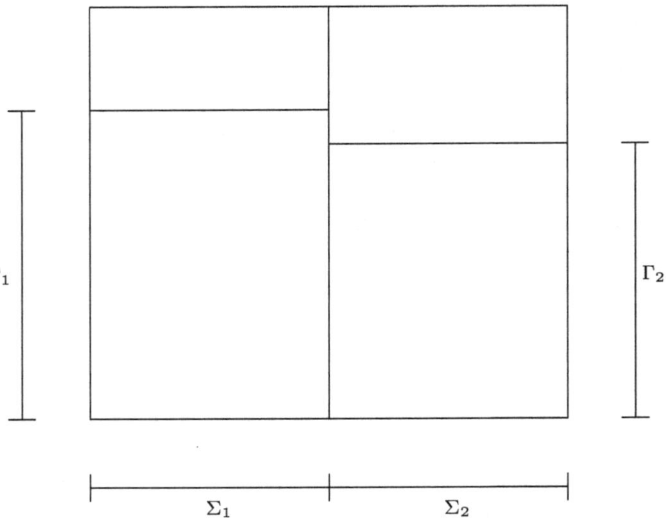

Scenario 2

In this scenario, Σ_i are sets of sequences, (see Diagram 5.3.2) corresponding, intuitively, to a set of models in language $\mathcal{L}_i$; Σ_i will be the set of α_i-models and the subsets Γ_i are to be seen as the "best" models where β_i will hold. The languages are supposed to be disjoint sublanguages of a common language $\mathcal{L}$. As the Σ_i have symmetrical roles, there is no intuitive reason for multiplication not to be commutative.

We can interpret the situation twofold:

First, we work separately in sublanguages $\mathcal{L}_1$ and $\mathcal{L}_2$, and, say, α_i and β_i are both defined in $\mathcal{L}_i$, we look at $\alpha_i \mid\!\sim \beta_i$ in the sublanguage $\mathcal{L}_i$ or we consider both α_i and β_i in the big language $\mathcal{L}$ and look at $\alpha_i \mid\!\sim \beta_i$ in $\mathcal{L}$. These two ways are a priori completely different. Speaking in preferential terms, it is not at all clear why the orderings on the submodels should have anything to do with the orderings on the whole models. It seems to be a very desirable property, but we have to postulate it, which we do now (an overview is given in Table 5.4). We first give informally a list of such rules, mainly to show the connection with the first scenario. Later (see Definition 5.3.1), we will introduce formally some rules for which we show a connection with interpolation. Here "(*big* $*$ *big* $\Rightarrow$ *big*)" stands for "if both factors are big, so will be the product"; this will be abbreviated by "$b * b \Rightarrow b$" in Table 5.4.

We have (*big* $* \, 1 \Rightarrow$ *big*) Let $\Gamma_1 \subseteq \Sigma_1$; if $\Gamma_1 \in \mathcal{F}(\Sigma_1)$, then $\Gamma_1 \times \Sigma_2 \in \mathcal{F}(\Sigma_1 \times \Sigma_2)$ (and we have the dual rule for Σ_2 and Γ_2).

This property preserves proportions, so it seems intuitively quite uncontested whenever we admit coherence over products. (Recall that there was nothing to show in the first scenario.)

When we reconsider the above case, and suppose $\alpha \mid\!\sim \beta$ is in the sublanguage, so $M(\beta) \in \mathcal{F}(M(\alpha))$ holds in the sublanguage, so by $(big * 1 \Rightarrow big)$, $M(\beta) \in \mathcal{F}(M(\alpha))$ in the big language $\mathcal{L}$.

We obtain the dual rule for small (and likewise, medium size) sets:

$(small * 1 \Rightarrow small)$ Let $\Gamma_1 \subseteq \Sigma_1$; if $\Gamma_1 \in \mathcal{I}(\Sigma_1)$, then $\Gamma_1 \times \Sigma_2 \in \mathcal{I}(\Sigma_1 \times \Sigma_2)$ (and we have the dual rule for Σ_2 and Γ_2), establishing $All = 1$ as the neutral element for multiplication.

We look now at other plausible rules:

$(small * x \Rightarrow small)$ $\Gamma_1 \in \mathcal{I}(\Sigma_1)$, $\Gamma_2 \subseteq \Sigma_2 \Rightarrow \Gamma_1 \times \Gamma_2 \in \mathcal{I}(\Sigma_1 \times \Sigma_2)$.

$(big * big \Rightarrow big)$ $\Gamma_1 \in \mathcal{F}(\Sigma_1)$, $\Gamma_2 \in \mathcal{F}(\Sigma_2) \Rightarrow \Gamma_1 \times \Gamma_2 \in \mathcal{F}(\Sigma_1 \times \Sigma_2)$.

$(big * medium \Rightarrow medium)$ $\Gamma_1 \in \mathcal{F}(\Sigma_1)$, $\Gamma_2 \in \mathcal{M}^+(\Sigma_2) \Rightarrow \Gamma_1 \times \Gamma_2 \in \mathcal{M}^+(\Sigma_1 \times \Sigma_2)$.

$(medium * medium \Rightarrow medium)$ $\Gamma_1 \in \mathcal{M}^+(\Sigma_1)$, $\Gamma_2 \in \mathcal{M}^+(\Sigma_2) \Rightarrow \Gamma_1 \times \Gamma_2 \in \mathcal{M}^+(\Sigma_1 \times \Sigma_2)$.

When we accept all the above rules, we can invert $(big * big \Rightarrow big)$, as a big product must be composed of big components. Likewise, at least one component of a small product has to be small; see Proposition 5.3.1.

We see that these properties give a lot of modularity. We can calculate the consequences of α and α' separately—provided α, α' use disjoint alphabets—and put the results together afterwards. Such properties are particularly interesting for classification purposes, where subclasses are defined with disjoint alphabets.

Recall that we work here with a notion of "big" and "small" subsets, which may be thought of as defined by a filter (ideal), though we usually will not need the full strength of a filter (ideal). But assume as usual that $A \subseteq B \subseteq C$ and $A \subseteq C$ is big together imply $B \subseteq C$ is big, and that $C \subseteq C$ is big, and define $A \subseteq B$ as small iff $(B - A) \subseteq B$ is big; call all subsets which are neither big nor small medium size. For an extensive discussion, see Gabbay and Schlechta (2016).

Let $X' \cup X'' = X$ be a disjoint cover, so $\Pi X = \Pi X' \times \Pi X''$. We consider subsets Σ, etc., of ΠX. If not stated otherwise, Σ, etc., need not be a product $\Sigma' \times \Sigma''$. We will sometimes write $\Pi' := \Pi X'$, $\Sigma'' := \Sigma \upharpoonright X''$. The roles of X' and X'' are interchangeable, e.g., instead of $\Gamma \upharpoonright X' \subseteq \Sigma \upharpoonright X'$, we may also write $\Gamma \upharpoonright X'' \subseteq \Sigma \upharpoonright X''$.

We consider here the following two sets of three finite product rules about size and μ. Both sets will be shown to be equivalent in Proposition 5.3.1.

Definition 5.3.1

(1) $(S * 1)$ $\Delta \subseteq \Sigma' \times \Sigma''$ is big iff there is $\Gamma = \Gamma' \times \Gamma'' \subseteq \Delta$ such that $\Gamma' \subseteq \Sigma'$ and $\Gamma'' \subseteq \Sigma''$ are big.

(2) $(S * 2)$ $\Gamma \subseteq \Sigma$ is big $\Rightarrow \Gamma \upharpoonright X' \subseteq \Sigma \upharpoonright X'$ is big, where Σ is not necessarily a product.

(3) $(S * 3)$ $A \subseteq \Sigma$ is big $\Rightarrow$ there is a $B \subseteq \Pi' \times \Sigma''$ big such that $B \upharpoonright X'' \subseteq A \upharpoonright X''$; again, Σ is not necessarily a product.

(4) $(\mu * 1)$ $\mu(\Sigma' \times \Sigma'') = \mu(\Sigma') \times \mu(\Sigma'')$.
Note that μ is relative to different sets. So, using above notation $X = X' \cup X''$, we should write more precisely:
$(\mu * 1)$ $\mu_X(\Sigma' \times \Sigma'') = \mu_{X'}(\Sigma') \times \mu_{X''}(\Sigma'')$, or, for $I = J \cup J'$ instead of $X = X' \cup X''$, as we will write in Chap. 9, $(\mu * 1)$ $\mu_I(\Sigma' \times \Sigma'') = \mu_J(\Sigma') \times \mu_{J'}(\Sigma'')$ etc.
(5) $(\mu * 2)$ $\mu(\Sigma) \subseteq \Gamma \Rightarrow \mu(\Sigma \upharpoonright X') \subseteq \Gamma \upharpoonright X'$.
(6) $(\mu * 3)$ $\mu(\Pi X' \times \Sigma'') \upharpoonright X'' \subseteq \mu(\Sigma) \upharpoonright X''$.
(7) $(s * s)$ Let $\Gamma_i \subseteq \Sigma_i$; then $\Gamma_1 \times \Gamma_2 \subseteq \Sigma_1 \times \Sigma_2$ is small iff $\Gamma_1 \subseteq \Sigma_1$ is small or $\Gamma_1 \subseteq \Sigma_1$ is small. A generalisation to more than two factors is obvious.
(8) One can also consider weakenings, e.g., $(S * 1')$ $\Gamma' \times \Sigma'' \subseteq \Sigma' \times \Sigma''$ is big iff $\Gamma' \subseteq \Sigma'$ is big.

Proposition 5.3.1
This is Proposition 5.2.1 of Gabbay and Schlechta (2011), which also contains the proof.

(1) Let $(S * 1)$ hold. Then
$\Gamma' \times \Gamma'' \subseteq \Sigma' \times \Sigma''$ is small iff $\Gamma' \subseteq \Sigma'$ or $\Gamma'' \subseteq \Sigma''$ is small.
(2) If the filters over A are principal filters, generated by $\mu(A)$, i.e., $B \subseteq A$ is big iff $\mu(A) \subseteq B \subseteq A$ for some $\mu(A) \subseteq A$, then
$(S * i)$ is equivalent to $(\mu * i)$, $i = 1, 2, 3$.
(3) Let the notion of size satisfy (Opt), (iM), and $(< \omega * s)$; see the tables "Rules on size" in Sect. 5.3.2. Then $(\mu * 1)$ and $(s * s)$ are equivalent.

Discussion
We compare these rules to probability defined size.
Let "big" be defined by "more than 50%". If $\Pi X'$ and $\Pi X''$ have three elements each, then subsets of $\Pi X'$ or $\Pi X''$ of *card* ≥ 2 are big. But taking the product may give $4/9 < 1/2$. So the product rule "*big* $*$ *big* $=$ *big*" will not hold there. One direction will hold, of course.
Next, we discuss the prerequisite $\Sigma = \Sigma' \times \Sigma''$. Consider the following example:

Example 5.3.1
Take a language of five propositional variables, with $X' := \{a, b, c\}$, $X'' := \{d, e\}$. Consider the model set $\Sigma := \{\pm a \pm b \pm cde,\ -a - b - c - d \pm e\}$, i.e., of eight models of de and two models of $-d$. The models of de are 8/10 of all elements of Σ, so it is reasonable to call them a big subset of Σ. But its projection on X'' is only 1/3 of Σ''.
So we have a potential *decrease* when going to the coordinates.
This shows that weakening the prerequisite about Σ as done in $(S * 2)$ is not innocent.

Remark 5.3.2
When we set small sets to 0, big sets to 1, we have the following Boolean rules for filters:

(1) $0 + 0 = 0$
(2) $1 + x = 1$
(3) $-0 = 1, \ -1 = 0$
(4) $0 * x = 0$
(5) $1 * 1 = 1$.

There are no such rules for medium size sets, as the union of two medium size sets may be big, but also stay medium.

Such multiplication rules capture the behaviour of Reiter defaults and of defeasible inheritance.

5.3.4 Hamming Relations and Distances

5.3.4.1 Hamming Relations and Multiplication of Size

We now define Hamming relations in various flavours, and then (see Proposition 5.3.4) show that (smooth) Hamming relations generate a notion of size which satisfies our conditions, defined in Definition 5.3.1. This is discussed in more detail in Gabbay and Schlechta (2011), Chap. 5.

We will conclude this section by showing that our conditions $(\mu * 1)$ and $(\mu * 2)$, see Definition 5.3.1, essentially characterise Hamming relations.

Definition 5.3.2
We abuse notation, and define a relation $\subseteq$ on $\Pi X'$, $\Pi X''$, and $\Pi X' \times \Pi X''$.

(1) Define $x \prec y :\Leftrightarrow x \preceq y$ and $x \neq y$; thus $\sigma \preceq \tau$ iff $\sigma \prec \tau$ or $\sigma = \tau$.

(2) We say that a relation $\preceq$ satisfies $(GH3)$ iff

$(GH3)$ $\sigma'\sigma'' \preceq \tau'\tau'' \Leftrightarrow \sigma' \preceq \tau'$ and $\sigma'' \preceq \tau''$.

(Thus, $\sigma'\sigma'' \prec \tau'\tau''$ iff $\sigma'\sigma'' \preceq \tau'\tau''$ and $(\sigma' \prec \tau'$ or $\sigma'' \prec \tau'')$.)

(3) Call a relation $\prec$ a *GH* (= general Hamming) relation iff the following two conditions hold:

$(GH1)$ $\sigma' \preceq \tau' \wedge \sigma'' \preceq \tau'' \wedge (\sigma' \prec \tau' \vee \sigma'' \prec \tau'') \Rightarrow \sigma'\sigma'' \prec \tau'\tau''$

(where $\sigma' \preceq \tau'$ iff $\sigma' \prec \tau'$ or $\sigma' = \tau'$),

$(GH2)$ $\sigma'\sigma'' \prec \tau'\tau'' \Rightarrow \sigma' \prec \tau' \vee \sigma'' \prec \tau''$.

$(GH2)$ means that some compensation is possible, e.g., $\tau' \prec \sigma'$ might be the case, but $\sigma'' \prec \tau''$ wins in the end, so $\sigma'\sigma'' \prec \tau'\tau''$.

We use (GH) for $(GH1) + (GH2)$.

Example 5.3.2
The circumscription relation satisfies $(GH3)$ with $\neg p \leq p$ and $\bigwedge \pm q_i \leq \bigwedge \pm q_i'$ iff $\forall i (\pm q_i \leq \pm q_i')$.

Remark 5.3.3

(1) The independence makes sense because the concept of models, and thus the usual interpolation for classical logic, rely on the independence of the assignments.

(2) This corresponds to social choice for many independent dimensions.

(3) We can also consider such factorisation as an approximation: we can do part of the reasoning independently.

Definition 5.3.3

Given a relation $\preceq$, define as usual a principal filter $\mathcal{F}(X)$ generated by the $\preceq$-minimal elements:

$\mu(X) := \{x \in X : \neg\exists x' \prec x.x' \in X\}$,

$\mathcal{F}(X) := \{A \subseteq X : \mu(X) \subseteq A\}$.

The following proposition summarises various properties for the different Hamming relations:

Proposition 5.3.4

This is Proposition 5.2.4 of Gabbay and Schlechta (2011), which also contains the proof.

Let $\Pi X = \Pi X' \times \Pi X''$, $\Sigma \subseteq \Pi X$.

(1) Let $\preceq$ *be a smooth relation satisfying* $(GH3)$. *Then* $(\mu * 2)$ *holds, and thus* $(S * 2)$ *by Proposition 5.3.1, (2).*

(2) Let again $\Sigma' := \Sigma \restriction X'$, $\Sigma'' := \Sigma \restriction X''$. *Let* $\preceq$ *be a smooth relation satisfying* $(GH3)$. *Then*

$\mu(\Sigma') \times \mu(\Sigma'') \subseteq \Sigma \Rightarrow \mu(\Sigma) = \mu(\Sigma') \times \mu(\Sigma'')$.

(Here $\Sigma = \Sigma' \times \Sigma''$ *will not necessarily hold.)*

(3) Let again $\Sigma' := \Sigma \restriction X'$, $\Sigma'' := \Sigma \restriction X''$. *Let* $\preceq$ *be a relation satisfying* $(GH3)$, *and* $\Sigma = \Sigma' \times \Sigma''$. *Then* $(\mu * 1)$ *holds, and thus, by Proposition 5.3.1, (2),* $(S * 1)$ *holds.*

(4) Let $\preceq$ *be a smooth relation satisfying* $(GH3)$; *then* $(\mu * 3)$ *holds, and thus by Proposition 5.3.1, (2),* $(S * 3)$ *holds.*

(5)$(\mu * 1)$ *and* $(\mu * 2)$ *and the usual axioms for smooth relations characterise smooth relations satisfying* $(GH3)$.

(6) Let $\sigma \prec \tau \Leftrightarrow \tau \notin \mu(\{\sigma, \tau\})$ *and* $\prec$ *be smooth. Then* μ *satisfies* $(\mu * 1)$ *(or, by Proposition 5.3.1, equivalently* $(s * s)$*) iff* $\prec$ *is a GH relation.*

(7) Let $\Gamma' \subseteq \Sigma'$, $\Gamma'' \subseteq \Sigma''$, $\Gamma' \times \Gamma'' \subseteq \Sigma' \times \Sigma''$ *be small; let* $(GH2)$ *hold; then* $\Gamma' \subseteq \Sigma'$ *is small or* $\Gamma'' \subseteq \Sigma''$ *is small.*

(8) Let $\Gamma' \subseteq \Sigma'$ *be small,* $\Gamma'' \subseteq \Sigma''$; *let* $(GH1)$ *hold; then* $\Gamma' \times \Gamma'' \subseteq \Sigma' \times \Sigma''$ *is small.*

Example 5.3.3

Even for smooth relations satisfying $(GH3)$, the converse of $(\mu * 2)$ is not necessarily true:

Let $\sigma' \prec \tau'$, $\tau'' \prec \sigma''$, $\Sigma := \{\sigma, \tau\}$; then $\mu(\Sigma) = \Sigma$, but $\mu(\Sigma') = \{\sigma'\}$, $\mu(\Sigma'') = \{\tau''\}$, so $\mu(\Sigma) \neq \mu(\Sigma') \times \mu(\Sigma'')$.

We need the additional assumption that $\mu(\Sigma') \times \mu(\Sigma'') \subseteq \Sigma$; see Proposition 5.3.4 (2).

Example 5.3.4

The following are examples of *GH* relations:

Define on all components X_i a relation $\prec_i$.

(1) The set variant Hamming relation:

Let the relation $\prec$ be defined on $\Pi\{X_i : i \in I\}$ by $\sigma \prec \tau$ iff for all j $\sigma_j \preceq_j \tau_j$, and there is at least one i such that $\sigma_i \prec_i \tau_i$.

(2) The counting variant Hamming relation:

Let the relation $\prec$ be defined on $\Pi\{X_i : i \in I\}$ by $\sigma \prec \tau$ iff the number of i such that $\sigma_i \prec_i \tau_i$ is bigger than the number of i such that $\tau_i \prec_i \sigma_i$.

(3) The weighed counting Hamming relation:

Like the counting relation, but we give different (numerical) importance to different i, e.g., $\sigma_1 \prec \tau_1$ may count 1, $\sigma_2 \prec \tau_2$ may count 2, etc.

□

Note

Note that $(\mu * 1)$ results in a strong independence result in the second scenario: Let $\sigma\rho' \prec \tau\rho'$; then $\sigma\rho'' \prec \tau\rho''$ for all ρ''. Thus, whether $\{\rho''\}$ is small or medium size (i.e., $\rho'' \in \mu(\Sigma')$), the behaviour of $\Sigma \times \{\rho''\}$ is the same. This we do not have in the first scenario, as small sets may behave very differently from medium size sets. (But, still, their internal structure is the same; only the minimal elements change.) When $(\mu * 2)$ holds, if $\sigma\sigma' \prec \tau\tau'$ and $\sigma \neq \tau$, then $\sigma \prec \tau$, i.e., we need not have $\sigma' = \tau'$.

5.3.4.2 Hamming Distances and Revision

This short section is mainly intended to put our work in a broader perspective, by showing a connection of Hamming distances to modular revision as introduced by Parikh and his co-authors. The main result here is Corollary 5.3.7. We will not go into details of motivation here, and refer the reader to Parikh (1999) for further discussion. An introduction to theory revision is given in Chap. 7.

Thus, we have modular distances and relations, i.e., Hamming distances and relations, we have modular revision as described below, and we have modular logic, which has the (semantical) interpolation property. We want to point out here in particular this cross-reference from modular revision to modular logic, i.e., logic with interpolation.

We recall the following:

Definition 5.3.4
Given a distance d, define for two sets X, Y

$X \mid Y := \{y \in Y : \exists x \in X(\neg \exists x' \in X, y' \in Y.d(x', y') < d(x, y))\}$.

We assume that $X \mid Y \neq \emptyset$ if $X, Y \neq \emptyset$. Note that this is related to the consistency axiom of AGM theory revision: revising by a consistent formula gives a consistent result. The assumption may be wrong due to infinite descending chains of distances.

Definition 5.3.5
Given $\mid$, we can define an AGM revision operator $*$ as follows:

$T * \phi := Th(M(T) \mid M(\phi))$,

where T is a theory, and $Th(X)$ is the set of formulas which hold in all $x \in X$.

It was shown in Lehmann et al. (2001) that a revision operator thus defined satisfies the AGM revision postulates.

Definition 5.3.6
Let d be an abstract distance on some product space $X \times Y$ and its components. (We require distances only that they be comparable, that $d(x, y) = 0$ iff $x = y$, and that $d(x, y) \geq 0$.)

d is called a generalised Hamming distance (GHD) iff it satisfies the following two properties:

($GHD1$) $d(\sigma, \tau) \leq d(\alpha, \beta)$ and $d(\sigma', \tau') \leq d(\alpha', \beta')$ and $(d(\sigma, \tau) < d(\alpha, \beta)$ or $d(\sigma', \tau') < d(\alpha', \beta')) \Rightarrow d(\sigma\sigma', \tau\tau') < d(\alpha\alpha', \beta\beta')$,

($GHD2$) $d(\sigma\sigma', \tau\tau') < d(\alpha\alpha', \beta\beta') \Rightarrow d(\sigma, \tau) < d(\alpha, \beta)$ or $d(\sigma', \tau') < d(\alpha', \beta')$.

(Compare this definition to Definition 5.3.2.)

We have a result analogous to the relation case:

Fact 5.3.5
This is Fact 5.2.5 of Gabbay and Schlechta (2011), which also contains the proof.

Let $\mid$ be defined by a generalised Hamming distance; then $\mid$ satisfies

(1) $(\mid *)$ $(\Sigma_1 \times \Sigma_1') \mid (\Sigma_2 \times \Sigma_2') = (\Sigma_1 \mid \Sigma_2) \times (\Sigma_1' \mid \Sigma_2')$,

(2) $(\Sigma_1' \mid \Sigma_2') \times (\Sigma_1'' \mid \Sigma_2'') \subseteq \Sigma_2$ and $(\Sigma_2' \mid \Sigma_1') \times (\Sigma_2'' \mid \Sigma_1'') \subseteq \Sigma_1 \Rightarrow (\Sigma_1) \mid (\Sigma_2) = (\Sigma_1' \mid \Sigma_2') \times (\Sigma_1'' \mid \Sigma_2'')$ if the distance is symmetric

(where Σ_i is not necessarily $\Sigma_i' \times \Sigma_i''$, etc.).

These properties translate to logic as follows:

Corollary 5.3.6
If ϕ and ψ are defined on a separate language from that of ϕ' and ψ', and the distance satisfies ($GHD1$) and ($GHD2$), then for revision the following holds:

$(\phi \wedge \phi') * (\psi \wedge \psi') = (\phi * \psi) \wedge (\phi' * \psi')$.

Corollary 5.3.7
By Corollary 5.3.6, Hamming distances generate decomposable revision operators a la Parikh (see Parikh (1999)) also in the generalised form of variable K and ϕ.

We conclude with a small result on partial (semantical) revision:

Fact 5.3.8
This is Fact 5.2.8 of Gabbay and Schlechta (2011), which also contains the proof.
Let $|$ be defined by a Hamming distance, then $\Pi X \mid \Sigma \subseteq \Gamma \Rightarrow \Pi X' \mid (\Sigma \upharpoonright X') \subseteq \Gamma \upharpoonright X'$.
(Recall that Π' is the restriction of Π to X'.)

5.3.4.3 Discussion of Representation

It would be nice to have a representation result like the one for Hamming relations; see Proposition 5.3.4, (5). But this is impossible, for the following reason:

In constructing the representing distance from revision results, we made arbitrary choices (see the proofs in Lehmann et al. (2001) or Schlechta (2004)). That is, we choose sometimes arbitrarily $d(x, y) \leq d(x', y')$ when we do not have enough information to decide. (This is an example of the fact that the problem of "losing ignorance" should not be underestimated; see, e.g., Gabbay and Schlechta (2009).) As we do not follow the same procedure for all cases, there is no guarantee that the different representations will fit together.

Of course, it might be possible to come to a uniform choice, and one could then attempt a representation result. This is left as an open problem.

5.3.5 *Summary of Properties*

We summarise in this section properties related to multiplicative laws.

They are collected in Table 5.4.

$pr(b) = b$ means the projection of a big set on one of its coordinates is big again.

Note that $A \times B \subseteq X \times Y$ big $\Rightarrow A \subseteq X$ big is intuitively better justified than the other direction, as the proportion might increase in the latter, and decrease in the former. See the table "Rules on size", Sect. 5.3.2, "increasing proportions".

Table 5.4 Multiplication laws

Multiplication laws

Multiplication law	Scenario 1 (see Diagram 5.3.1)			Scenario 2 (* symmetrical, only one side shown) (see Diagram 5.3.2)				
	Corresponding algebraic addition property	Logical property	Relation property	Algebraic property ($\Gamma_i \subseteq \Sigma_i$)	Logical property (α, β in L_1, α', β' in L_2 $L = L_1 \cup L_2$ (disjoint))	Interpolation		
						Multiplic. law	Relation property	Interpolation
Non-monotonic logic								
$x * 1 \Rightarrow x$	Trivial			$\Gamma_1 \in \mathcal{F}(\Sigma_1) \Rightarrow$ $\Gamma_1 \times \Sigma_2 \in \mathcal{F}(\Sigma_1 \times \Sigma_2)$	$\alpha \mid\!\sim_{L_i} \beta \Rightarrow \alpha \mid\!\sim_L \beta$			
$1 * x \Rightarrow x$	Trivial							
$x * s \Rightarrow s$	(iM) $A \subseteq B \in \mathcal{I}(X) \Rightarrow A \in \mathcal{I}(X)$	$\alpha \mid\!\sim \neg\beta \Rightarrow$ $\alpha \mid\!\sim \neg\beta \vee \gamma$	–	$\Gamma_1 \in I(\Sigma_1) \Rightarrow$ $\Gamma_1 \times \Gamma_2 \in I(\Sigma_1 \times \Sigma_2)$	$\alpha \mid\!\sim_{L_1} \beta$, $\beta' \vdash_{L_2} \alpha' \Rightarrow$ $\alpha \wedge \alpha' \mid\!\sim (\beta \wedge \alpha') \vee (\alpha \wedge \beta')$			
$s * x \Rightarrow s$	$(eM\mathcal{I})$ $X \subseteq Y \Rightarrow \mathcal{I}(X) \subseteq \mathcal{I}(Y)$, $X \subseteq Y \Rightarrow$ $\mathcal{F}(Y) \cap \mathcal{P}(X) \subseteq \mathcal{F}(X)$	$\alpha \wedge \beta \mid\!\sim \neg\gamma \Rightarrow$ $\alpha \mid\!\sim \neg\beta \vee \neg\gamma$	–					
$b * b \Rightarrow b$ $(\mu * 1)$	$(< \omega * s)$, $(\mathcal{M}^+_\omega)$ (3) $A \in \mathcal{F}(X), X \in \mathcal{F}(Y) \Rightarrow$ $A \in \mathcal{F}(Y)$	$\alpha \mid\!\sim \beta$, $\alpha \wedge \beta \mid\!\sim \gamma$ $\Rightarrow \alpha \mid\!\sim \gamma$	– (Filter)	$\Gamma_1 \in \mathcal{F}(\Sigma_1), \Gamma_2 \in \mathcal{F}(\Sigma_2) \Rightarrow$ $\Gamma_1 \times \Gamma_2 \in \mathcal{F}(\Sigma_1 \times \Sigma_2)$	$\alpha \mid\!\sim_{L_1} \beta$, $\alpha' \mid\!\sim_{L_2} \beta' \Rightarrow$ $\alpha \wedge \alpha' \mid\!\sim_L \beta \wedge \beta'$	$b * b \Leftrightarrow b$ $(\mu * 1)$	(GH)	$\mid\!\sim \circ \mid\!\sim$

$b * m \Rightarrow m$	$(< \omega * s)$, $(\mathcal{M}^{+}_{\omega})$ (2) $A \in \mathcal{M}^{+}(X), X \in \mathcal{F}(Y) \Rightarrow$ $A \in \mathcal{M}^{+}(Y)$	$\alpha \mid\!\sim \beta, \alpha \wedge \beta \not\mid\!\sim \neg\gamma$ $\Rightarrow \alpha \not\mid\!\sim \neg\beta \vee \neg\gamma$	– (Filter)	$\Gamma_1 \in \mathcal{F}(\Sigma_1), \Gamma_2 \in \mathcal{M}^{+}(\Sigma_2) \Rightarrow$ $\Gamma_1 \times \Gamma_2 \in \mathcal{M}^{+}(\Sigma_1 \times \Sigma_2)$	$\alpha \not\mid\!\sim_{L_1} \neg\beta, \alpha' \mid\!\sim_{L_2} \beta' \Rightarrow$ $\alpha \wedge \alpha' \not\mid\!\sim_{L} \neg(\beta \wedge \beta')$			
$m * b \Rightarrow m$	$(< \omega * s)$, $(\mathcal{M}^{+}_{\omega})$ (1) $A \in \mathcal{F}(X), X \in \mathcal{M}^{+}(Y) \Rightarrow$ $A \in \mathcal{M}^{+}(Y)$	$\alpha \not\mid\!\sim \neg\beta, \alpha \wedge \beta \mid\!\sim \gamma$ $\Rightarrow \alpha \not\mid\!\sim \neg\beta \vee \neg\gamma$	– (Filter)					
$m * m \Rightarrow m$	$(\mathcal{M}^{++})$ $A \in \mathcal{M}^{+}(X), X \in \mathcal{M}^{+}(Y)$ $\Rightarrow A \in \mathcal{M}^{+}(Y)$	Rational monotony	Ranked	$\Gamma_1 \in \mathcal{M}^{+}(\Sigma_1), \Gamma_2 \in \mathcal{M}^{+}(\Sigma_2)$ $\Rightarrow$ $\Gamma_1 \times \Gamma_2 \in \mathcal{M}^{+}(\Sigma_1 \times \Sigma_2)$	$\alpha \not\mid\!\sim_{L_1} \neg\beta, \alpha' \not\mid\!\sim_{L_2} \neg\beta' \Rightarrow$ $\alpha \wedge \alpha' \not\mid\!\sim_{L} \neg(\beta \wedge \beta')$			
$b * b \Leftrightarrow b$, $pr(b) = b$ $(\mu * 2)$					$\alpha \mid\!\sim \beta \Rightarrow \alpha \restriction L_1 \mid\!\sim \beta \restriction L_1$ and $\alpha \mid\!\sim_{L_1} \beta, \alpha' \mid\!\sim_{L_2} \beta' \Rightarrow$ $\alpha \wedge \alpha' \mid\!\sim_{L} \beta \wedge \beta'$	$(\mu * 1)$ + $(\mu * 2)$	(GH3)	$\vdash \circ \mid\!\sim$
J' small					$\alpha \wedge \alpha' \mid\!\sim \beta \wedge \beta' \Leftrightarrow$ $\alpha \mid\!\sim \beta, \alpha' \vdash \beta'$		*forget*(J')	-
Theory revision								
				$(\mid *)$: $(\Sigma_1 \times \Sigma'_1) \mid (\Sigma_2 \times \Sigma'_2) =$ $(\Sigma_1 \mid \Sigma_2) \times (\Sigma'_1 \mid \Sigma'_2)$	$(\phi \wedge \phi') * (\psi \wedge \psi') =$ $(\phi * \psi) \wedge (\phi' * \psi')$		*(GHD)*	$(\phi \wedge \phi') * (\psi \wedge \psi') \vdash \rho \Rightarrow$ $\phi' * \psi' \vdash \rho\ \phi$, ψ in J, ϕ', ψ', ρ in $L - J$

5.4 A Short Discussion of Their Abstract Properties in Our Context

Preferential structures are not only well examined, but they also have nice logical properties, like “or”. They have a “core”, the ideal situation—unless there are no minimal models, and then everything collapses, just as with inconsistent classical theories—but they do not treat subideal situations. It is all or nothing.

Part of this is due to the fact that general preferential structures are purely algebraic, the relation is not controlled by the language. This points already to a solution, as adopted in Chap. 11. We do not consider arbitrary preferential relations, but construct special ones, respecting the language. In addition, working without copies makes the structure more flexible, offering graceful degradation. (Copies provide, in principle, an all-or-nothing result: we need the whole set to minimize the non-minimal elements.)

Chapter 6
Deontic Logic, Contrary-to-Duty Obligations

6.1 Summary

We mainly concentrate on contrary-to-duty obligations (*CTD*'s), a special form of obligation. CTDs correspond to groups of situations, where all members of group 1 are "better" than all members of group 2, better than all members of group 3, etc. Intuitively, if you cannot satisfy obligation 1, thus cannot create a situation in group 1, then you should at least satisfy obligation 2, and if not even this, then obligation 3, etc.

We discuss

- $\mathcal{A}$-ranked preferential structures, formalising above groups, in several variants,
- semantic representation results for those structures,
- syntactic representation results,
- add an accessibility relation to such structures in Sect. 6.5 and
- give a representation result for the full structure with the accessibility relation in Sect. 6.5.4.

6.2 Deontic Logic

Deontic logic is the logic of obligations. In situation X, one should do y.

Deontic logic was plagued with paradoxa. For instance, obligations are not closed under "or" with arbitrary other commands. It is an obligation that one should not steal, but not that (you should not steal or kill the neighbour's cat). This is Ross' paradox. (The situation is different when the second command is also a reasonable obligation.) If you should buy a cake, and you should buy bread, then you should buy a cake or bread, this is better than nothing.

D.M. Gabbay and K. Schlechta, *A New Perspective on Nonmonotonic Logics*,
DOI 10.1007/978-3-319-46817-4_6

A semantics for obligations was given by Hansson, see Hansson (1971), by preferential models, before they were re-discovered in the context of nonmonotonic logic.

For more on the subject, see e.g. Meyer et al. (1994), Gabbay and Schlechta (2016).

6.3 Contrary-to-Duty Obligations

A much discussed problem is the analysis and modelling of contray-to-duty obligations. Suppose you are not to have a fence in your neighbourhood. If you do not oblige, e.g. you have a dog, and do not want it to stray, then you should at least paint it white, etc. So the best situation is where there is no fence, the second best, where there is a fence, but it is painted white, etc. This is discussed in detail in Gabbay and Schlechta (2008f), and we present another solution in Chap. 13. Closure conditions for obligations are discussed in Gabbay and Schlechta (2008f).

Our solution to contrary-to-duty (CTD) obligations in Gabbay and Schlechta (2008f), $\mathcal{A}$-ranked structures, can be seen as a precursor to our much more complex approach in Chap. 11. For this reason, and for the convenience of the reader, we repeat this solution here in detail, but without proofs. The approach in Chap. 11 is given without representation results. But the full development of the CTD situation is probably enough to point towards representation results and techniques for the situation in Chap. 11. The solution of the CTD case is a special preferential structure, which is organised in layers. The lowest layer corresponds to the ideal obligation (no fence in our above example), the second layer from below to the second best obligation (fence, but painted white), etc. We think that this is a very natural and simple solution to the question of contrary-to-duty obligations.

We refine now the above example:

Example 6.3.1
Consider the following hierarchy of obligations making fences as unobtrusive as possible, involving contrary-to-duty obligations.

(1) You should have no fence (main duty).
(2) If this is impossible (e.g. you have a dog which might invade neighbours' property), it should be less than 3-feet high (contrary to duty, but second best choice).
(3) If this is impossible too (e.g. your dog might jump over it), it should be white (even more contrary to duty, but still better than nothing).
(4) If all is impossible, you should get the neighbours' consent (etc.).

Before we describe our approach to the CTD problem, we introduce the necessary formal machinery.

6.4 $\mathcal{A}$-Ranked Structures

We go into more detail here for the following two reasons:

(1) the present construction is in many aspect simpler than the one carried out in Chap. 11,
(2) Chapter 13 will see a very different treatment of contrary-to-duty obligations, so it seemed good to present our approach of Gabbay and Schlechta (2008f) here in more detail.

For proofs and more detail and discussion, the reader is referred to Gabbay and Schlechta (2008f).

Definition 6.4.1
Let $\boldsymbol{A}$ be a fixed set, and $\mathcal{A}$ a finite, totally ordered (by $<$) disjoint cover by non-empty subsets of $\boldsymbol{A}$.

For $x \in \boldsymbol{A}$, let $rg(x)$ the unique $A \in \mathcal{A}$ such that $x \in A$, so $rg(x) < rg(y)$ is defined in the natural way.

A preferential structure $\langle \mathcal{X}, \prec \rangle$ ($\mathcal{X}$ a set of pairs $\langle x, i \rangle$) is called $\mathcal{A}$-ranked iff for all x, x' $rg(x) < rg(x')$ implies $\langle x, i \rangle \prec \langle x', i' \rangle$ for all $\langle x, i \rangle, \langle x', i' \rangle \in \mathcal{X}$.

6.4.1 *Representation Results for $\mathcal{A}$-Ranked Structures*

6.4.1.1 Discussion

The not necessarily smooth and the smooth case will be treated differently.

Strangely, the smooth case is simpler, as an added new layer in the proof settles it. Yet, this is not surprising when looking closer, as minimal elements never have higher rank, and we know from (μCUM) that minimising by minimal elements suffices. All we have to add that any element in the minimal layer minimises any element higher up.

In the simple, not necessarily smooth case, we have to go deeper into the original proof to obtain the result.

The following idea, inspired by the treatment of the smooth case, will not work: Instead of minimising by arbitrary elements, minimise only by elements of minimal rank, as the following example shows. If it worked, we might add just another layer to the original proof without $(\mu\mathcal{A})$, (see Definition 6.4.2), as in the smooth case.

Example 6.4.1
Consider the base set $\{a, b, c\}$, $\mu(\{a, b, c\}) = \{b\}$, $\mu(\{a, b\}) = \{a, b\}$, $\mu(\{a, c\}) = \emptyset$, $\mu(\{b, c\}) = \{b\}$, $\mathcal{A}$ defined by $\{a, b\} < \{c\}$.

Obviously, $(\mu\mathcal{A})$ is satisfied. μ can be represented by the (not transitive!) relation $a \prec c \prec a$, $b \prec c$, which is $\mathcal{A}$-ranked.

But trying to minimise a in $\{a, b, c\}$ in the minimal layer will lead to $b \prec a$, and thus $a \notin \mu(\{a, b\})$, which is wrong. □ .

The proofs of the general and transitive general case are (minor) adaptations of the proofs for general preferential structures. For the smooth case, we only have to add a supplementary layer in the end which will make the construction $\mathcal{A}$-ranked.

6.4.1.2 $\mathcal{A}$-Ranked General and Transitive Structures

Introduction

We will show here the following representation result:

Let $\mathcal{A}$ be given.

An operation $\mu : \mathcal{Y} \to \mathcal{P}(Z)$ is representable by an $\mathcal{A}-$ranked preferential structure iff μ satisfies $(\mu \subseteq)$, (μPR), $(\mu\mathcal{A})$ (Proposition 6.4.1), and, moreover, the structure can be chosen transitive (Proposition 6.4.2).

Note that we carefully avoid any unnecessary assumptions about the domain $\mathcal{Y} \subseteq \mathcal{P}(Z)$ of the function μ.

Definition 6.4.2
We define a new condition:

Let $\mathcal{A}$ be given as defined in Definition 6.4.1.

$(\mu\mathcal{A})$ If $X \in \mathcal{Y}$, $A, A' \in \mathcal{A}$, $A < A'$, $X \cap A \neq \emptyset$, $X \cap A' \neq \emptyset$ then $\mu(X) \cap A' = \emptyset$.

This new condition will be central for the modified representation.

The Basic, Not Necessarily Transitive, Case

Proposition 6.4.1
Let $\mathcal{A}$ be given.

An operation $\mu : \mathcal{Y} \to \mathcal{P}(Z)$ is representable by an $\mathcal{A}-$ranked preferential structure iff μ satisfies $(\mu \subseteq)$, (μPR),$(\mu\mathcal{A})$.

The Transitive Case

Proposition 6.4.2
Let $\mathcal{A}$ be given.

An operation $\mu : \mathcal{Y} \to \mathcal{P}(Z)$ is representable by an $\mathcal{A}-$ranked transitive preferential structure iff μ satisfies $(\mu \subseteq)$, (μPR), $(\mu\mathcal{A})$.

6.4.1.3 $\mathcal{A}$-Ranked Smooth Structures

All smooth cases have a simple solution. We use one of our existing proofs for the not necessarily $\mathcal{A}$-ranked case, and add one litte result:

Fact 6.4.3
Let $(\mu\mathcal{A})$ hold, and let $\mathcal{Z} = \langle \mathcal{X}, \prec \rangle$ be a smooth preferential structure representing μ, i.e. $\mu = \mu_{\mathcal{Z}}$.

Suppose that
$\langle x, i \rangle \prec \langle y, j \rangle$ implies $rg(x) \leq rg(y)$.

Define $\mathcal{Z}' := \langle \mathcal{X}, \sqsubset \rangle$ where $\langle x, i \rangle \sqsubset \langle y, j \rangle$ iff $\langle x, i \rangle \prec \langle y, j \rangle$ or $rg(x) < rg(y)$.

Then $\mathcal{Z}'$ is $\mathcal{A}$-ranked.
$\mathcal{Z}'$ is smooth, too, and $\mu_{\mathcal{Z}} = \mu_{\mathcal{Z}'} =: \mu'$.

In addition, if $\prec$ is free from cycles, so is $\sqsubset$, if $\prec$ is transitive, so is $\sqsubset$.

All that remains to show then is that our constructions of smooth and of smooth and transitive structures satisfy the condition
$\langle x, i \rangle \prec \langle y, j \rangle$ implies $rg(x) \leq rg(y)$.

Proposition 6.4.4
Let $\mathcal{A}$ be given.

Let - for simplicity - $\mathcal{Y}$ be closed under finite unions, and $\mu : \mathcal{Y} \rightarrow \mathcal{P}(Z)$. Then there is a $\mathcal{Y}$-smooth $\mathcal{A}$-ranked preferential structure $\mathcal{Z}$, s.t. for all $X \in \mathcal{Y}$ $\mu(X) = \mu_{\mathcal{Z}}(X)$ iff μ satisfies $(\mu \subseteq)$, (μPR), (μCUM), $(\mu\mathcal{A})$.

Proposition 6.4.5
Let $\mathcal{A}$ be given.

Let - for simplicity - $\mathcal{Y}$ be closed under finite unions, and $\mu : \mathcal{Y} \rightarrow \mathcal{P}(Z)$. Then there is a $\mathcal{Y}$-smooth $\mathcal{A}$-ranked transitive preferential structure $\mathcal{Z}$, s.t. for all $X \in \mathcal{Y}$ $\mu(X) = \mu_{\mathcal{Z}}(X)$ iff μ satisfies $(\mu \subseteq)$, (μPR), (μCUM), $(\mu\mathcal{A})$.

6.4.1.4 The Logical Properties with Definability Preservation

Proposition 6.4.6
Let $\mid\sim$ be a logic for $\mathcal{L}$. Set $T^{\mathcal{M}} := Th(\mu_{\mathcal{M}}(M(T)))$, where $\mathcal{M}$ is a preferential structure.

(1) Then there is a (transitive) definability preserving classical preferential model $\mathcal{M}$ s.t. $\overline{\overline{T}} = T^{\mathcal{M}}$ iff (LLE), (CCL), (SC), (PR) hold for all $T, T' \subseteq \mathcal{L}$.
(2) The structure can be chosen smooth, iff, in addition (CUM) holds.
(3) The structure can be chosen $\mathcal{A}$-ranked, iff, in addition
$(\mathcal{A}$-min) $T \not\vdash \neg\alpha_i$ and $T \not\vdash \neg\alpha_j$, $i < j$ implies $\overline{\overline{T}} \vdash \neg\alpha_j$ holds.

Proposition 6.4.7
Consider for a logic $\mid\sim$ on $\mathcal{L}$ the properties
(LLE), (CCL), (SC), (PR), (CUM), $(\mathcal{A}$-min) hold for all $T, T' \subseteq \mathcal{L}$,
and for a function $\mu : \boldsymbol{D}_{\mathcal{L}} \rightarrow \mathcal{P}(M_{\mathcal{L}})$ the properties

(μdp) μ is definability preserving, i.e. $\mu(M(T)) = M(T')$ for some T'
$(\mu \subseteq)$, (μPR), (μCUM), $(\mu \mathcal{A})$
for all $X, Y \in \boldsymbol{D}_{\mathcal{L}}$.

It then holds:

(a) If μ satisfies (μdp), $(\mu \subseteq)$, (μPR), then $\mid\!\sim$ defined by $\overline{\overline{T}} := T^{\mu} := Th(\mu(M(T)))$ satisfies (LLE), (CCL), (SC), (PR). If μ satisfies in addition (μCUM), then (CUM) will hold, too. If μ satisfies in addition $(\mu \mathcal{A})$, then $(\mathcal{A}$-min) will hold, too.
(b) If $\mid\!\sim$ satisfies (LLE), (CCL), (SC), (PR), then there is $\mu : \boldsymbol{D}_{\mathcal{L}} \to \mathcal{P}(M_{\mathcal{L}})$ s.t. $\overline{\overline{T}} = T^{\mu}$ for all $T \subseteq \mathcal{L}$ and μ satisfies (μdp), $(\mu \subseteq)$, (μPR). If, in addition, (CUM) holds, then (μCUM) will hold, too. If, in addition, $(\mathcal{A}$-min) holds, then $(\mu \mathcal{A})$ will hold, too.

6.5 Application of $\mathcal{A}$-Ranked Structures to Contrary-to-Duty Conditionals

6.5.1 Outline of the Solution

Example 6.3.1 can be modelled as follows ($\mu(x)$ is the minimal models of x):
Layer 1: $\mu(True)$: all best models have no fence.
Layer 2: $\mu(fence)$: all best models with a fence are less than 3 ft. high.
Layer 3: μ(*fence* and more than 3 ft. high): all best models with a tall fence have a white fence.
Layer 4: μ(*fence* and non-white and $\geq$ 3 ft): in all best models with a non-white fence taller than 3 feet, you have permission
Layer 5: all the rest
This will be modelled by a corresponding $\mathcal{A}$-ranked structure.

In summary:

(1) We have a hierarchy of situations, where one group (e.g. preventing accidents) is strictly better than another group (e.g. helping victims).
(2) Within each group, preferences are not so clear (first help person A, or person B, first call ambulance, etc.?).
(3) We have a subset of situations which are attainable, this can be modelled by an accessibility relation which tells us which situations are possible or can be reached.

Diagram 6.5.1

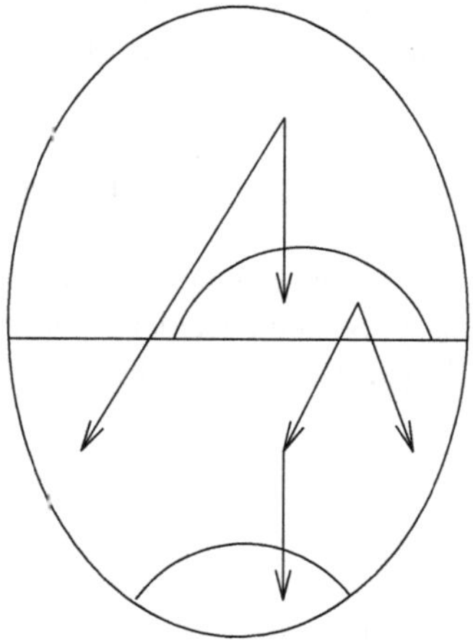

Each layer behaves inside like any preferential structure.

Amongst each other, layers behave like ranked structures.

$\mathcal{A}-$ **ranked structure**

We combine all three ideas, consider what we call $\mathcal{A}$-ranked structures, structures which are organised in levels A_1, A_2, A_3, etc., where all elements of A_1 are better than any element of A_2 — this is basically rankedness —, and where inside each A_i we have an arbitrary relation of preference. Thus, an $\mathcal{A}$-ranked structure is between a simple preferential structure and a fully ranked structure.

See Diagram 6.5.1.

Remark: It is not at all necessary that the rankedness relation between the different layers and the relation inside the layers express the same concept. For instance, rankedness may express deontic preference, whereas the inside relation expresses normality or some usualness.

In addition, we have an accessibility relation R, which tells us which situations are reachable.

It is perhaps easiest to motivate the precise choice of modelling by layered (or contrary to duty) obligations.

For any point t, let $R(t) := \{s : tRs\}$, the set of R-reachable points from t. Given a preferential structure $\mathcal{X} := \langle X, \prec\rangle$, we can relativize $\mathcal{X}$ by considering only those points in X, which are reachable from t.

Let $X' \subseteq X$, and $\mu(X')$ the minimal points of X, we will now consider $\mu(X') \cap R(t)$ - attention, not: $\mu(X' \cap R(t))$! This choice is motivated by the following: norms are universal, and do not depend on one's situation t.

If $\mathcal{X}$ describes a simple obligation, then we are obliged to Y iff $\mu(X') \cap R(t) \neq \emptyset$, and $\mu(X') \cap R(t) \subseteq Y$. The first clause excludes obligations to the unattainable. We can write this as follows, supposing that X' is the set of models of ϕ', and Y is the set of models of ψ :
$m \models \phi' > \psi$.
Thus, we put the usual consequence relation $\mid\sim$ into the object language as $>$, and relativize to the attainable (from m).

If an $\mathcal{A}$-ranked structure has two or more layers, then we are, if possible, obliged to fulfill the lower obligation, e.g. prevent an accident, but if this is impossible, we are obliged to fulfill the upper obligation, e.g. help the victims, etc.

See Diagram 6.5.2.

Diagram 6.5.2

The overall structure is visible from t

Only the inside of the circle is visible from s

Half-circles are the sets of minimal elements of layers

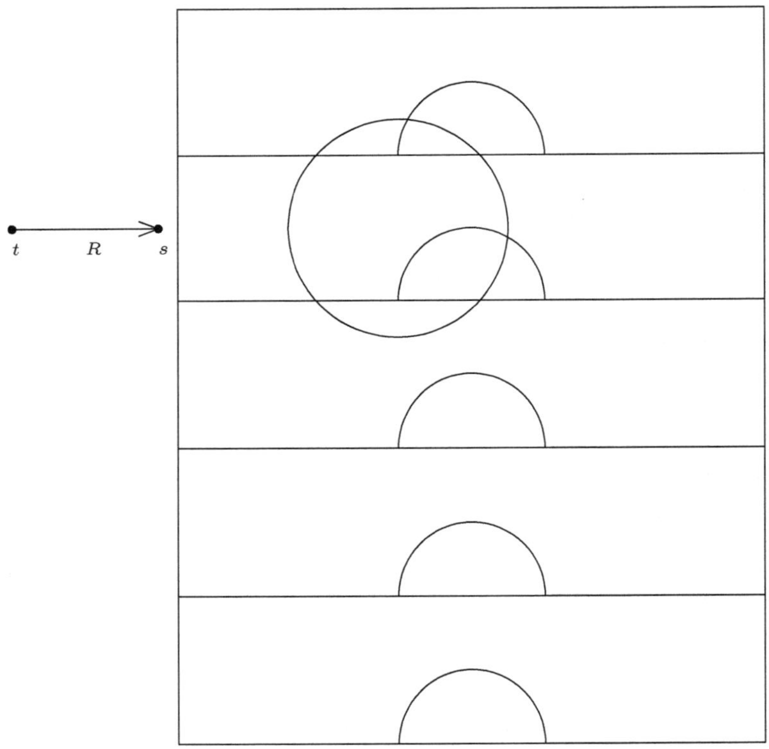

$\mathcal{A}-$ **ranked structure and accessibility**

Let now, for simplicity, $\boldsymbol{B}$ be a subset of the union of all layers A, and let $\boldsymbol{B}$ be the set of models of β. This can be done, as the individual subset can be found by considering $A \cap \boldsymbol{B}$, and call the whole structure $\langle \mathcal{A}, \boldsymbol{B} \rangle$.

Then we say that m satisfies $\langle \mathcal{A}, \boldsymbol{B} \rangle$ iff in the lowest layer A where $\mu(A) \cap R(m) \neq \emptyset$ $\mu(A) \cap R(m) \subseteq \boldsymbol{B}$.

When we want a terminology closer to usual conditionals, we may write, e.g. $(A_1 > B_1; A_2 > B_2; \ldots.)$ expressing that the best is A_1, and then B_1 should hold, the second best is A_2, then B_2 should hold, etc. (The B_i are just $A_i \cap \boldsymbol{B}$.) See Diagram 6.5.3.

6.5.2 Formal Modelling and Summary of Results

We started with an investigation of "best fulfillment" of abstract requirements, and contrary-to-duty obligations. - See also Gabbay (2008) and Gabbay (2008a).

It soon became evident that semi-ranked preferential structures give a natural semantics to contrary-to-duty obligations, just as simple preferential structures give a natural semantics to simple obligations — the latter goes back to Hansson Hansson (1971).

A semi-ranked - or $\mathcal{A}$-ranked preferential structure, as we will call them later, as they are based on a system of sets $\mathcal{A}$ - has a finite number of layers, which amongst them are totally ordered by a ranking, but the internal ordering is just any (binary) relation. It thus has stronger properties than a simple preferential structure, but not as strong ones as a (totally) ranked structure.

The idea is to put the (cases of the) strongest obligation at the bottom, and the weaker ones more towards the top. Then, fulfillment of a strong obligation makes the whole obligation automatically satisfied, and the weaker ones are forgotten.

Beyond giving a natural semantics to contrary to duty obligations, semi-ranked structures seem very useful for other questions of knowledge representation. For instance, any blackbird might seem a more normal bird than any penguin, but we might not be so sure within each set of birds.

Thus, this generalisation of preferential semantics seems very natural and welcome.

The second point of this chapter is to make some, but not necessarily all, situations accessible to each point of departure. Thus, if we imagine agent a to be at point p, some fulfillments of the obligation, which are reachable to agent a' from point p' might just be impossible to reach for him. Thus, we introduce a second relation, of accessibility in the intuitive sense, denoting situations which can be reached. If this relation is transitive, then we have restrictions on the set of reachable situations: if p is accessible from p', and p can access situation s, then so can p', but not necessarily the other way round.

On the formal side, we characterise:

(1) $\mathcal{A}$-ranked structures,
(2) satisfaction of an $\mathcal{A}$-ranked conditional once an accessibility relation between the points p, p', etc. is given.

For the convience of the reader, we now state the main formal results of this section — together with the more unusual definitions.

On (1):

Let $\boldsymbol{A}$ be a fixed set, and $\mathcal{A}$ a finite, totally ordered (by $<$) disjoint cover by non-empty subsets of $\boldsymbol{A}$.

For $x \in \boldsymbol{A}$, let $rg(x)$ the unique $A \in \mathcal{A}$ such that $x \in A$, so $rg(x) < rg(y)$ is defined in the natural way.

A preferential structure $\langle \mathcal{X}, \prec \rangle$ ($\mathcal{X}$ a set of pairs $\langle x, i \rangle$) is called $\mathcal{A}$-ranked iff for all x, x' $rg(x) < rg(x')$ implies $\langle x, i \rangle \prec \langle x', i' \rangle$ for all $\langle x, i \rangle, \langle x', i' \rangle \in \mathcal{X}$. See Definition 5.2.1 for the definition of preferential structures, and Diagram 6.5.1 for an illustration.

We then have:

Let $\mid\sim$ be a logic for $\mathcal{L}$. Set $T^{\mathcal{M}} := Th(\mu_{\mathcal{M}}(M(T)))$, and $\overline{\overline{T}} := \{\phi : T \mid\sim \phi\}$. where $\mathcal{M}$ is a preferential structure.

(1) Then there is a (transitive) definability preserving classical preferential model $\mathcal{M}$ s.t. $\overline{\overline{T}} = T^{\mathcal{M}}$ iff (LLE), (CCL), (SC), (PR) hold for all $T, T' \subseteq \mathcal{L}$.
(2) The structure can be chosen smooth, iff, in addition (CUM) holds.
(3) The structure can be chosen $\mathcal{A}$-ranked, iff, in addition
    ($\mathcal{A}$-min) $T \not\vdash \neg\alpha_i$ and $T \not\vdash \neg\alpha_j$, $i < j$ implies $\overline{\overline{T}} \vdash \neg\alpha_j$ holds.

See Definition 5.2.2 for the logic defined by a preferential structure, Definition 2.2.6 for the logical conditions, Definition 5.2.3 for smoothness.

On (2)

Given a transitive accessibility relation R, $R(m) := \{x : mRx\}$.

Given $\mathcal{A}$ as above, let $\boldsymbol{B} \subseteq \boldsymbol{A}$ be the set of "good" points in $\boldsymbol{A}$, and set $\mathcal{C} := \langle \mathcal{A}, \boldsymbol{B} \rangle$.

We define:

(1) $\mu(\mathcal{A}) := \bigcup\{\mu(A_i) : i \in I\}$
    (warning: this is NOT $\mu(\boldsymbol{A})$)
(2) $\mathcal{A}_m := R(m) \cap \boldsymbol{A}$,
(3) $\mu(\mathcal{A}_m) := \bigcup\{\mu(A_i) \cap R(m) : i \in I\}$
(3a) $\nu(\mathcal{A}_m) := \mu(\mu(\mathcal{A}_m))$
    (thus $\nu(\mathcal{A}_m) = \{a \in \boldsymbol{A} : \exists A \in \mathcal{A}(a \in \mu(A), a \in R(m)$, and $\neg\exists a'(\exists A' \in \mathcal{A}(a' \in \mu(A'), a' \in R(m), a' \prec a\}$.
(4) $m \models \mathcal{C} :\leftrightarrow \nu(\mathcal{A}_m)) \subseteq \boldsymbol{B}$.

See Diagram 6.5.3

Then the following hold:

Let $m, m' \in M$, $A, A' \in \mathcal{A}$, $\boldsymbol{A}$ be the set of models of α.

(1) $m \models \Box\neg\alpha$, mRm' $\Rightarrow m' \models \Box\neg\alpha$
(2) mRm', $\nu(\mathcal{A}_m) \cap A \neq \emptyset$, $\nu(\mathcal{A}_{m'}) \cap A' \neq \emptyset$, $\Rightarrow A \leq A'$ (in the ranking)
(3) mRm', $\nu(\mathcal{A}_m) \cap A \neq \emptyset$, $\nu(\mathcal{A}_{m'}) \cap A' \neq \emptyset$, $m \models \mathcal{C}$, $m' \not\models \mathcal{C}$, $\Rightarrow A < A'$

Conversely, these conditions suffice to construct an accessibility relation between M and $\boldsymbol{A}$ satisfying them, so they are sound and complete.

6.5.3 Overview

Next, we will give special definitions for our framework.

We then start the main formal part, and give representation results for $\mathcal{A}$-ranked structures, first for the general case, then for the smooth case.

Finally, we characterise changes due to restricted accessibility.

Definition 6.5.1

We have the usual framework of preferential structures, i.e. either a set with a possibly non-injective labelling function, or, equivalently, a set of possible worlds with copies. The relation of the preferential structure will be fixed, and will not depend on the point m from where we look at it.

Next, we have a set $\boldsymbol{A}$, and a finite, disjoint cover $A_i : i < n$ of $\boldsymbol{A}$, with a relation "of quality" $<$, $\mathcal{A}$ will denote the A_i (and thus $\boldsymbol{A}$), and $<$, i.e. $\mathcal{A} = \langle\{A_i : i \in I\}, <\rangle$.

(We may assume that all A_i are described by a formula.)

Finally, we have $\boldsymbol{B} \subseteq \boldsymbol{A}$, the subset of "good" elements of $\boldsymbol{A}$ - which we also assume to be described by a formula.

In addition, we have a binary relation of accessibility, R, which we assume transitive - modal operators will be defined relative to R. R determines which part of the preferential structure is visible.

Let $R(s) := \{t : sRt\}$.

Note that automatically for $X \subseteq \boldsymbol{A}$, $\mu(X) \subseteq A_j$ when j is the smallest i s.t. $X \cap A_i \neq \emptyset$.

The idea is now to make the A_i the layers, and "trigger" the first layer A_j s.t. $\mu(A_j) \cap R(x) \neq \emptyset$, and check whether $\mu(A_j) \cap R(x) \subseteq B_j$. A suitable ranked structure will automatically find this A_j.

More definitions and results for such $\mathcal{A}$ and $\mathcal{C}$ will be found in Sect. 6.5.4.

The representation results for $\mathcal{A}$-ranked structures were given already in Sect. 6.4, so we can turn immediately to Sect. 7.3.5 of Gabbay and Schlechta (2008f), which we quote for the convenience of the reader.

6.5.4 Formal Results and Representation for Hierarchical Conditionals

We look here at the following problem:
Given

(1.1) a finite, ordered partition $\mathcal{A}$ of $\boldsymbol{A}$, $\mathcal{A} = \langle\{A_i : i \in I\}, <\rangle$
(1.2) a normality relation $\prec$, which is an $\mathcal{A}$-ranking, defining a choice function μ on subsets of $\boldsymbol{A}$, (so, obviously, $A < A'$ iff $\mu(A \cup A') \cap A' = \emptyset$),
(1.3) a subset $\boldsymbol{B} \subseteq \boldsymbol{A}$, and we set $\mathcal{C} := \langle\mathcal{A}, \boldsymbol{B}\rangle$ (thus, the B_i are just $A_i \cap \boldsymbol{B}$, this way of writing saves a little notation),
(2.1) a set of models M,
(2.2) an accessibility relation R on M, with some finite upper bound on R-chains,
(2.3) an unknown extension of R to pairs (m, a), $m \in M$, $a \in \boldsymbol{A}$,
(3.1) a notion of validity $m \models \mathcal{C}$, for $m \in M$, defined by $m \models \mathcal{C}$ iff $\{a \in \boldsymbol{A} : \exists A \in \mathcal{A}(a \in \mu(A), a \in R(m)$, and

$$\neg\exists a'(\exists A' \in \mathcal{A}(a' \in \mu(A'), a' \in R(m), a' \prec a\} \subseteq \boldsymbol{B},$$

(3.2) a subset M' of M
give a criterion which decides whether it is possible to construct the extension of R to pairs (m, a) s.t. $\forall m \in M.(m \in M' \Leftrightarrow m \models \mathcal{C})$.

We first show some elementary facts on the situation, and give the criterion in Proposition 6.5.3, together with the proof that it does what is wanted.

Fact 6.5.1
Reachability for a transitive relation is characterised by
$y \in R(x) \rightarrow R(y) \subseteq R(x)$.

Let now S be a set with an accessibility relation R', generated by transitive closure from the intransitive subrelation R. All modal notation will be relative to this R.

Let $\boldsymbol{A} = M(\alpha)$, $A_i = M(\alpha_i)$.

Definition 6.5.2

(1) $\mu(\mathcal{A}) := \bigcup\{\mu(A_i) : i \in I\}$

(warning: this is NOT $\mu(\boldsymbol{A})$)
(2) $\mathcal{A}_m := R(m) \cap \boldsymbol{A}$,
(3) $\mu(\mathcal{A}_m) := \bigcup\{\mu(A_i) \cap R(m) : i \in I\}$
(3a) $\nu(\mathcal{A}_m) := \mu(\mu(\mathcal{A}_m))$
(thus $\nu(\mathcal{A}_m) = \{a \in \boldsymbol{A} : \exists A \in \mathcal{A}(a \in \mu(A), a \in R(m)$, and
$\neg\exists a'(\exists A' \in \mathcal{A}(a' \in \mu(A'), a' \in R(m), a' \prec a\}$.
(4) $m \models \mathcal{C} :\leftrightarrow \nu(\mathcal{A}_m)) \subseteq \boldsymbol{B}$.

See Diagram 6.5.3

Diagram 6.5.3

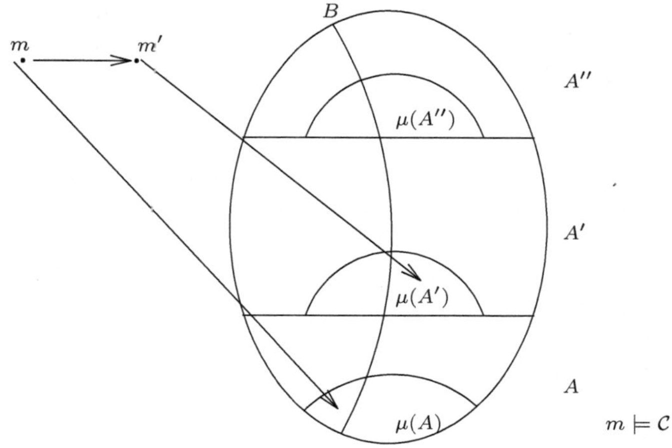

Here, the "best" element m sees is in B, so $\mathcal{C}$ holds in m.
The "best" element m' sees is not in B, so $\mathcal{C}$ does not hold in m'.

Validity of $\mathcal{C}$ from m and m'

We have the following Fact for $m \models \mathcal{C}$:

Fact 6.5.2
Let $m, m' \in M$, $A, A' \in \mathcal{A}$.

(1) $m \models \Box\neg\alpha,\ mRm' \Rightarrow m' \models \Box\neg\alpha$
(2) $mRm',\ \nu(\mathcal{A}_m) \cap A \neq \emptyset,\ \nu(\mathcal{A}_{m'}) \cap A' \neq \emptyset, \Rightarrow A \leq A'$
(3) $mRm',\ \nu(\mathcal{A}_m) \cap A \neq \emptyset,\ \nu(\mathcal{A}_{m'}) \cap A' \neq \emptyset,\ m \models \mathcal{C},\ m' \not\models \mathcal{C}, \Rightarrow A < A'$

We solve now the representation task described at the beginning of Sect. 6.5.4, all we need are the properties shown in Fact 6.5.2.

(Note that constructing R between the different m, m' is trivial: we could just choose the empty relation.)

Proposition 6.5.3
If the properties of Fact 6.5.2 hold, we can extend R to solve the representation problem described at the beginning of this Sect. 6.5.4.

Construction 6.5.1
We choose now elements as possible, which ones are chosen exactly do not matter.
$X_i := \{b_i, c_i\}$ iff $\mu(A_i) \cap \boldsymbol{B} \neq \emptyset$ and $\mu(A_i) - \boldsymbol{B} \neq \emptyset$, $b_i \in \mu(A_i) \cap \boldsymbol{B}$, $c_i \in \mu(A_i) - \boldsymbol{B}$.
$X_i := \{c_i\}$ iff $\mu(A_i) \cap \boldsymbol{B} = \emptyset$ and $\mu(A_i) - \boldsymbol{B} \neq \emptyset$, $c_i \in \mu(A_i) - \boldsymbol{B}$
$X_i := \{b_i\}$ iff $\mu(A_i) \cap \boldsymbol{B} \neq \emptyset$ and $\mu(A_i) - \boldsymbol{B} = \emptyset$, $b_i \in \mu(A_i) \cap \boldsymbol{B}$,
$X_i := \emptyset$ iff $\mu(A_i) = \emptyset$.

Case 1:
Let m be R-minimal and $m \models \mathcal{C}$. Let i_0 be the first i s.t. $b_i \in X_i$, make $\gamma(m) := i_0$, and make $R(m) := \{b_{i_0}\} \cup \bigcup\{X_i : i > i_0\}$. This makes $\mathcal{C}$ hold. (This leaves us as many possibilities open as possible — remember we have to decrease the set of reachable elements now.)
Case 2:
Let m be R-minimal and $m \not\models \mathcal{C}$. Let i_0 be the first i s.t. $c_i \in X_i$, make $\gamma(m) := i_0$, and make $R(m) := \bigcup\{X_i : i \geq i_0\}$. This makes $\mathcal{C}$ false.

Let all R-predecessors of m be determined, and $i := max\{\gamma(m') : m'Rm\}$.
Case 3: $m \models \mathcal{C}$. Let j be the smallest $i' \geq i$ with $\mu(A_{i'}) \cap \boldsymbol{B} \neq \emptyset$. Let $R(m) := \{b_j\} \cup \bigcup\{X_k : k > j\}$, and $\gamma(m) := j$.
Case 4: $m \not\models \mathcal{C}$.
Case 4.1: For all $m'Rm$ with $i = \gamma(m')$ $m' \not\models \mathcal{C}$.

Take one such m' and set $R(m) := R(m')$, $\gamma(m) := i$.
Case 4.2: There is $m'Rm$ with $i = \beta(m')$ $m' \models \mathcal{C}$.

Let j be the smallest $i' > i$ with $\mu(A_{i'}) - \boldsymbol{B} \neq \emptyset$. Let $R(m) := \bigcup\{X_k : k \geq j\}$. (Remark: To go from $\models$ to $\not\models$, we have to go higher in the hierarchy.)

Obviously, validity is done as it should be. It remains to show that the sets of reachable elements decrease with R.

Fact 6.5.4
In above construction, if mRm', then $R(m') \subseteq R(m)$.

We consider an example for illustration.

Example 6.5.1
Let $a_1Ra_2RcRc_1$, $b_1Rb_2Rb_3RcRd_1Rd_2$.

Let $\mathcal{C} = (A_1 > B_1, \ldots, A_n > B_n)$ with the C_i consistency with $\mu(A_i)$.

Let $\mu(A_2) \cap B_2 = \emptyset$, $\mu(A_3) \subseteq B_3$, and for the other i hold neither of these two.

Let $a_1, a_2, b_2, c_1, d_2 \models \mathcal{C}$, the others $\not\models \mathcal{C}$.

Let $\mu(A_1) = \{a_{1,1}, a_{1,2}\}$, with $a_{1,1} \in B_1$, $a_{1,2} \notin B_1$,

$\mu(A_2) = \{a_{2,1}\}$, $\mu(A_3) = \{a_{3,1}\}$ (there is no reason to differentiate), and the others like $\mu(A_1)$. Let $\mu A := \bigcup\{\mu(A_i) : i \leq n\}$.

We have to start at a_1 and b_1, and make $R(x)$ progressively smaller.

Let $R(a_1) := \mu A - \{a_{1,2}\}$, so $a_1 \models \mathcal{C}$. Let $R(a_2) = R(a_1)$, so again $a_2 \models \mathcal{C}$.

Let $R(b_1) := \mu A - \{a_{1,1}\}$, so $b_1 \not\models \mathcal{C}$. We now have to take $a_{1,2}$ away, but $a_{2,1}$ too to be able to change. So let $R(b_2) := R(b_1) - \{a_{1,2}, a_{2,1}\}$, so we begin at $\mu(A_3)$, which is a (positive) singleton. Then let $R(b_3) := R(b_2) - \{a_{3,1}\}$.

We can choose $R(c) := R(b_3)$, as $R(b_3) \subseteq R(a_2)$.

Let $R(c_1) := R(c) - \{a_{4,2}\}$ to make $\mathcal{C}$ hold again. Let $R(d_1) := R(c)$, and $R(d_2) := R(c_1)$. $\square$

Chapter 7
Theory Revision, Theory Contraction and Conditionals

7.1 Summary

We present here some basic ideas and results for theory revision, the closely related theory contraction, and, finally counterfactual conditionals. The semantics for conditionals and update is very similar to that for theory revision, both are based on a notion of minimal distance, which codes minimal change. Again, we see here the idea of minimal or limited influence change.

In somewhat more detail.

We first introduce AGM theory revision, theory contraction and epistemic entrenchment, in the syntactic and the semantic variant. We then discuss distance semantics for theory revision, representation results, and definability preservation problems. Diagram 7.2.1 shows that the language of theory revision is relatively poor, this has consequences for representation, see Proposition 7.2.8.

We then mention safe contraction in Sect. 7.3, which will be used in Chap. 11 to choose consistent subsets of defaults.

We conclude with a comment on conditionals and update.

7.2 Theory Revision

We begin with a very succinct introduction into AGM theory revision, and the subsequent results by the second author as published in Schlechta (2004). It is not supposed to be a self-contained introduction, but to help the reader recollect the situation.

Recall from the introduction that theory revision was invented in order to "fuse" together two separately consistent, but together inconsistent theories or formulas to a consistent result. The by far best known approach is that by Alchourron, Gardenfors, and Makinson, and know as the AGM approach, see Alchourron et al. (1985). They formulated "rationality postulates" for various variants of theory revision, which we give now in a very succinct form. Lehmann, Magidor, Schlechta, see Lehmann et al.

D.M. Gabbay and K. Schlechta, *A New Perspective on Nonmonotonic Logics*,
DOI 10.1007/978-3-319-46817-4_7

(2001), gave a distance semantics for theory revision, this is further elaborated in Schlechta (2004), and presented here in very brief outline, too.

For more on the subject, see e.g. Makinson (2009), Booth et al. (2004), Schlechta (1991).

Definition 7.2.1

We present in parallel the logical and the semantical (or purely algebraic) side. For the latter, we work in some fixed universe U, and the intuition is $U = M_{\mathcal{L}}$, $X = M(K)$, etc. so, e.g. $A \in K$ becomes $X \subseteq B$.

(For reasons of readability, we omit most caveats about definability.)

$K_{\perp}$ will denote the inconsistent theory.

We consider two functions, –and $*$, taking a deductively closed theory and a formula as arguments, and returning a (deductively closed) theory on the logics side. The algebraic counterparts work on definable model sets. It is obvious that $(K-1)$, $(K*1)$, $(K-6)$, $(K*6)$ have vacuously true counterparts on the semantical side. Note that K (X) will never change; everything is relative to fixed K (X). $K * \phi$ is the result of revising K with ϕ. $K - \phi$ is the result of subtracting enough from K to be able to add $\neg\phi$ in a reasonable way, called contraction.

Moreover, let $\leq_K$ be a relation on the formulas relative to a deductively closed theory K on the formulas of $\mathcal{L}$, and let $\leq_X$ be a relation on $\mathcal{P}(U)$ or a suitable subset of $\mathcal{P}(U)$ relative to a fixed X. When the context is clear, we simply write $\leq$. $\leq_K$ $(\leq_X)$ is called a relation of epistemic entrenchment for K (X).

Table 7.1, "AGM theory revision", presents "rationality postulates" for contraction (–), rationality postulates revision $(*)$ and epistemic entrenchment. In AGM tradition, K will be a deductively closed theory, ϕ, ψ formulas. Accordingly, X will be the set of models of a theory, A, B the model sets of formulas.

In the further development, formulas ϕ, etc. may sometimes also be full theories. As the transcription to this case is evident, we will not go into details.

Remark 7.2.1

(1) Note that $(X \mid 7)$ and $(X \mid 8)$ express a central condition for ranked structures: If we note $X \mid .$ by $f_X(.)$, we then have $f_X(A) \cap B \neq \emptyset \Rightarrow f_X(A \cap B) = f_X(A) \cap B$.
(2) It is trivial to see that AGM revision cannot be defined by an individual distance (see Definition 7.2.3) Suppose $X \mid Y := \{y \in Y : \exists x_y \in X(\forall y' \in Y.d(x_y, y) \leq d(x_y, y'))\}$. Consider a, b, c. $\{a, b\} \mid \{b, c\} = \{b\}$ by $(X \mid 3)$ and $(X \mid 4)$, so $d(a, b) < d(a, c)$. But on the other hand $\{a, c\} \mid \{b, c\} = \{c\}$, so $d(a, b) > d(a, c)$, a contradiction.

Proposition 7.2.2

We refer here to Table 7.2, "AGM interdefinability". Contraction, revision and epistemic entrenchment are interdefinable by the following equations, i.e. if the defining side has the respective properties, so will the defined side. (See Alchourron et al. (1985).)

Speaking in terms of distance-defined revision, $X \mid A$ is the set of those $a \in A$ which are closest to X, and $X \ominus A$ is the set of y which are either in X or in $\boldsymbol{C}(A)$ and closest to X among those in $\boldsymbol{C}(A)$.

Table 7.1 AGM theory revision

AGM theory revision			
Contraction, $K - \phi$			
$(K-1)$	$K-\phi$ is deductively closed		
$(K-2)$	$K-\phi \subseteq K$	$(X\ominus 2)$	$X \subseteq X \ominus A$
$(K-3)$	$\phi \notin K \Rightarrow K-\phi = K$	$(X\ominus 3)$	$X \not\subseteq A \Rightarrow X\ominus A = X$
$(K-4)$	$\not\vdash \phi \Rightarrow \phi \notin K-\phi$	$(X\ominus 4)$	$A \neq U \Rightarrow X\ominus A \not\subseteq A$
$(K-5)$	$K \subseteq \overline{(K-\phi)\cup\{\phi\}}$	$(X\ominus 5)$	$(X\ominus A)\cap A \subseteq X$
$(K-6)$	$\vdash \phi \leftrightarrow \psi \Rightarrow K-\phi = K-\psi$		
$(K-7)$	$(K-\phi)\cap(K-\psi) \subseteq$ $K-(\phi\wedge\psi)$	$(X\ominus 7)$	$X\ominus(A\cap B) \subseteq$ $(X\ominus A)\cup(X\ominus B)$
$(K-8)$	$\phi \notin K-(\phi\wedge\psi) \Rightarrow$ $K-(\phi\wedge\psi) \subseteq K-\phi$	$(X\ominus 8)$	$X\ominus(A\cap B) \not\subseteq A \Rightarrow$ $X\ominus A \subseteq X\ominus(A\cap B)$
Revision, $K * \phi$			
$(K*1)$	$K*\phi$ is deductively closed	–	
$(K*2)$	$\phi \in K*\phi$	$(X \mid 2)$	$X \mid A \subseteq A$
$(K*3)$	$K*\phi \subseteq \overline{K\cup\{\phi\}}$	$(X \mid 3)$	$X\cap A \subseteq X \mid A$
$(K*4)$	$\neg\phi \notin K \Rightarrow$ $\overline{K\cup\{\phi\}} \subseteq K*\phi$	$(X \mid 4)$	$X\cap A \neq \emptyset \Rightarrow$ $X \mid A \subseteq X\cap A$
$(K*5)$	$K*\phi = K_\perp \Rightarrow \vdash \neg\phi$	$(X \mid 5)$	$X \mid A = \emptyset \Rightarrow A = \emptyset$
$(K*6)$	$\vdash \phi\leftrightarrow\psi \Rightarrow K*\phi = K*\psi$	-	
$(K*7)$	$K*(\phi\wedge\psi) \subseteq$ $\overline{(K*\phi)\cup\{\psi\}}$	$(X \mid 7)$	$(X \mid A)\cap B \subseteq$ $X \mid (A\cap B)$
$(K*8)$	$\neg\psi \notin K*\phi \Rightarrow$ $\overline{(K*\phi)\cup\{\psi\}} \subseteq K*(\phi\wedge\psi)$	$(X \mid 8)$	$(X \mid A)\cap B \neq \emptyset \Rightarrow$ $X \mid (A\cap B) \subseteq (X \mid A)\cap B$
Epistemic entrenchment			
$(EE1)$	$\leq_K$ is transitive	$(EE1)$	$\leq_X$ is transitive
$(EE2)$	$\phi \vdash \psi \Rightarrow \phi \leq_K \psi$	$(EE2)$	$A \subseteq B \Rightarrow A \leq_X B$
$(EE3)$	$\forall\phi,\psi$ $(\phi \leq_K \phi\wedge\psi$ or $\psi \leq_K \phi\wedge\psi)$	$(EE3)$	$\forall A, B$ $(A \leq_X A\cap B$ or $B \leq_X A\cap B)$
$(EE4)$	$K \neq K_\perp \Rightarrow$ $(\phi \notin K$ iff $\forall\psi.\phi \leq_K \psi)$	$(EE4)$	$X \neq \emptyset \Rightarrow$ $(X \not\subseteq A$ iff $\forall B.A \leq_X B)$
$(EE5)$	$\forall\psi.\psi \leq_K \phi \Rightarrow \vdash \phi$	$(EE5)$	$\forall B.B \leq_X A \Rightarrow A = U$

A Remark on Intuition The idea of epistemic entrenchment is that ϕ is more entrenched than ψ (relative to K) iff $M(\neg\psi)$ is closer to $M(K)$ than $M(\neg\phi)$ is to $M(K)$. In shorthand, the more we can twiggle K without reaching $\neg\phi$, the more is ϕ entrenched. Truth is maximally entrenched—no twiggling whatever will reach falsity. The more ϕ is entrenched, the more we are certain about it. Seen this way, the properties of epistemic entrenchment relations are very natural (and trivial): As only the closest points of $M(\neg\phi)$ count (seen from $M(K)$), ϕ or ψ will be as entrenched as $\phi\wedge\psi$, and there is a logically strongest ϕ' which is as entrenched as ϕ—this is just the sphere around $M(K)$ with radius $d(M(K), M(\neg\phi))$.

Table 7.2 AGM interdefinability

AGM interdefinability	
$K * \phi := \overline{(K - \neg\phi)} \cup \phi$	$X \mid A := (X \ominus \boldsymbol{C}(A)) \cap A$
$K - \phi := K \cap (K * \neg\phi)$	$X \ominus A := X \cup (X \mid \boldsymbol{C}(A))$
$K - \phi := \{\psi \in K : (\phi <_K \phi \vee \psi \text{ or} \vdash \phi)\}$	$X \ominus A := \begin{cases} X \; iff \; A = U, & \\ \bigcap\{B : X \subseteq B \subseteq U, A <_X A \cup B\} & otherwise \end{cases}$
$\phi \leq_K \psi :\leftrightarrow \begin{cases} \vdash \phi \wedge \psi \\ or \\ \phi \notin K - (\phi \wedge \psi) \end{cases}$	$A \leq_X B :\leftrightarrow \begin{cases} A, B = U \\ or \\ X \ominus (A \cap B) \not\subseteq A \end{cases}$

Definition 7.2.2
$d : U \times U \to Z$ is called a pseudo-distance on U iff (d1) holds
(d1) Z is totally ordered by a relation $<$.

If, in addition, Z has a $<$-smallest element 0, and (d2) holds, we say that d respects identity
(d2) $d(a, b) = 0$ iff $a = b$.

If, in addition, (d3) holds, then d is called symmetric
(d3) $d(a, b) = d(b, a)$
(for any $a, b \in U$).

Note that we can force the triangle inequality to hold trivially (if we can choose the values in the real numbers): It suffices to choose the values in the set $\{0\} \cup [0.5, 1]$, i.e., in the interval from 0.5 to 1, or as 0.

Definition 7.2.3
We define the collective and the individual variant of choosing the closest elements in the second operand by two operators, $|, \uparrow: \mathcal{P}(U) \times \mathcal{P}(U) \to \mathcal{P}(U)$:
Let d be a distance or pseudo-distance.
$X \mid Y := \{y \in Y : \exists x_y \in X.\forall x' \in X, \forall y' \in Y(d(x_y, y) \leq d(x', y')\}$
(the collective variant, used in theory revision)
and
$X \uparrow Y := \{y \in Y : \exists x_y \in X.\forall y' \in Y(d(x_y, y) \leq d(x_y, y')\}$
(the individual variant, used for counterfactual conditionals and theory update).

Thus, $A \mid_d B$ is the subset of B consisting of all $b \in B$ that are closest to A. Note that, if A or B is infinite, $A \mid_d B$ may be empty, even if A and B are not empty. A condition assuring nonemptiness will be imposed when necessary.

Definition 7.2.4
An operation $|: \mathcal{P}(U) \times \mathcal{P}(U) \to \mathcal{P}(U)$ is representable iff there is a pseudo-distance $d : U \times U \to Z$ such that
$A \mid B = A \mid_d B := \{b \in B : \exists a_b \in A \forall a' \in A \forall b' \in B(d(a_b, b) \leq d(a', b'))\}$.

The following is the central definition, it describes the way a revision $*_d$ is attached to a pseudo-distance d on the set of models.

Definition 7.2.5
$T *_d T' := Th(M(T) \mid_d M(T'))$.

$*$ is called representable iff there is a pseudo-distance d on the set of models s.t. $T * T' = Th(M(T) \mid_d M(T'))$.

Fact 7.2.3
This is Fact 8.2.1 in Gabbay and Schlechta (2009), where a proof can be found.

A distance based revision satisfies the AGM postulates provided

(1) it respects identity, i.e. $d(a, a) < d(a, b)$ for all $a \neq b$,
(2) it satisfies a limit condition: minima exist,
(3) it is definability preserving.

(It is trivial to see that the first two are necessary, and Example 7.2.1 (2) below shows the necessity of (3). In particular, (2) and (3) will hold for finite languages.)

Definition 7.2.6
For $X, Y \neq \emptyset$, set $U_Y(X) := \{z : d(X, z) \leq d(X, Y)\}$.

Fact 7.2.4
This is Fact 8.2.2 in Gabbay and Schlechta (2009), where a proof can be found.

Let $X, Y, Z \neq \emptyset$. Then

(1) $U_Y(X) \cap Z \neq \emptyset$ iff $(X \mid (Y \cup Z)) \cap Z \neq \emptyset$,
(2) $U_Y(X) \cap Z \neq \emptyset$ iff $\boldsymbol{C}Z \leq_X \boldsymbol{C}Y$ - where $\leq_X$ is epistemic entrenchment relative to X.

Definition 7.2.7
We refer here to Table 7.3, “Distance representation and revision”. It defines the loop condition, and shows the correspondence between the semantical and the syntactic sides.

The prerequisites are as follows:

Let $U \neq \emptyset$, $\mathcal{Y} \subseteq \mathcal{P}(U)$ satisfy $(\cap)$, $(\cup)$, $\emptyset \notin \mathcal{Y}$.

Let $A, B, X_i \in \mathcal{Y}$, $\mid: \mathcal{Y} \times \mathcal{Y} \to \mathcal{P}(U)$.

Let $*$ be a revision function defined for arbitrary consistent theories on both sides. (This is thus a slight extension of the AGM framework, as AGM works with formulas only on the right of $*$.)

Proposition 7.2.5
This is Proposition 8.2.2 in Gabbay and Schlechta (2009), where a proof can be found.

The following connections between the logical and the algebraic side might be the most interesting ones. We will consider in all cases also the variant with full theories. It is shown in Table 7.4, “Revision and definability preservation”.

Given $*$ which respects logical equivalence, let $M(T) \mid M(T') := M(T * T')$, conversely, given $\mid$, let $T * T' := Th(M(T) \mid M(T'))$. We then have

Table 7.3 Distance representation and revision

Distance representation and revision		
		$(*Equiv)$ $\models T \leftrightarrow S, \models T' \leftrightarrow S', \Rightarrow T * T' = S * S'$
		$(*CCL)$ $T * T'$ is a consistent, deductively closed theory
	$(\mid Succ)$ $A \mid B \subseteq B$	$(*Succ)$ $T' \subseteq T * T',$
	$(\mid Con)$ $A \cap B \neq \emptyset \Rightarrow A \mid B = A \cap B$	$(*Con)$ $Con(T \cup T') \Rightarrow T * T' = \overline{T \cup T'},$
Intuitively, Using symmetry $d(X_0, X_1) \leq d(X_1, X_2),$ $d(X_1, X_2) \leq d(X_2, X_3),$ $d(X_2, X_3) \leq d(X_3, X_4)$ $\ldots$ $d(X_{k-1}, X_k) \leq d(X_0, X_k)$ $\Rightarrow$ $d(X_0, X_1) \leq d(X_0, X_k),$ i.e., transitivity, or absence of loops involving $<$	$(\mid Loop)$ $(X_1 \mid (X_0 \cup X_2)) \cap X_0 \neq \emptyset,$ $(X_2 \mid (X_1 \cup X_3)) \cap X_1 \neq \emptyset,$ $(X_3 \mid (X_2 \cup X_4)) \cap X_2 \neq \emptyset,$ $\ldots$ $(X_k \mid (X_{k-1} \cup X_0)) \cap X_{k-1} \neq \emptyset$ $\Rightarrow$ $(X_0 \mid (X_k \cup X_1)) \cap X_1 \neq \emptyset$	$(*Loop)$ $Con(T_0, T_1 * (T_0 \vee T_2)),$ $Con(T_1, T_2 * (T_1 \vee T_3)),$ $Con(T_2, T_3 * (T_2 \vee T_4))$ $\ldots$ $Con(T_{k-1}, T_k * (T_{k-1} \vee T_0))$ $\Rightarrow$ $Con(T_1, T_0 * (T_k \vee T_1))$

Table 7.4 Revision and definability preservation

Revision and definability preservation			
(1.1)	$(K * 7)$	$\Rightarrow$	$(X \mid 7)$
(1.2)		$\Leftarrow (\mu dp)$	
(1.3)		$\Leftarrow$ B is the model set for some ϕ	
(1.4)		$\not\Leftarrow$ in general	
(2.1)	$(*Loop)$	$\Rightarrow$	$(\mid Loop)$
(2.2)		$\Leftarrow (\mu dp)$	
(2.3)		$\Leftarrow$ all X_i are the model sets for some ϕ_i	
(2.4)		$\not\Leftarrow$ in general	

Table 7.5, "Distance representation", summarises representation of theory revision functions by structures with a distance.

By "pseudo-distance" we mean here a pseudo-distance which respects identity, and is symmetrical.

$(\mid \emptyset)$ means that if $X, Y \neq \emptyset$, then $X \mid_d Y \neq \emptyset$.

The following Example 7.2.1 shows that, in general, a revision operation defined on models via a pseudo-distance by $T * T' := Th(M(T) \mid_d M(T'))$ might not satisfy $(*Loop)$ or $(K * 7)$, unless we require $\mid_d$ to preserve definability.

Example 7.2.1
Consider an infinite propositional language $\mathcal{L}$.

Let X be an infinite set of models, m, m_1, m_2 be models for $\mathcal{L}$. Arrange the models of $\mathcal{L}$ in the real plane s.t. all $x \in X$ have the same distance < 2 (in the real plane) from m, m_2 has distance 2 from m, and m_1 has distance 3 from m.

Let T, T_1, T_2 be complete (consistent) theories, T' a theory with infinitely many models, $M(T) = \{m\}$, $M(T_1) = \{m_1\}$, $M(T_2) = \{m_2\}$. The two variants diverge now slightly

(1) $M(T') = X \cup \{m_1\}$. T, T', T_2 will be pairwise inconsistent.
(2) $M(T') = X \cup \{m_1, m_2\}$, $M(T'') = \{m_1, m_2\}$.

Assume in both cases $Th(X) = T'$, so X will not be definable by a theory.

Now for the results.

Then $M(T) \mid M(T') = X$, but $T * T' = Th(X) = T'$.
(1) We easily verify $Con(T, T_2 * (T \vee T))$, $Con(T_2, T * (T_2 \vee T_1))$, $Con(T, T_1 * (T \vee T))$, $Con(T_1, T * (T_1 \vee T'))$, $Con(T, T' * (T \vee T))$, and conclude by Loop (i.e., $(*Loop)$) $Con(T_2, T * (T' \vee T_2))$, which is wrong.
(2) So $T * T'$ is consistent with T'', and $\overline{(T * T') \cup T''} = T''$. But $T' \cup T'' = T''$, and $T * (T' \cup T'') = T_2 \neq T''$, contradicting $(K * 7)$. □

Proposition 7.2.6
Let $U \neq \emptyset$, $\mathcal{Y} \subseteq \mathcal{P}(U)$ be closed under finite $\cap$ and finite $\cup$, $\emptyset \notin \mathcal{Y}$.
(a) $\mid$ is representable by a symmetric pseudo-distance $d : U \times U \to Z$ iff $\mid$ satisfies $(\mid Succ)$ and $(\mid Loop)$ in Definition 7.2.7.
(b) $\mid$ is representable by an identity respecting symmetric pseudo-distance $d : U \times U \to Z$ iff $\mid$ satisfies $(\mid Succ)$, $(\mid Con)$, and $(\mid Loop)$ in Definition 7.2.7.
See Lehmann et al. (2001) or Schlechta (2004).

Proposition 7.2.7
Let $\mathcal{L}$ be a propositional language.
(a) A revision operation $*$ is representable by a symmetric consistency and definability preserving pseudo-distance iff $*$ satisfies $(*Equiv)$, $(*CCL)$, $(*Succ)$, $(*Loop)$.

Table 7.5 Distance representation

Distance representation				
$\mid -$ function		Distance Structure		$*-$ function
$(\mid Succ) + (\mid Con) + (\mid Loop)$	$\Leftrightarrow$, $(\cup) + (\cap)$ Proposition 7.2.6	pseudo-distance	$\Leftrightarrow$ $(\mu dp) + (\mid \emptyset)$ Proposition 7.2.7	$(*Equiv) + (*CCL) + (*Succc)+$ $(*Con) + (*Loop)$
any finite characterization	$\not\Leftrightarrow$ Proposition 7.2.8		$\not\Rightarrow$ without (μdp) Example 7.2.1	

(b) A revision operation $*$ *is representable by a symmetric consistency and definability preserving, identity respecting pseudo-distance iff* $*$ *satisfies* $(*Equiv)$, $(*CCL)$, $(*Succ)$, $(*Con)$, $(*Loop)$.
See Lehmann et al. (2001) or Schlechta (2004).

Example 7.2.2
This example shows the expressive weakness of revision based on distance: not all distance relations can be reconstructed from the revision operator. Thus, a revision operator does not allow to "observe" all distance relations, so transitivity of $\leq$ cannot necessarily be captured in a short condition, requiring arbitrarily long conditions, see Proposition 7.2.8.

Note that even when the pseudo-distance is a real distance, the resulting revision operator $|_d$ does not always permit to reconstruct the relations of the distances: revision is a coarse instrument to investigate distances.

Distances with common start (or end, by symmetry) can always be compared by looking at the result of revision

$a \mid_d \{b, b'\} = b$ iff $d(a, b) < d(a, b')$,
$a \mid_d \{b, b'\} = b'$ iff $d(a, b) > d(a, b')$,
$a \mid_d \{b, b'\} = \{b, b'\}$ iff $d(a, b) = d(a, b')$.

This is not the case with arbitrary distances $d(x, y)$ and $d(a, b)$, as this example will show.

We work in the real plane, with the standard distance, the angles have 120°. a' is closer to y than x is to y, a is closer to b than x is to y, but a' is farther away from b' than x is from y. Similarly for b, b'. But we cannot distinguish the situation $\{a, b, x, y\}$ and the situation $\{a', b', x, y\}$ through $|_d$. (See Diagram 7.2.1):

Seen from a, the distances are in that order: y, b, x.
Seen from a', the distances are in that order: y, b', x.
Seen from b, the distances are in that order: y, a, x.
Seen from b', the distances are in that order: y, a', x.
Seen from y, the distances are in that order: $a/b, x$.
Seen from y, the distances are in that order: $a'/b', x$.
Seen from x, the distances are in that order: $y, a/b$.
Seen from x, the distances are in that order: $y, a'/b'$.

Thus, any $c \mid_d C$ will be the same in both situations (with a interchanged with a', b with b'). The same holds for any $X \mid_d C$ where X has two elements.

Thus, any $C \mid_d D$ will be the same in both situations, when we interchange a with a', and b with b'. So we cannot determine by $|_d$ whether $d(x, y) > d(a, b)$ or not. □

Diagram 7.2.1

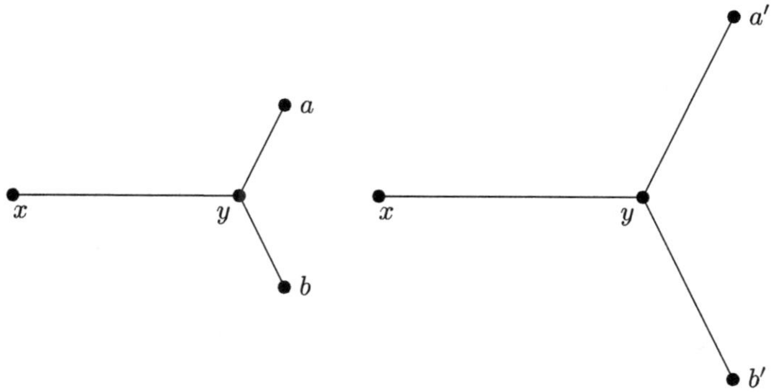

Indiscernible by revision

Proposition 7.2.8
There is no finite characterization of distance-based |-operators.

(Attention: this is, of course, false when we fix the left-hand side: the AGM axioms give a finite characterization. So this also shows the strength of being able to change the left-hand side.)

See Schlechta (2004).

7.3 Theory Contraction

7.3.1 Introduction

Our way to obtain consistent default theories is similar to Safe Contraction (see Sect. 7.3), but also has some important differences.

The theory of Safe Contraction seems quite difficult, due to a multitude of possibilities to consider. An excellent overview is in Rott and Hansson (2014).

First, we present the basics of Safe Contraction in a somewhat more general manner than usual. A translation to the semantical version is trivial, and omitted. We then point to the important differences.

7.3.2 Some of the AGM Axioms for Theory Contraction

In the following, K is a set of formulas, ϕ etc. a formula, and a theory is a deductively (under classical logic) closed set of formulas. The idea is to "subtract" from K enough formulas so one cannot deduce ϕ any more. The result of this "subtraction" is written $K \dot{-} \phi$.

$(K \dot{-} 1)$ If K is a theory, then so is $K \dot{-} \phi$

$(K \dot{-} 2)$ $K \dot{-} \phi \subseteq K$
$(K \dot{-} 3)$ If $\phi \notin K$, then $K \dot{-} \phi = K$
$(K \dot{-} 4)$ If $\phi \notin \overline{\emptyset}$, then $\phi \notin \underline{K \dot{-} \phi}$
$(K \dot{-} 5)$ $K \subseteq \overline{(K \dot{-} \phi) \cup \{\phi\}}$
$(K \dot{-} 6)$ If $\overline{\phi} = \overline{\psi}$, then $K \dot{-} \phi = K \dot{-} \psi$
$(K \dot{-} 7)$ $(K \dot{-} \phi) \cap (K \dot{-} \psi) \subseteq K \dot{-} (\phi \wedge \psi)$.

7.3.3 The General Picture of Safe Contraction

7.3.3.1 The Formula Version

K is a set of formulas.

Suppose $K \vdash \phi$.

Let $K \coprod \phi := \{K' \subseteq K : K' \vdash \phi$ and for all $K'' \subset K',\ K'' \not\vdash \phi\}$ - (the minimal $K' \subseteq K$ which allow to deduce ϕ)

$K' \in K \coprod \phi$ is called a ϕ-kernel of K.

Let μ be a choice function on $\mathcal{P}(\mathcal{K})$ s.t. for $K' \in K \coprod \phi$, $\emptyset \neq \mu(K') \subseteq K'$. μ chooses elements to be removed from K'.

Let $K \ominus \phi := K - \bigcup\{\mu(K') : K' \in K \coprod \phi\}$.

Then $K \ominus \phi \not\vdash \phi$. (Proof obvious.)

Let $K \dot{-} \phi := \overline{K \ominus \phi}$.

We may define μ by a relation $\prec$ on K : $\mu(K') := \{\psi \in K' : \neg\exists\psi' \in K'.\psi' \prec \psi\}$, then $\mu(K')$ is the set of $\prec$-minimal elements of K'.

In traditional Safe Contraction, this is the case, and, in addition, $\prec$ is supposed free from cycles, and invariant under logical equivalence on both sides. Additional axioms for $\prec$, which are not always required, are
$(H2\uparrow)$ If $\phi \prec \psi$ and $\psi \vdash \gamma$, then $\phi \prec \gamma$
$(H2\downarrow)$ If $\phi \prec \psi$ and $\gamma \vdash \phi$, then $\gamma \prec \psi$
In many cases, K is assumed to be deductively closed.

7.3.4 A Modification

By modifying the conditions on $\prec$ slightly, we can bring idea and semantics of safe contraction closer to other forms of theory revision.

This is as follows: In Lehmann et al. (2001), we gave a distance semantics to theory revision. There, the distance had two arguments, of course, K and ϕ, or, rather, $M(K)$ and $M(\phi)$. In safe contraction, $\prec$ is to be made independent from K. So, let us order individual models by $\prec$, and extend $\prec$ to general model sets, by taking the inf of the models in the set—we suppose that this exists. We then eliminate the smallest suitable models. We have not pursued this idea.

7.3.5 *The Differences to Our Situation*

First, our approach is axiom based, the set of defaults is not deductively closed. So, the $\mathcal{K}$ is relatively poor.

Second, the order $\prec$ is not given on the default formulas, but on the formulas, to which they are attached. In particular, e.g. "normally ϕ" may occur attached to several places, and may be eliminated in one place, but upheld in a more specific one.

7.4 Conditionals and Update

The basic idea of the semantics for conditionals and update is the same as for theory revision: minimal change is coded by distance between models (worlds). Of course, this is again the assumption that influence changes rare. Now, we know that something has changed or has to change, but we limit changes.

The difference is that the distance is now measured individually, and not to the whole set of "old" worlds.

In theory revision, we considered $\{b \in B : \exists a \in A \forall a' \in A, b' \in B. d(a, b) \leq d(a'b')\}$, now we consider $\{b \in B : \exists a \in A \forall b' \in B. d(a, b) \leq d(a, b')\}$.

The sentence "if it were raining, I would open the umbrella" is interpreted as "if it were raining and all other conditions stay the same, as far as possible, I would open the umbrella", meaning that we consider the worlds closest to the actual world, but in which it rains.

The semantics of counterfactual conditionals was first given in abstract terms by R. Stalnaker, see Stalnaker (1968), and in terms of distance by D. Lewis, see Lewis (1973). For early work on the semantics of update, see Katsuno and Mendelzon (1990).

As most of the basic tenets of conditionals and update are very close to or the same as those for theory revision, we will not go into details.

For more on the subject, see e.g. Katsuno and Mendelzon (1990), Lewis (1973), Schlechta and Makinson (1994), Stalnaker (1968).

Chapter 8
Neurology

8.1 Introduction and Summary

We concentrate on the work by G. Edelman. As he came from the outside (research on the immune system, awarded by a Nobel Prize), he had a fresh look on the subject. He tries to give a global view on development and functioning of the brain. This is somewhat in contrast to much of present research in neuroscience, which tends to concentrate more on the functioning of single neurons, or small groups of neurons. His books are very interesting, but not very reader friendly. Thus, this Chapter can also be seen as a service to potential readers of his work.

To put his work somewhat in perspective, and complement it, we begin with remarks inspired by other books on the subject.

8.1.1 *Summary*

We discuss

- the differences between a computer and the brain,
- the organisation of the brain into different areas, "experts, which discuss among each other",
- the flexibility of the brain,
- very important: the recursive or reentrant organisation of the brain,
- the absence of a strict hierarchy, there is no homunculus,
- the constructive nature of the brain, perceptions are constructed,
- the role of language as a product, not a method of the brain, but its effects through recursion,
- prototypes: the brain seems to work with prototypes, rather than with statements,
- the role of adaptation in species and individual evolution,
- the dynamic structure of processing information (and producing consciousness),

D.M. Gabbay and K. Schlechta, *A New Perspective on Nonmonotonic Logics*,
DOI 10.1007/978-3-319-46817-4_8

- the binding problem, and, finally
- a short remark on abstract constructions and how they may come about.

8.2 General Remarks on the Functioning of the Brain

8.2.1 Difference Between the Brain and Computers

John von Neumann argued already in his 1958 lectures (Neumann 1958) that the brain and computers work differently. The brain is a very massively parallel organ, and with relatively low computational depth. The first is obvious, the second follows from the low speed of neurons, and their low precision. In deeper computations, low precision results in unreliable results - errors can compensate or aggravate each other.

The robustness and flexibility of the brain are further arguments that the brain does not work like a computer. There is no central part where information comes together and actions are decided. The organisation is more complex. Moreover, the brain does not seem to have a central clock, like a computer. (This remark is due to F. Crick.) And, obviously, we cannot access our memory by addresses.

8.2.2 Further Remarks

These remarks are based on or influenced by Churchland (2007), Churchland (1986), Churchland (1989), Roth (1996), and conversations with Ch.v.d. Malsburg, FIAS, Frankfurt/Germany.

8.2.2.1 The Structure of the Brain

- There is no clear boundary between memory and processing unit of the brain.
- The brain is not a huge amorphous mass of neurons and synapses, but is structured into different centres, which have their own internal and individual organisation, and communicate with each other—but not every centre with every other centre either. This inter-centre communication can be (mainly) unidirectional, or bidirectional.

  This raises the obvious questions

  - which computational differences result from the structural differences of the different brain centres?
  - what are the computational advantages of the global organisation of the brain into different centres, of "experts talking to each other", and their various ways of communication?

Is the human utterance "Let me think!" an indication, shutting off "lateral" disturbances, which might occur otherwise, if the brain were an amorphous mass?

It seems that not much is known about these questions.

- The power of the human brain (in comparison to animal brains) is not the power of new software running on old hardware, but is the power of new hardware, a new structure of the brain.
- The states of the brain can best be described as (very large) vectors, and its working as vector transformation. The central learning mechanism is the modification of synaptic weights.
- The brain is highly flexible, but not totally so. In some cases, the work of one brain center can be done by a different one, e.g. in case of lesions.
- The brain is highly recursive, it does not have a linear processing order from input (senses) to output (muscles).

8.2.2.2 The Constructive Nature of the Brain

The brain *constructs* our perceptions, as well as our ideas of "self" etc. "Reality" is a construction of the brain. Due to the recursive nature of the (higher) brain, in a strict sense, every observation is "theory" laden, there is no pure observation. This recursive nature of the brain allows "re-processing" of information, manipulation of ideas, abstractions, etc.

8.2.2.3 Language and the Brain

- Language is not the form which the brain uses to work. Language is an "epiphenomenon", iself a creation of the brain, but it can influence the work of the brain through recursiveness. It can make thoughts more precise, and help new combinations by use of the corresponding particles (prepositions, etc.). We can influence the brain activity of other people through communication.
  The complexity of information processing of the brain exceeds the complexity of language. Likewise, new information, also scientific theories, often are richer than their linguistic expression.
  Language is a product of the brain, not a primary tool of the brain. Language can be a secondary tool of the brain, as it can help to sharpen and combine our ideas. (A pebble is a primary tool, the stone axe made with the help of a pebble is a secondary tool.)
  The (higher) brain does not necessarily work in one go, but it can correct itself. In this sense, J.v. Neumann was overly pessimistic. We can do long reasoning chains, but we need correction mechanisms. Language and logic, secondary tools, are such mechanisms.
  A good example is defeasible inheritance. Principled cautious reasoning can lead

to daring conclusions, as some objections may be excluded out of caution. So, to be really cautious, we need a correction mechanism.

- The working of the brain is not necessarily synthetic, like language or logic, building up more complicated entities from elementary ones. Example, a face is not built from eyes, etc., but first seen as a whole. This is illustrated by the fact that we often learn to structure global perceptions - for instance when learning about music or wine.
- The brain seems to work with (and memorise) prototypes, and similarity relations between prototypes, and not with statements, semantic atoms, and their combinations. (Note the similarity to the Stalnaker/Lewis counterfactual conditionals semantics with possible worlds and distances between them, see also Chap. 7.)

8.2.2.4 The Brain and Evolution

The brain seems to have a somewhat peculiar role in evolution. It does not seem to be tied as closely as other organs towards improving performance, e.g. Roth (1996), p. 348. Thus, evolution has only prevented the brain from seriously damaging our chances of survival. But we cannot use the evolutionary argument to show that our ideas of the world are "true", they are only sufficiently good and efficient to enable our survival, or not leading to disaster. There is no way to achieve validity of our perceptions and thoughts through an evolutionary argument.

8.3 Summary of Work by Edelman et al.

This is a summary of work by Gerald Edelman and co-authors, Edelman (1989), Edelman (2000), and Edelman (2004). Edelman (1989) seems the clearest book of all three.

8.3.1 Edelman's Research History

Gerald Edelman shared the 1972 Nobel Price for Physiology or Medicine for his work on antibodies. He subsequently turned to research on consciousness, and saw parallels between the immune and the neural system. Both are adaptive, and learn during evolution and the individual history of the animal.

8.3.2 Terminology

Unfortunately, his terminology is not always very clear, and sometimes used in a not very common way. We make some introductory remarks here, and elaborate on some terms (in particular: "category", "concept", "value-category memory", "remembered present") later in the text.

(1) Map
"Map" is used in the mathematical sense of "function", or even "relation", not in the cartographic sense. The latter reading just does not make much sense.
A map of one brain region A into another region B is thus just a number of neurons which connect the neurons of A to neurons of B. There are also higher maps, which connect map M to map M'. There do not seem to be any restrictions in the sense that, if neuron $n \in M$ is connected to $n' \in M$, and n is connected to $m \in M'$ and n' to $m' \in M'$, then m is also connected to m' (not necessarily a homomorphism), or so.

- Global map:
  A global map is a map between maps, or even higher, a map between global maps, etc.

(2) Reentrant connection
E. insists that a reentrant connection is more than a connection with feedback. For example: Suppose we have two regions, A and A', with 5 neurons each, a, b, c, d, e and a', b', c', d', e', and neurons going from a to a', b' to b, c to c', d' to d, e to e', then A and A' are connected in a reentrant way. (We think E. would call the above example a connection with feedback, if all but one connection were to go in the same direction, here, e.g., if there was no connection from d' to d, but one from d to d'. But this is just guessing.) These are toy examples, and such areas and connections may involve thousands of neurons, and connections may be flexible.
Examples of many types of reentrant connections are given in Edelman (1989), p. 48, 66.

(3) Degeneracy
A system is degenerated if different causes can have the same effect—abstractly, it is a non-injective function. For instance, the same proteine can be coded by different genes, or different (groups of) neurons can be involved in reactions with the same result. Degeneracy is a reason for the enormous flexibility and resilience of the nervous system, and fundamental for the robustness of living beings against injury etc. Graceful degradation is the consequence of a more general form of degeneracy, fully the same effect cannot be achieved any more, but at least some approximation to it. A nice example of degeneracy in meteorology is the following: A cloud in windy weather may sit stable on a mountain top, but the participating molecules change all the time.
There is also the converse, which has no special name in E.'s work, however: the same neuron or group of neurons can be active in very different processes.

Thus, different neurons can be active in the same process, and the same neuron can be active in different processes.

8.3.3 *The Main Ideas*

We summarise here his, probably, main ideas.

(1) The Theory of Neuronal Group Selection (TNGS). The brain is subject to several selection mechanisms (in the sense of Darwin)

  (1.1) It was formed and selected in evolution according to the animals' needs, see, however, Sect. 8.2.2, at the end.
  (1.2) During the individual's development, environment and behaviour formed or weakened new or existing connections.
  (1.3) On the scale of short term developments, connections can be strengthened or weakened, and firing be synchronised, in particular through reentrant connections.

(2) Degeneracy:
The fact that, often, various neuron groups can perform the same task, and, conversely, the same neuron group different tasks, gives the brain its plasticity, in development (TNGS) and functioning—see above.
(3) Reentrant connections:
Reentrant connections allow a re-processing of combined old ("stored") and new information, in several centres of the brain—see above.
(4) Maps and recursive (global) maps:
Maps, in particular reentrant maps, connect and integrate parts of the brain within and between different centres, and their functions, thus, e.g. between different sensory modalities (colour, lines, etc.). Maps between maps (global maps) allow higher constructions—without any upper limit. They are the basis of consciousness, there is no fixed superior center ("homunculus"). No single dominant high-level area has been found.
(5) "The remembered present":
Present sensory input is interpreted on the background of past experience and attributed values (positive, negative values of an experience, in the value memory system), and also modifies the latter through synaptic changes. In this sense, there is no strict difference between past and present, they influence each other. This is achieved by reentrant maps.
There is no memory in the sense that a computer has a memory, memory is rather the tendency to interpret and categorise the world in a certain way.
(6) The Thalamo Cortical System (TCS), functional clusters, and the dynamic core:
During the evolution of reptiles to birds and mammals, new, reentrant and degenerate connections formed between different areas of the cortex and between cor-

tex and thalamus, forming the Thalamo Cortical System (TCS). Its neurons talk mainly to themselves, i.e. active neurons originating in the TCS, mostly end in the TCS, too, they form a "functional cluster". This cluster is composed of several sub-clusters. The process of synchronised firing of neurons in the TCS is called the "dynamic core", and results in consciousness. Thus, the dynamic core is defined by the interactions of neurons, or groups of neurons, and not bound to particular neurons.

My remark: It seems that one has to consider the brain on two very different levels. On the first, the basic entity is the neuron - an object. On the second, the basic entity is the firing of a neuron over time—a process. The basic entities compose to groups of neurons and areas of the brain on the one hand, and synchronised firing groups of neurons on the other hand.

(7) The role of consciousness:

Consciousness has an evolutionary advantage, and is thus favoured by selection.

(8) Implicitly: The overall organisation of the brain:

The brain is extremely plastic. The role of individual neurons and neuron groups changes on longer and shorter time scales, often, different neuron (groups) can do the same function (degeneracy). Mostly, there is no fixed task assigned to one neuron.

The brain consists of several neuronal groups, which have different internal structures (e.g. more or less parallel connections, etc.), and are also connected to each other in different ways, e.g. directly, indirectly, in loops or linear ways, etc.

Such connections are made by "maps", which are often massively parallel, sometimes, and especially in higher brains, in both directions (reentrant), directly or in longer loops. This allows, e.g., the combined treatment of new sensory information with "stored" information.

Such maps exist on several levels, higher brains also have maps between maps (global maps), there is no fixed upper limit on this recursive "meta" construction.

The plasticity, variation of centres, reentrant connections, and recursive meta construction are the basis for the power of our brain.

We turn to a more detailed discussion.

8.3.4 Edelman's Theory in More Detail

8.3.4.1 Selection and Degeneracy, Theory of Neuronal Group Selection (TNGS)

(1) A selection mechanism needs two things

(1.1) a variable population from which to chose
(1.2) a choice mechanism

(2) In evolution, genetic variability provides the variation, reproduction (or not) the choice mechanism. In addition, different genes may generate the same protein, this is an additional variability.
(3) Recall that Edelman calls this phenomenon (different possible causes, same effect) degeneracy. This is one of his main concepts. In the theory of the brain, degeneracy occurs as different neurons, different areas of the brain may have the same effect in the functioning of the brain. This is one reason for the enormous plasticity of the brain (coping with lesions, late manifestation of Alzheimer's, etc.).

8.3.4.2 Reentry

(1) The discovery that higher brains have reentrant connections is fundamental. For instance, past experience can reinforce connections between sensory input neurons (e.g. by a hypothesis what is seen), which, in turn, results in a new experience.
(2) Reentrant connections thus provide a selection mechanism for connections, and they can synchronise neuron activity. The latter is not a static selection, but a selection between different possible processes (synchronise A with B or with C?).
(3) In Edelman's work, the existence of reentrant connections is the single most important basic property of the higher brain.

8.3.4.3 The Thalamo Cortical System (TCS)

The human brain has three basic connection systems.

(1) The thalamo cortical system (TCS) is a system of dense reentrant connections between different areas of the cortex and between the cortex and the thalamus.
(2) The system of long loops connecting the cortex with subcortical structures like basal ganglia and the thalamus. They form loops, but are less flexible, more parallel, and not reentrant. They are more direct maps from one area to the other, thus less flexible. They connect the processes of the TCS to unconscious, automatic routines. The TCS's processes are cut off from the outside during dreaming.
(3) The value system. It originates from the locus coeruleus, and reaches fan-like over the whole brain. It becomes (globally) active, when "interesting" things happen (reward, punishment, change …). Its activity makes changes in neuron connections easier, and thus facilitates learning. (In Edelman (2000), p. 105, there is a very striking example: People remembered years later exactly what they did at the moment they heard about Kennedy's assassination. The value system became very active, and "burned" the circumstances into the memory.) To emphasise the (internal) interactions of the TCS, in particular the memory and

the value system, E. calls the central memory system value-category memory. (Edelman (2004), German: p. 62, English: p. 53). By degeneracy, the memory is not organised like a computer memory, where every bit of information has its (fixed for the moment) place.

8.3.4.4 The Binding Problem

See, e.g. Malsburg (1994, 1999).

(1) The human cortex has 10^9 neurons, 10^{15} synapses. There are more than 30 specialised areas for visual perception (e.g. colour, movement, spatial direction) distributed all over the brain, but there is no superior area to collect this information. Likewise, there are hundreds of different memory systems in the brain, for vision, touch, movement, etc. Neurons, which correspond to neighbouring areas of the visual field are more strongly connected. This is a "basic hypothesis" about the world: Things in the same area belong together. Of course, things are not always so simple, the cow's mouth is very close to the grass, but they do not (directly) belong together.
(2) The different properties (colour, form, etc.) of an object have to be "bound" together, this is the binding problem. It is solved through reentrant connections between the different areas of the brain, without any superior area. The binding results in synchronous firing of the corresponding areas of the brain. This was simulated in a model.

8.3.4.5 Categories and Concepts, Global Maps

(1) Classification couple (Edelman (1989), p. 48):
(My) Example: We see and touch an object, e.g. a pen.
Neurons of the visual system map (locally) to an area V of the cortex, neurons of touch detection map (locally) to an area T of the cortex. The reentrant connections between V and T create a classification couple, a higher order map, a map between two local maps (or, here better, between the two images of the local maps). The simultaneous activation of V and T strengthens the reentrant connections between V and T. Thus, certain patterns in V will be associated with certain pattern in T. Suppose we see now a different pen, with the same tactile properties, we associate a new pattern in V with the same pattern in T, and thus form a generalisation (same T, different V). Degeneracy provides another means of generalisation, as different neuron (groups) can be associated with the same effect—they behave like a non-injective function. In reality, more than a pair of sensor channels might be associated in a higher order map, it might be a classification n-tupel.

(2) Categories:
A classification n-tupel with the generalisations as described results in a category, e.g. the category "mouse" for a cat is a generalisation of an n-tupel of seeing, hearing, etc., (and finally taste, if the cat is lucky).
(3) Global mappings (Edelman (1989), p. 54 f):
Global mappings map multiple reentrant local sensor and motor maps to each other and to simple (i.e. not maps) areas of the brain, like brain stem, hippocampus, etc. They are, among other things, the basis for perceptual categories.
(4) Concepts (Edelman (1989), p. 140–145):
"Of all the subjects to be considered in formulating an adequate theory of consciousness, that of concepts is the most daunting - concepts are difficult to define, and their existence and neural bases must be inferred indirectly from experimental evidence." (Edelman (1989), p. 140). The following is an attempt to summarise some of his ideas, unfortunately, they are not very clear. The best explanation is perhaps in Edelman (1989), p. 140 ff, see also Edelman (2004), German: p. 57 ff, English: p. 48 ff. In particular in his more popular books, the problems are rather glossed over. In E's terminology, "concept" is more general than "category". A concept must include relations, he gives the example of a cat which forms the concept of its own forward motion. This involves an idea of "self", a comparison (relation) of past and present, etc. (Edelman (2004) German: p. 58, English: p. 49). E. conjectures that global mappings are the neurological basis of concepts, as they can and do map very different lower maps (sensory, motor, and arbitrary maps of the brain's own activity) to each other in a reentrant manner. Concepts formation categorises the brain's own global mappings (Edelman (1989), p. 99). (It seems that the concept of a "concept" (sorry!) was developped in neuropsychological experience with traumatised humans, showing a difference between category and concept formation (Edelman (1989), p. 295).)

8.3.4.6 Consciousness

(1) The reentrant connections of the TCS are highly degenerate, and there is no direct correspondence between conscious representation and activity of specific neurons. Thus, the brain is not organised like a computer. This activity of the TCS generates primary consciousness, a combination of present perception and past memories, which E. calls "remembered present". The combination of many modalities (visual, etc.) creates the perception of a "scene". The activities have to persist several hundreds of milliseconds to create a conscious experience. As said, the idea of the dynamic core and its importance for consciousness is another central idea of E's work.
(2) Conscious activities result in much stronger synchronisation of the brain than unconscious processes do. The pattern of synchronisation can vary widely from individuum to individuum.

In an ideal gas, molecules are totally independent, in a crystal, they are totally dependent. In both cases, the complexity is low (measured suitably). The brain has local centres, which "speak" a lot to each other, like experts do, the resulting complexity is high.
Different types of synchronisation correspond to different states of the brain. In epileptic seizures (petit mal), the whole brain fires in synchrony, in deep sleep, the brain fires slowly in big amplitudes. In both cases, the complexity of the process is low. In consciousness and REM sleep, the brain fires in local synchrony and changing patterns, and the complexity of the process is high.

(3) Conscious experience is integrated, it has creative and constructive properties. It is intentional, i.e. directed towards the (outer and inner) world. It allows to go back and forth between present and past, and frees the animal from its ties to the present. It necessitates a distinction between self and nonself.

8.3.4.7 Higher Consciousness and Language

(This part is sketched only in his books.)
The primate brain has the Broca and Wernicke areas, and new, highly reentrant connections between these and the value-category memory. This results in new possibilities to form concepts, in symbolic representation, the combinatorial possibilities of language, and higher order consciousness. Semantics, as concepts, preceded language in evolution. Language allows to stabilise thoughts and make them more precise (Edelman (1989), p. 263).

And, a side remark, but important: Deduction and selection (through reentrant brain connections, and development of the brain) seem to be the only ways of thinking, the latter is more creative. (See Edelman (2000), p. 214.)

8.3.5 Comments, and Our Own Ideas

(1) E's books seem difficult to read. This is certainly partly due to the complexity of the subject, but it also seems to us that he is not always very clear. We would have preferred a neurological definition of his concepts, even if they might be too coarse or slightly wrong. A major question seems to us why "concept" - contrary to "category" - is so difficult to define. Is there a gradual way from "category" to "concept"? Or are our words not adapted to discuss the difference? Has it something to do with the plasticity of the brain? These questions might be naive, still, one should pose them—or, preferably, better ones.
(2) Self-organisation is common, in nature, e.g. crystals organise themselves.
(3) It seems to us that the brain is partly highly compartmentalised, partly very little so. Basic routines are—roughly—happening in the compartments. The flexibility of the brain is, of course, due to the part which is not dedicated to a specific activity. Thus, the same neuron can participate in one activity, and

the next moment in a meta-activity about the former activity. This is, what we understood.

(4) Of course, when the participating neurons change, it has to be clear that the new neurons replace the old ones, that they do the same job. When the old neurons participate in a new activity, the brain has to "re-interpret itself". For example, there are no neurons for the object level, others for the meta-level, etc.

(5) One has to distinguish clearly between the structure of the brain, and the processes happening in the brain. Many important phenomena seem to be processes, roles, carried out by varying actors. The basic entity when thinking about the brain and its functioning is probably a neuron (or group of neurons) which fire (a process). We then build more complex processes by recursively - partly also with loops - connecting simpler structures.

(6) Reentrant connections seem to indicate how we form abstractions a (simplified) picture is sent back to the area of the original data, and subjected to the same processes again. Thus, operations on abstracta are basically the same operations as those on the percepts. They avoid the "homunculus" too: it suffices to memorise the level homunculus (e.g. via "pointers" in computer science language) we work on currently, to code meta-information.

(7) We think it is important to distinguish a view on the brain "from the inside", and one "from the outside". It seems safe to say that the human brain is equivalent to an immensely complicated Turing machine, as the brain is finite, and its possible states (functions of the synapses etc.) can safely be said to be finite, too.

(To go a bit more into details: For growing nerve calls and new connections, we can introduce sufficiently many "virtual" cells and connections, which become "real", as well as the inverse for dying cells, etc. As we consider the molecular level, there might also be some chance involved. A chance reaction may be considered a transition of the brain to a different state, in function of that state. So, all in all, we have finite processor (brain), a finite tape with finitely many possible symbols (states of the world), and with a second dimension of the tape (chance), also with finitely many possible symbols. So all is finite, and even deterministic, and we have a Turing machine.)

(8) Animals can also form categories, but they cannot manipulate them as we do.

(9) There seems to be an enormous step from the "darkness" of unconscious neuronal processes to the "light" of conscious ones. Is the description of the difference in his books sufficient to explain the difference? Or, are we just not used to see the enormity of the difference on the neuronal level?

(10) The "evolutionary" theory of consciousness means that consciousness has evolved according to "Darwinian" principles, like other properties of living beings, see, however, Sect. 8.2.2, at the end. In particular, consciousness is "useful" (Edelman (1989), p. 92), or, at least not harmful. If it is stable and evolving over longer periods, it is probably useful.

The "physical" theory of consciousness means that the neurological substrate influences what we "see", but not the other way round (Edelman (2004) German: p. 82 ff, English: p. 76 ff). It is like watching TV, what we see might influence us, but the actors will not see and hear us. As the TV example shows, this is possible, but a real theory of consciousness has to explain why the influence works only in one direction.

If we see correctly, both together have a problem with what we experience. When we live useful things (eating good food, having a new insight, etc.), we have pleasurable experiences. When we live bad things (being hurt, etc.) we have unpleasant experiences. Why? It could be just the other way round, as the experience does not matter. Eating good food might be accompanied by a feeling of intense pain (though the pain in the neurological sense is not there), etc. As the feeling does not matter (no connection the other way round), we would still go for good food, though feeling intense pain. As this is not so, there should be a reason why evolution has associated positive experience with positive feeling. But then we need some kind of influence from feeling to experience.

A way out could be as follows: our brain necessarily produces consciousness, one cannot have higher neuronal activity without consciousness, it is like current which generates a magnetic field, it is a necessary "side effect".

(11) Why do we need so many neurons to produce thoughts? And how does this mess of billions of firing neurons result in clear thoughts and concepts? Is this an effect of recursiveness, too, resulting in ever sharper thoughts? In this context, it might be interesting to see what happens in the brain, when we learn to distinguish fine differences, e.g. of taste, learning to appreciate wine, becoming a connaisseur.

(12) Does consciousness appear all of a sudden? If so, why and when? Or is there graded consciousness, with "lesser" consciousness in simpler structured or smaller brains? How can we find out?

(13) A general comment on neuroscience: Describing thoughts as certain firing patterns of certain groups of neurons does not make the concept of thought obsolete, less precise, etc. It is a different level of description. Speaking in terms of quarks does not make chemistry superfluous. The fascinating question is how one generates the other, how firing patterns result in precise ideas, logical thinking, language, etc. But we should not see one as an obsolete way of thinking and speaking, to be replaced by the other.

(14) This is a general comment on the literature trying to explain animal behaviour, e.g. Menzel and Fischer (2011):

It seems that there are ongoing conflicts between different schools in comparative psychology, on how to interpret animal behaviour. Some take a more anthropomorphic view, seeing planning, etc. in animals, others a more computational view, etc. In this context, the Darwinian approach—we, our behaviour, our brain developed from animals, so there is a continuum from animals to humans—is sometimes criticised as favouring an anthropomorphic view. But, do we have any competing theory of development? We think that any tradition of thought can be stretched too far. It might well be justified in some cases, but not in others. Often, there is probably no "experimentum crucis", but one has to use good sense to know when to stop. Of course, new data may overturn a view, so one should keep the door open for re-interpreting observations.

(15) Language
Language allows faster processing, as we can easily neglect inessential aspects. ("Zoomability" in Edelman (1989), p. 33.) In addition, the operators of human language allow easy manipulation of information (connection via "AND" etc.).

(16) Fast categorization
Animals seem to be able to do fast categorizations and generalisations on the basis of only a few examples (see Edelman (1989), p. 110). Why is this so? Is this a hint for the adaption of the brain to the world? Or do we chose "natural" categories is such experiments, categories we share with the animals?

8.4 Abstract Constructions—Another Part of Human Reasoning

We put this very tentative and abstract section here, as we think it should be closely linked to studies of the brain. We think that the problem is very important, as it might elucidate successes and perhaps failures of human discovery. The process of constructing abstract entities is part of reasoning, but not a subject of traditional logics. A good example of an immensely successful abstract construction is the atom. It was constructed in thought experiments by Leucippus and Democritus in the 5th century BC. The solution to the binding problem shows that already the animal brain is able to construct entities from sensory input. Humans tend to think that events in the world are caused by something, and, when we know the reasons for events to happen, we hope to avoid negative ones. This is basic for our survival. The brain constructs the lion from sensory input through binding, and we may have constructed a supernatural giant to explain thunder, earthquakes, etc. This is an old tradition. The atom is different, as it lacks the properties usual objects and entities have. So, can we conjecture how the brain forms abstract constructions?

We think that the reentrant structure of the brain allows to work on abstractions in the same way as it works on "real objects" and "real processes". We think that forming the abstract entity of a circle can be done by analogy to forming a round bowl from a piece of clay, a geometrical point by analogy to sharpening the tip of an arrow. The concrete processes and their results are familiar, and applying the analogue process to an entity of thought will give the result. The natural playfulness of the brain makes it possible. The question seems to be whether, in principle, we (have the leisure to) take the problem and result seriously, or abandon them as useless phantasies. Of course, the resulting construction may be useless, lead to alchemy or superstitions, checking their meaningfulness is a different question. We should not forget that our categories are constructions, too, they are not "out there". They are similar to abstract objects, like atoms, abstract forces, like gravity. Categories, concepts, language, are better explored than abstract objects (from an abstract point of view, of course). Are there any common laws for all such constructions? Or for the way they are created? Obviously, past experience in such creation is useful. This is all guesswork. A verification would have to come from neuroscience, observing the brain working on abstract constructions.

8.5 Some Speculative Remarks About Prototypes

First, a word of caution. These remarks are guesses, and might be very wrong. They are more questions than statements, and continue considerations in Sect. 8.2.2.

It seems that the brain works with prototypes, rather than with statements, pre-linguistic or not. This is the origin of these comments.

We already mentioned that "finding" the right prototype is similar to finding the models in counterfactual reasoning in the Stalnaker/Lewis semantics. It might be more complicated, as the number of close prototypes might be considered, too, not only their distance.

The brain has to do more with prototypes than just find the right ones.

- For instance, it seems highly unlikely that our penguin prototype has all the information about birds, too, a bird prototype will have. A penguin prototype will have an incomplete picture only. Children may have a penguin prototype, without knowing that they are birds. For reasons of economy, it seems likely that a full penguin prototype is constructed "on the fly" (no pun intended) from the penguin prototype, bird prototype, animal prototype, even object prototype.
  This construction process may have some similarity to theory revision, defeasible inheritance, etc.
- Prototypes probably cannot only be combined, but also de-composed into smaller entities, which are not pre-defined (these operations are thus not synthetic, as language and logic operations are), and modified in other, perhaps more subtle ways.
  For instance, an animal has to recognise a pattern corresponding to an instinctual image in an actually perceived pattern.
- It seems an interesting research project to find out, among other things:

(1) How such prototypes are coded in the brain.
(2) The differences between prototypes of animals and human beings.
(3) The various operations done on the prototypes, and which are easier or faster to execute than others, and how those operations are realised.
(4) Does working with prototypes—and the ways these operations are done—lead naturally to certain desirable properties of reasoning, e.g. corresponding to homogenousness, specificity (and perhaps other properties), which have to be imposed artificially in logic?
    Does working with prototypes instead of sets of defining properties explain our difficulties to "define" natural kinds like human beings, etc.?
(5) How does language interact with prototypes?

(6) It seems unlikely that the combination of prototypes, e.g. of penguin and bird prototypes, corresponds to pure upward or pure downward inheritance, things are probably less "pure".
(7) Are prototypes "rich" as conscious scenes are?

Finally, one may ask if mathematical intuition is the construction and use of suitable prototypes.

Part III
New Results

Chapter 9
Independence and Interpolation

9.1 Summary

We first repeat the essential ideas on monotone and antitone interpolation in Gabbay and Schlechta (2011), see Sect. 9.2, in particular Proposition 9.2.1. This is done to emphasise the connection to independent relations as discussed in Chap. 10, and to put the question of (semantic) interpolation more in perspective.

We then turn to interpolation for nonmonotonic logic, and its connection to multiplication properties of abstract size in Sect. 9.3. One of the main results there, Proposition 9.3.23 was shown already in Gabbay and Schlechta (2011), but we proceed here in a more systematic way.

We begin with preparatory definitions and remarks:

(1) We first introduce three variants of interpolation and abstract size properties in Sect. 9.3.1. (Most of this material was presented already in Gabbay and Schlechta (2011).)
(2) We discuss some general and combinatorial results in Sect. 9.3.2, which will be used later.
(3) We turn to conditions and results for abstract multiplication and generating (preferential) relations, see Sect. 9.3.3, in further preparation of our results.
(4) We discuss several examples of simple relations in Sect. 9.3.4.

We finally turn to the interpolation results for nonmonotonic logics themselves

(1) We show that $(\mu * 1)$ entails interpolation, Proposition 9.3.23, but also that the condition is not necessary, in Sect. 9.3.5. These results were already shown in Gabbay and Schlechta (2011).
(2) D. Makinson posed the question to one of the authors if there is also a (single) interpolant for both directions when the formulas are equivalent, i.e. if $\phi \mid\sim \psi$ and $\psi \mid\sim \phi$, is there some α such that $\phi \mid\sim \alpha \mid\sim \psi$ and $\psi \mid\sim \alpha \mid\sim \phi$? We show in Sect. 9.3.6, Proposition 9.3.25, that this is the case, but we need in our proof, in addition to $(\mu * 1)$, smoothness of the generating relation, more precisely its consequence

D.M. Gabbay and K. Schlechta, *A New Perspective on Nonmonotonic Logics*,
DOI 10.1007/978-3-319-46817-4_9

$\mu(X) \subseteq Y,\ \mu(Y) \subseteq X \Rightarrow \mu(X) = \mu(Y)$
which is shown in Fact 9.3.4.
We do not know if the smoothness property is necessary to show the result.

(3) Finally, we show in Sect. 9.3.7, Proposition 9.3.26, that interpolation is also a consequence of condition $(\mu * 4)$. Condition $(\mu * 4)$ is conceptually simpler than condition $(\mu * 1)$, as it does not involve a change of the sets where the sizes are calculated. The price to pay is a more complicated notation.

For more on the subject, see e.g. Craig (1957), Gabbay and Maksimova (2005), Gabbay and Schlechta (2010), Amir (2002).

9.2 Monotone and Antitone Interpolation

9.2.1 Overview and Background

This section contains no new results about interpolation, and serves only to recall the problem and solution. All details are to be found in Gabbay and Schlechta (2011).

We recall here that (logical) interpolation is a consequence of a particular form of independence, and thus of limited influence change. It is thus a well-examined case of the general pattern—that rare influence change, together with suitable concepts, is basic for "reasonable" reasoning. For this reason, it is important to recall the main ideas in the present book, where the fundamental concept is rarity of influence change.

9.2.2 Monotone and Antitone Semantic Interpolation

The following interpolation results can be read upward (monotonic logic) or downward (the core of nonmonotonic logic, in the following sense: γ is the theory of the minimal models of α, and not just any formula which holds in the set of minimal models, which would be downward and then upward again in the sense of model set inclusion); in the latter case we have to be careful: we usually cannot go upward again, so we have the sharpest possible case in mind. The case of mixed movement—down and then up—as in full nonmonotonic logic is treated in Sect. 9.3.

We consider here the semantical property of monotony or antitony in the following sense (in the two-valued case; the generalisation to the many-valued case is straightforward):

Let $\vdash$ be some logic such that $\phi \vdash \psi$ implies $M(\phi) \subseteq M(\psi)$ (the monotone case) or $M(\psi) \subseteq M(\phi)$ (the antitone case). In the many-valued case, the corresponding property is that $\rightarrow$ (or $\vdash$) respects $\leq$, the order on the truth values.

The problem (for simplicity, for the two-valued case) reads now as follows:

If $M(\phi) \subseteq M(\psi)$ (or, symmetrically, $M(\psi) \subseteq M(\phi)$), is there a "simple" model set A such that $M(\phi) \subseteq A \subseteq M(\psi)$ or $M(\psi) \subseteq A \subseteq M(\phi)$. Obviously, the problem is the same in both cases. We will see that such an A will always exist, so all such logics have semantical interpolation (but not necessarily syntactic interpolation).

The main conceptual problem is to define "simple model set". We have to look at the syntactic problem for guidance. Suppose ϕ is defined using propositional variables p and q, ψ using q and r. α has to be defined using only q. What are the models of α? By the very definition of validity in classical logic, neither p nor r have any influence on whether m is a model of α or not. Thus, if m is a model of α, we can modify m on p and r, and it will still be a model. Classical models are best seen as functions from the set of propositional variables to $\{TRUE, FALSE\}$, $\{t, f\}$, etc. In this terminology, any m with $m \models \alpha$ is "free" to choose the value for p and r, and we can write the model set A of α as $\{t, f\} \times M_q \times \{t, f\}$, where M_q, the set of values for q α-models, may have ($\emptyset$, $\{t\}$, $\{f\}$, $\{t, f\}$).

So, the semantical interpolation problem is to find sets which may be restricted on the common variables, but are simply the Cartesian product of the possible values for the other variables. To summarise: Let two model sets X and Y be given, where X itself is restricted on variables $\{p_1, \ldots, p_m\}$ (i.e. the Cartesian product for the rest), Y is restricted on $\{r_1, \ldots, r_n\}$, then we have to find a model set A which is restricted only on $\{p_1, \ldots, p_m\} \cap \{r_1, \ldots, r_n\}$; and such that $X \subseteq A \subseteq Y$, of course.

Formulated this way, our approach, the problem and its solution, has two trivial generalisations

- for multi-valued logics we take the Cartesian product of more than just $\{t, f\}$.
- ϕ may be the hypothesis, and ψ the consequence, but also vice versa; there is no direction in the problem. Thus, any result for classical logic carries over to the core part of, e.g. preferential logics.

The main result for the situation with $X \subseteq Y$ is that there is always such a semantical interpolant A as described above (see Proposition 9.2.1). (Our proof works also for "parallel interpolation", a concept introduced by Kourousias and Makinson (2007), see Gabbay and Schlechta (2011).)

9.2.2.1 The Result for the Many-Valued Case

We assume here that the max and min of arbitrary sets of truth values will always exist.

Recall that we do not work with sets of models or sequences any more, but with arbitrary functions $f, g, h : M \to V$, where each $m \in M$ is a function $m : L \to V$, where, intuitively, L stands for the set of propositional variables, V for the set of truth values, M for the set of many-valued models, and f, etc. for functions (intuitively, $f = f_\phi$, etc.) assigning each m a value, intuitively, the value of ϕ in m. Again, we will consider $f \leq g$ and look for some h with $f \leq h \leq g$, where $I(f) \cup I(g) \subseteq I(h)$.

Definition 9.2.1

(1) Let $J \subseteq L, f : M \to V$. Define
$f^+(m, J) := max\{f(m') : m \restriction J = m' \restriction J\}$ and
$f^-(m, J) := min\{f(m') : m \restriction J = m' \restriction J\}$.

(Similarly, if m is defined only on J, the condition is $m' \restriction J = m$, instead of $m \restriction J = m' \restriction J$.)

(2) Call M rich iff for all $m, m' \in M$, $J \subseteq L$, $(m \restriction J) \cup (m' \restriction (L - J)) \in M$ (i.e. we may cut and paste models).

This assumption is usually given; it is mainly here to remind the reader that it is not trivial, and we have to make sure it really holds. A problem might, e.g. arise when we consider only subsets of all models, i.e. some $M' \subseteq M$, and not the full M.

Note that the possibility of arbitrary combinations of models is also an aspect of independence.

(3) A reminder: Call $f : M \to V$ insensitive to $J \subseteq L$ iff for all m, n: $m \restriction (L - J) = n \restriction (L - J)$ implies $f(m) = f(n)$, i.e. the values of m on J have no importance for f.

(4) We will sometimes write m_J for $m \restriction J$.

Let $L = J \cup J' \cup J''$ be a disjoint union. If $f : M \to V$ is insensitive to $J \cup J''$, we can define $f(m, J')$ as any $f(m')$ such that $m' \restriction J' = m_{J'}$.

Proposition 9.2.1

This is Proposition 4.2.3 in Gabbay and Schlechta (2011).

Let M be rich, $f, g : M \to V$, $f(m) \leq g(m)$ for all $m \in M$. Let $L = J \cup J' \cup J''$; let f be insensitive to J, g be insensitive to J''.

Then $f^+(m, J') \leq g^-(m, J')$ for all $m \in M$, and any $h : M \to V$ which is insensitive to $J \cup J''$ is an interpolant iff

$f^+(m, J') \leq h(m_J m_{J'} m_{J''}) = h(m \restriction J') \leq g^-(m, J')$ for all $m \in M$.

(h can be extended to the full M in a unique way, as it is insensitive to $J \cup J''$; so it does not really matter whether we define h on L or on J'.)

See Diagram 9.2.1.

Diagram 9.2.1

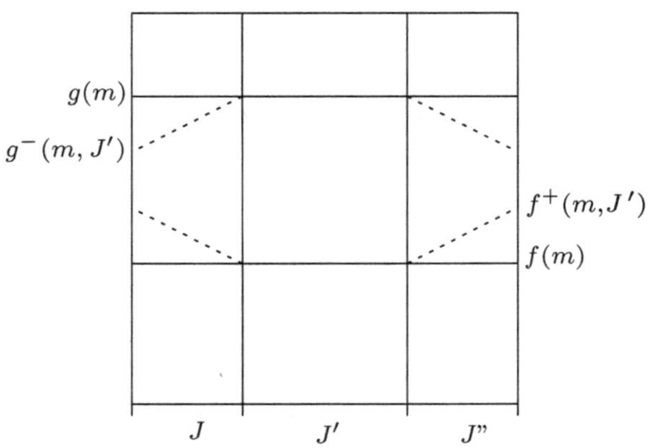

Semantic interpolation, $f \leq g$

Definition 9.2.2
It is thus justifiable in the above situation to call $f^{+}(m, J')$ and $g^{-}(m, J')$ the standard interpolants; more precisely, we call h such that $h(m) = f^{+}(m, J')$ or $h(m) = g^{-}(m, J')$ with $R(h) \subseteq J'$ a standard interpolant.

Our result has an additional reading: it defines an interval of interpolants, with lower bound $f^{+}(m, J')$ and upper bound $g^{-}(m, J')$. But these interpolants have a particular form. If they exist, i.e. iff $f \leq g$, then $f^{+}(m, J')$ depends only on f and J' (and m), but *not* on g, and $g^{-}(m, J')$ depends only on g and J', *not* on f. Thus, they are universal, as we have to look only at one function and the set of common variables.

Remark 9.2.2
We can now consider the triple $\langle f, h, g\rangle$, where h is an interpolant for f and g, and $\langle F, H, G\rangle$, where (for fixed J, J', J'') each $h \in H$ is an interpolant for all $f \in F$, $g \in G$.

We leave it to the reader to explore these relation similarly to Chap. 10 in detail, but wanted to give enough information to see the connection.

9.3 Interpolation for Nonmonotonic Logic and Size

9.3.1 Introduction

The connection between nonmonotonic logic and the abstract concept of size was investigated in Gabbay and Schlechta (2016); see also Gabbay and Schlechta (2009). There, we looked among other things at abstract addition of size. Here, we will show a connection to abstract multiplication of size. Our semantical approach used decomposition of set-theoretical products. An important step was to write a set of models Σ as a product of some set Σ' (which was a restriction of Σ) and some full Cartesian product. So, when we speak about size, we will have (slightly simplified) some big subset Σ_1 of one product Π_1 and some big subset Σ_2 of another product Π_2, and will now check whether $\Sigma_1 \times \Sigma_2$ is a big subset of $\Pi_1 \times \Pi_2$, in shorthand, whether "*big* $*$ *big* = *big*". (See Definition 5.3.1 for precise definitions.) Such conditions are called coherence conditions, as they do not concern the notion of size itself, but the way the sizes defined for different base sets are connected. Our main result here says that if the logic under investigation is defined from a notion of size which satisfies sufficiently many conditions, then this logic will have interpolation of type 3 or even 2 (type 3 is the weakest form discussed here).

Consider now some set product $X \times X'$. (Intuitively, X and X' are model sets on sublanguages J and J' of the whole language L.) When we have now a rule such as, if Y is a big subset of X, and Y' a big subset of X', then $Y \times Y'$ is a big subset of $X \times X'$, and conversely, we can calculate consequences separately in the sublanguages and put them together to have the overall consequences. But this is the principle behind interpolation: we can work with independent parts.

This is made precise in Definition 5.3.1, in particular by the rule

$(\mu * 1)$: $\mu(X \times X') = \mu(X) \times \mu(X')$.

Recall that μ is defined here on different sets. So, if $I = J \cup J'$, $J \cap J' = \emptyset$, and $X \subseteq \Pi J$, $X' \subseteq \Pi J'$, so $X \times X' \subseteq \Pi I$, then we can write more precisely, as we will do in more complicated situations.

$(\mu * 1)$: $\mu_I(X \times X') = \mu_J(X) \times \mu_{J'}(X')$.

(Note that the conditions $(\mu * i)$ and $(S * i)$ are equivalent, as shown in Proposition 5.3.1 (for principal filters).)

A variant of $(\mu * 1)$ is $(\mu * 4)$:

$(\mu * 4)$: $\mu(X \times X') = \mu(X \times \Pi J') \upharpoonright J \times \mu(\Pi J \times X') \upharpoonright J'$

Here, μ is always μ_I, so it is not necessary to distinguish different μ's.

The main result is that the multiplicative size rules $(\mu * 1)$, and similarly $(\mu * 4)$, entail nonmonotonic interpolation of the form $\phi \mid\!\sim \alpha \mid\!\sim \psi$; see Propositions 9.3.23 and 9.3.26.

We now take a closer look at interpolation for nonmonotonic logic.

9.3.1.1 The Three Variants of Interpolation

Consider preferential logic, a rule like $\phi \mid\!\sim \psi$. This means that $\mu(\phi) \subseteq M(\psi)$. So we go from $M(\phi)$ to $\mu(\phi)$, the minimal models of ϕ, and then to $M(\psi)$, and, abstractly, we have $M(\phi) \supseteq \mu(\phi) \subseteq M(\psi)$, so we have neither necessarily $M(\phi) \subseteq M(\psi)$, nor $M(\phi) \supseteq M(\psi)$; the relation between $M(\phi)$ and $M(\psi)$ may be more complicated. Thus, we have neither the monotone, nor the antitone case. For this reason, our general results for monotone or antitone logics do not hold any more.

But we also see here that classical logic is used, too. Suppose that there is a ϕ' which describes exactly $\mu(\phi)$; then we can write $\phi \mid\!\sim \phi' \vdash \psi$.

So we can split preferential logic into a core part—going from ϕ to its minimal models—and a second part, which is just classical logic. (Similar decompositions are also natural for other nonmonotonic logics.) Thus, preferential logic can be seen as a combination of two logics, the nonmonotonic core, and classical logic. It is thus natural to consider variants of the interpolation problem, where $\mid\!\sim$ denotes again preferential logic, and $\vdash$ as usual classical logic

Given $\phi \mid\!\sim \psi$, is there "simple" α such that

(1) $\phi \mid\!\sim \alpha \vdash \psi$, or
(2) $\phi \vdash \alpha \mid\!\sim \psi$, or
(3) $\phi \mid\!\sim \alpha \mid\!\sim \psi$?

For interpolation of the forms $\phi \vdash \alpha \mid\!\sim \psi$ and $\phi \mid\!\sim \alpha \vdash \psi$, the reader is referred to Gabbay and Schlechta (2011), Chap. 5, we will concentrate here on interpolation of the form $\phi \mid\!\sim \alpha \mid\!\sim \psi$.

9.3.1.2 Discussion

The conditions are natural, and not too strong, and the connection between those multiplicative properties and interpolation gives quite deep insights into the basics of nonmonotonic logics.

We first quote the result from Gabbay and Schlechta (2011), the discussion of the converse, and of the extreme interpolants, and then turn to the new result with full equivalence, this is the main result of this chapter.

It seems that interpolation is too weak and too strong for common sense reasoning. Too strong, as it does not allow *birds* $\mid\!\sim$ *fly*, *penguins* $\mid\!\sim \neg$*fly* together. Given *birds* $\mid\!\sim$ *fly*, we can only have *penguins* $\not\mid\!\sim$ *fly*, see Fact 9.3.17. And it does not allow to conclude from *birds* $\mid\!\sim$ *fly* to *blackbirds* $\mid\!\sim$ *fly*, so it is too weak. We would rather like to have a strong form of interpolation, which allows to conclude from birds to blackbirds, but only in a default form, to allow for exceptions, like penguins.

Our construction in Chap. 11 permits exactly this, and is a default version of relation properties ($< Abs$) and (R1).

It seems difficult to find a condition equivalent to the existence of interpolants. One of the problems is the following: An interpolant α can be defined using only variables from J' (in above notation), and $\mu(\phi) \restriction J' \subseteq \alpha \restriction J'$ is necessary. But it is not required that $\alpha \restriction J' \subseteq \mu \restriction J'$, and we may need such a model to minimise a new model to be considered as we have "liberated" the variables from J.

The independent treatment of parts of the language can also be seen in work by R. Parikh and co-authors, see e.g. Chopra and Parikh (2000). It is useful in other contexts, too, e.g. analogical and inductive reasoning, progressive reasoning, etc. This is subject to further research.

9.3.1.3 Some Notation

We will consider here products and subsets of products.

We will have $I = J \cup J'$ or $I = J \cup J' \cup J''$, the J's pairwise disjoint, and consider $\Pi I := \Pi\{U_i : i \in I\}$, $\Pi J := \Pi\{U_i : i \in J\}$, etc., where, intuitively, I is the set of propositional variables of the language, $U_i = \{0, 1\}$, and $\sigma \in \Pi I$ will be a total model, $\sigma_J \in \Pi J$ a partial model, etc. In particular, all U_i will be non-empty.

9.3.2 *Some General and Combinatorial Results*

These facts are elementary, but useful, also for the intuition.

Fact 9.3.1

Let $I = J \cup J' \cup J''$, $A, B \subseteq \Pi I$. Then

$((A \restriction J \cup J') \times B \restriction J'') \restriction J' \cup J'' = A \restriction J' \times B \restriction J''$.

Proof
"$\supseteq$": Let $\phi'\psi'' \in A \upharpoonright J' \times B \upharpoonright J''$, then there are $\phi\phi'\phi'' \in A$, $\psi\psi'\psi'' \in B$, and $\phi\phi' \in A \upharpoonright J \cup J'$, $\phi\phi'\psi'' \in (A \upharpoonright J \cup J') \times (B \upharpoonright J'')$ and $\phi'\psi'' \in ((A \upharpoonright J \cup J') \times (B \upharpoonright J'')) \upharpoonright J' \cup J''$.

"$\subseteq$": Conversely, let $\phi'\psi'' \in ((A \upharpoonright J \cup J') \times (B \upharpoonright J'')) \upharpoonright J' \cup J''$, then there are ϕ such that $\phi\phi'\psi'' \in (A \upharpoonright J \cup J') \times (B \upharpoonright J'')$, and ϕ'', ψ, ψ' such that $\phi\phi'\phi'' \in A$, $\psi\psi'\psi'' \in B$, thus $\phi' \in A \upharpoonright J'$, $\psi'' \in B \upharpoonright J''$ and $\phi'\psi'' \in A \upharpoonright J' \times B \upharpoonright J''$. □

Fact 9.3.2
Let $I = J \cup J'$ - except in (8), where $I = J \cup J' \cup J''$ - with all J's pairwise disjoint.

(1) $A \subseteq B \rightarrow A \upharpoonright J \subseteq B \upharpoonright J$
(2) $A \subset B \not\rightarrow A \upharpoonright J \subset B \upharpoonright J$
(3) $A \upharpoonright J = B \upharpoonright J \not\rightarrow A \subseteq B$
(4) $A \upharpoonright J \subset B \upharpoonright J \not\rightarrow A \subseteq B$
(5) $A \cup B \upharpoonright J = A \upharpoonright J \cup B \upharpoonright J$
(6) $A \cap B \upharpoonright J \subseteq A \upharpoonright J \cap B \upharpoonright J$
(7) It is not always true that $A \upharpoonright J \cap B \upharpoonright J \subseteq A \cap B \upharpoonright J$
(8) Let $I = J \cup J' \cup J''$, and $A = \Pi J \times A \upharpoonright (J' \cup J'')$, $B = B \upharpoonright (J \cup J') \times \Pi J''$, then $A \upharpoonright J' \cap B \upharpoonright J' \subseteq A \cap B \upharpoonright J'$ holds, too.

Proof
J will be the first coordinate.

(1) Trivial.
(2) Consider $A = \{00\}$, $B = \{00, 01\}$.
(3) Consider $B = \{00\}$, $A = \{00, 01\}$.
(4) Consider $A = \{00, 01\}$, $B = \{00, 11\}$. Then $A \upharpoonright J = \{0\}$, $B \upharpoonright J = \{0, 1\}$.
(5) "$\subseteq$": Let $\sigma_J \in A \cup B \upharpoonright J$, so $\exists \tau \in A \cup B.\tau \upharpoonright J = \sigma_J$, so $\exists \tau \in A.\tau \upharpoonright J = \sigma_J$, or $\exists \tau \in B.\tau \upharpoonright J = \sigma_J$, so $\sigma_J \in A \upharpoonright J \cup B \upharpoonright J$.
"$\supseteq$" follows from (1).
(6) By (1).
(7) $A = \{01\}$, $B = \{00\}$, then $A \upharpoonright J = B \upharpoonright J = \{0\}$, but $A \cap B = \emptyset$, so $A \cap B \upharpoonright J = \emptyset$.
(8) Let $\sigma_{J'} \in A \upharpoonright J' \cap B \upharpoonright J'$, then there are $\sigma_J^A \sigma_{J'} \sigma_{J''}^A \in A$, $\sigma_J^B \sigma_{J'} \sigma_{J''}^B \in B$, so $\sigma_J^B \sigma_{J'} \sigma_{J''}^A \in A \cap B$, and $\sigma_{J'} \in (A \cap B) \upharpoonright J'$

□

The following fact and the argument are rather trivial, it mainly serves to illustrate a central idea, which is used several times.

Fact 9.3.3
Let $I = J \cup J' \cup J''$.

Let $A = (A \upharpoonright J \cup J') \times (A \upharpoonright J'')$, $B = (B \upharpoonright J) \times (B \upharpoonright J' \cup J'')$, $A \subseteq B$.

Let $\sigma = \sigma_J \sigma_{J'} \sigma_{J''}$ such that $\sigma_J \in B \upharpoonright J$, $\sigma_{J'} \in A \upharpoonright J'$, $\sigma_{J''} \in A \upharpoonright J''$.

Then $\sigma \in B$.

Proof
$\sigma_{J'} \in A \upharpoonright J'$, so there is $\sigma'_J \sigma_{J'} \sigma'_{J''} \in A$, so (by $\sigma_{J''} \in A \upharpoonright J''$) $\sigma'_J \sigma_{J'} \sigma_{J''} \in A \subseteq B$, so (by $\sigma_J \in B \upharpoonright J$) $\sigma = \sigma_J \sigma_{J'} \sigma_{J''} \in B$. □

Fact 9.3.4
If μ is generated by a smooth relation, $\mu(A) \subseteq B$, $\mu(B) \subseteq A$, then $\mu(A) = \mu(B)$.

Proof
Let $a \in \mu(A) - \mu(B)$. As $\mu(A) \subseteq B$, $a \in B$. So there is $b \prec a$, $b \in \mu(B) \subseteq A$, contradiction. □

Fact 9.3.5
Let $\mu(A) \subseteq B$, $\mu(B) \subseteq A$, μ generated by a smooth relation, then $\mu(A \cap B) \subseteq \mu(A) \cap \mu(B)$.

Proof
Suppose not, let $a \in \mu(A \cap B) \subseteq A \cap B \subseteq A$, but $a \notin \mu(A)$. So there must be $b \prec a$, $b \in \mu(A) \subseteq A$. By $\mu(A) \subseteq B$, $b \in B$, so $b \in A \cap B$, contradiction. □

Fact 9.3.6
Let $\mu(A) \subseteq B$, $\mu(B) \subseteq A$, then $\mu(A \cup B) \subseteq \mu(A) \cap \mu(B)$.

Proof
We show $a \in \mu(A)$, $a \in \mu(B)$ is analogous.

Suppose $a \in \mu(A \cup B) - \mu(A)$.

By $\mu(A \cup B) \subseteq A \cup B$, $a \in A$ or $a \in B$.

If $a \in A$, then there must be $b \in A.b \prec a$, so $a \notin \mu(A \cup B)$, a contradiction.

So suppose $a \in B - A$. If $a \notin \mu(B)$, then there is $b \in B.b \prec a$, so $a \notin \mu(A \cup B)$, a contradiction. So $a \in \mu(B) \subseteq A$, contradiction. □

9.3.3 *Conditions for Abstract Multiplication and Generating Relations*

9.3.3.1 Conditions for Abstract Multiplication

Definition 9.3.1
Let $\Sigma \subseteq \Pi J$, $\Gamma \subseteq \Pi J'$.

We suppose abstract measures are given on ΠI, ΠJ, $\Pi J'$, with e.g. $\mu_I(\Delta)$ denoting the smallest big subset of $\Delta \subseteq \Pi I$, intuitively the set of minimal elements of Δ, when μ_I is generated by a preferential relation, similar for μ_J and $\mu_{J'}$. When the context is clear, we might omit the indices for better readability.

We define

(1) $(\mu * 1)$:
$\mu(\Sigma \times \Gamma) = \mu(\Sigma) \times \mu(\Gamma)$
or, more precisely,
$\mu_I(\Sigma \times \Gamma) = \mu_J(\Sigma) \times \mu_{J'}(\Gamma)$
(2) $(\mu * 4)$:
$\mu(\Sigma \times \Gamma) = \mu(\Sigma \times \Pi J') \upharpoonright J \times \mu(\Pi J \times \Gamma) \upharpoonright J'$.

Remark 9.3.7
There is a multitude of possible modifications in this context, which may all be interesting for special situations, e.g. resulting in restricted, but not full, interpolation.

(1) We can restrict the J's for which the conditions hold.
(2) We can restrict the Σ's and Γ's for which the conditions hold.
(3) We can examine validity for μ generated by relations with various properties: smooth versus not necessarily smooth, transitive versus not necessarily transitive, etc.

Remark 9.3.8
This is trivial, but helpful

When checking $(\mu * 4)$, and $\Sigma = \{\sigma\}$, a singleton, $\Gamma \neq \emptyset$, and $\mu(X) = \emptyset \rightarrow X = \emptyset$, then $\mu(\Sigma \times \Pi J') \upharpoonright J = \mu(\Sigma \times \Gamma) \upharpoonright J = \{\sigma\}$. This simplifies the condition and its verification considerably. Likewise for $\Gamma = \{\gamma\}$, of course.

Example 9.3.1
$(\mu * 1)$ does not imply any of the following:

(1) $\mu_{J'}(X \upharpoonright J') \subseteq \mu_I(X) \upharpoonright J'$, even if the generating relations are smooth.
(2) transitivity or smoothness of the generating relation (even if each element has size 2).
(3) $\mu(A) \subseteq B$, $\mu(B) \subseteq A \Rightarrow \mu(A) = \mu(B)$, so a consequence of smoothness of the generating relation may fail.

Proof

(1) Let $I = J \cup J'$, in that order, i.e. J is the first, J' the second coordinate. Consider $\{00, 01, 10, 11\}$, with the order $00 \prec 01 \prec 10 \prec 11$, closed under transitivity, thus, it is smooth. This is the order in I. In J and J' set $0 \prec 1$.
This order satisfies $(\mu * 1)$, as we show now.
The possible products for this set are:
(1) $1 \times \{0, 1\}$ and $0 \times \{0, 1\}$,
(2) $\{1, 0\} \times 0$ and $\{1, 0\} \times 1$,
(3) $\{1, 0\} \times \{0, 1\}$.
We have for (1): $\mu_I(1 \times \{0, 1\}) = \{10\} = \mu_J\{1) \times \mu_{J'}\{0, 1\} = 10$, likewise for $0 \times \{0, 1\}$.
We have for (2): $\mu_I(\{1, 0\} \times 0) = 00 = \mu_J\{1, 0\} * \mu_{J'}(0) = 00$, likewise for $\{1, 0\} \times 1$.

We have for (3): $\mu_I(\{1,0\} \times \{0,1\}) = 00 = \mu_J(1,0) \times \mu_{J'}(0,1) = 00$.
Set now $X = \{10, 01\}$, then $\mu_I(X) = \{01\}$, $\mu_I(X) \restriction J' = \{1\}$, but $X \restriction J' = \{0,1\}$, so $\mu_{J'}(X \restriction J') = \{0\}$. Thus, $\mu_{J'}(X \restriction J') \not\subseteq \mu_I(X) \restriction J'$.

(2) Consider $\{00, 01, 10, 11\}$, with the order $00 \prec 01 \prec 10 \prec 11$, $00 \prec 10$, $01 \prec 11$, but without transitivity, i.e. $00 \not\prec 11$. Let μ_I defined by this relation. The set $\{00, 10, 11\}$ shows that smoothness fails. In J and J' set $0 \prec 1$.
This order satisfies $(\mu * 1)$, as we show now.
The possible products for this set are:
(1) $1 \times \{0,1\}$ and $0 \times \{0,1\}$,
(2) $\{1,0\} \times 0$ and $\{1,0\} \times 1$,
(3) $\{1,0\} \times \{0,1\}$.
We have for (1): $\mu_I(1 \times \{0,1\}) = \{10\} = \mu_J\{1) \times \mu_{J'}\{0,1\} = 10$, likewise for $0 \times \{0,1\}$.
We have for (2): $\mu_I(\{1,0\} \times 0) = 00 = \mu_J\{1,0\} * \mu_{J'}(0) = 00$, likewise for $\{1,0\} \times 1$.
We have for (3): $\mu_I(\{1,0\} \times \{0,1\}) = 00 = \mu_J(1,0) \times \mu_{J'}(0,1) = 00$.

(3) Consider the example for (2), and $A := \{00, 10, 11\}$, $B := \{00, 11\}$. Then $\mu(A) = \{00\}$, $\mu(B) = B$, so $\mu(A) \subseteq B$, $\mu(B) \subseteq A$, but $\mu(A) \neq \mu(B)$.

□

Fact 9.3.9
Let $I = J \cup J' \cup J''$, $\Delta := \Sigma \restriction J'$ for some Σ. Then $(\mu * 4)$ entails
$\mu[\Pi J \times \Delta \times \Pi J''] = \Big(\mu(\Pi I) \restriction J\Big) \times \Big(\mu[\Pi J \times \Delta \times \Pi J''] \restriction J'\Big) \times \Big(\mu(\Pi I) \restriction J''\Big)$.

Proof
Consider
(1) $\mu[\Pi J \times \Delta \times \Pi J'']$.
By applying $(\mu * 4)$ to the first $\times$ in (1),
$\mu[\Pi J \times \Delta \times \Pi J''] =$
(2) $\Big(\mu(\Pi I) \restriction J\Big) \times \Big(\mu[\Pi J \times \Delta \times \Pi J''] \restriction J' \cup J''\Big)$.
By applying $(\mu * 4)$ to the second $\times$ in (1),
$\mu[\Pi J \times \Delta \times \Pi J''] =$
$\Big(\mu[\Pi J \times \Delta \times \Pi J''] \restriction J \cup J'\Big) \times \Big(\mu(\Pi I) \restriction J''\Big)$,
Thus, we have for the second part of (2)
$\mu[\Pi J \times \Delta \times \Pi J''] \restriction J' \cup J'' =$
$\Big(\Big(\mu[\Pi J \times \Delta \times \Pi J''] \restriction J \cup J'\Big) \times \Big(\mu(\Pi I) \restriction J''\Big)\Big) \restriction J' \cup J'' =$
$\Big(\mu[\Pi J \times \Delta \times \Pi J''] \restriction J'\Big) \times \Big(\mu(\Pi I) \restriction J''\Big)$,
the second equality holds by Fact 9.3.1.

So we finally have $\mu[\Pi J \times \Delta \times \Pi J''] = \Big(\mu(\Pi I) \restriction J\Big) \times \Big(\mu[\Pi J \times \Delta \times \Pi J''] \restriction J'\Big) \times \Big(\mu(\Pi I) \restriction J''\Big)$. □

9.3.3.2 Conditions for the Generating Relations

Let again $I = J \cup J'$ with $J \cap J' = \emptyset$. For each $i \in I$ $X_i \neq \emptyset$, let $\Pi I := \Pi\{X_i : i \in I\}$, ΠJ, $\Pi J'$ analogously.

We define first

Definition 9.3.2
Let $\Sigma \subseteq \Pi J$, $\Gamma \subseteq \Pi J'$.

Let $\prec$ be a (transitive) relation on ΠI, $\sigma, \sigma' \in \Pi J$, $\tau, \tau' \in \Pi J'$. We define the following absoluteness condition for $\prec$:

($<$-Abs):

$\sigma\tau \prec \sigma\tau'$ iff $\sigma'\tau \prec \sigma'\tau'$ (for all $\sigma, \tau \neq \emptyset$ etc.)

By abuse of language, we can then write $\tau \prec \tau'$, and $\tau \perp \tau'$ iff $\sigma\tau \perp \sigma\tau'$. When necessary, we may index the relation, $\prec_{J'}$, etc.

Definition 9.3.3
Let again $\Sigma \subseteq \Pi J$, $\Gamma \subseteq \Pi J'$. Let $\sigma, \tau \in \Pi J$, $\sigma', \tau' \in \Pi J'$, and let relations $\prec$ be defined on ΠI, ΠJ, ΠJ. (We will not distinguish them as $\prec_I$ etc., as the context will disambiguate.) As usual, $\preceq$ will denote $\prec$ or $=$.

Analogously, σ_i will be an element of X_i, and we suppose relations $\prec$ are defined on all X_i, and on arbitrary ΠJ for (R1) below.

(1) $(GH1)$: $\sigma \preceq \tau \wedge \sigma' \preceq \tau' \wedge (\sigma \prec \tau \vee \sigma' \prec \tau') \Rightarrow \sigma\sigma' \prec \tau\tau'$.
(2) $(GH2)$: $\sigma\sigma' \prec \tau\tau' \Rightarrow \sigma \prec \tau \vee \sigma' \prec \tau'$.
    $(GH2)$ means that some compensation is possible, e.g. $\tau' \prec \sigma'$ might be the case, but $\sigma'' \prec \tau''$ wins in the end, so $\sigma'\sigma'' \prec \tau'\tau''$.
(3) $(GH3)$: $\sigma'\sigma'' \preceq \tau'\tau'' \Leftrightarrow \sigma' \preceq \tau'$ and $\sigma'' \preceq \tau''$.
    (Thus, $\sigma'\sigma'' \prec \tau'\tau''$ iff $\sigma'\sigma'' \preceq \tau'\tau''$ and ($\sigma' \prec \tau'$ or $\sigma'' \prec \tau''$).)
(4) (R1) $\sigma \prec \sigma'$ iff $\forall i(\sigma_i = \sigma'_i$ or $\sigma_i \prec \sigma'_i)$ and $\exists i(\sigma_i \prec \sigma'_i)$.

(The (GHi) were defined in Gabbay and Schlechta (2011).)

Remark 9.3.10
There is a multitude of possible definitions from longer to shorter sequences and inversely. For instance, we might weaken ($<$-Abs), to $\sigma\tau \prec \sigma\tau' \Rightarrow \sigma'\tau' \not\prec \sigma'\tau$ for all σ, τ, etc. We can introduce lexicographic relations, context dependent relations, like *fly* $\prec$ $\neg$*fly* for birds, but $\neg$*fly* $\prec$ *fly* for penguins, determine the order between longer sequences by an (abstract or probabilistic) majority rule from the smaller sequences, etc. ad libitum.

Fact 9.3.11
If (R1) holds for $\prec$, then so does $<$-Abs.

(This follows also from Facts 9.3.12 and 9.3.13.)

Proof
Trivial. □

Fact 9.3.12
$(\mu * 4)$ implies $(<Abs)$

Proof
Let σ be defined on J, τ, τ' on J'.

$\sigma\tau \prec \sigma\tau'$ iff $\sigma\tau' \notin \mu(\{\sigma\} \times \{\tau, \tau'\}) = \mu(\{\sigma\} \times \Pi J') \upharpoonright J \times \mu(\Pi J \times \{\tau, \tau'\}) \upharpoonright J' = \{\sigma\} \times \mu(\Pi J \times \{\tau, \tau'\}) \upharpoonright J'$, but the second part of this product is independent of σ. □

Fact 9.3.13
If (R1) holds for $\prec$, then so does $(\mu * 4)$.

Proof
Let J, J' be as in the definition of $(\mu * 4)$.

Let $\sigma\gamma \in (\Sigma \times \Gamma) - \mu(\Sigma \times \Gamma)$, then there ist $\sigma'\gamma' \in \Sigma \times \Gamma$, $\sigma'\gamma' \prec \sigma\gamma$. Let $K := \{i : (\sigma'\gamma')_i \prec (\sigma\gamma)_i\} \neq \emptyset$. Suppose $K \cap J \neq \emptyset$. Then by (R1) for all $\gamma'' \in \Pi J'$ $\sigma'\gamma'' \prec \sigma\gamma''$, so there is no $\rho \in \mu(\Sigma \times \Pi J')$ s.t. $\rho \upharpoonright J = \sigma$, so $\sigma\gamma \notin (\mu(\Sigma \times \Pi J') \upharpoonright J) \times (\mu(\Pi J \times \Gamma) \upharpoonright J')$. The case for $K \cap J' \neq \emptyset$ is analogous.

Conversely

Let $\prec$ be given on the $p_i/\neg p_i$. Define $\delta : I := J \cup J' \to \{0, 1\}$ as follows: if $p_i \prec \neg p_i$, then $\delta_i := 1$, if $p_i \succ \neg p_i$, then $\delta_i := 0$, otherwise δ_i is arbitrary, say 1. Then δ is minimal in ΠI. Let $\sigma\gamma \in \Sigma \times \Gamma$, but $\sigma\gamma \notin (\mu(\Sigma \times \Pi J') \upharpoonright J) \times (\mu(\Pi J \times \Gamma) \upharpoonright J')$. Then $\sigma \notin \mu(\Sigma \times \Pi J') \upharpoonright J$ or $\gamma \notin \mu(\Pi J \times \Gamma) \upharpoonright J'$. Suppose $\sigma \notin \mu(\Sigma \times \Pi J') \upharpoonright J$. Consider $\sigma(\delta \upharpoonright J')$, the concatenation of σ with $\delta \upharpoonright J'$. As $\sigma \notin \mu(\Sigma \times \Pi J') \upharpoonright J$, there must be $\tau\tau' \in \Sigma \times \Pi J'$, $\tau\tau' \prec \sigma(\delta \upharpoonright J')$, $\tau \neq \sigma$, and by choice of δ, $\tau \prec \sigma$ (by abuse of language, $\prec$ in J), so $\tau\gamma \prec \sigma\gamma$, thus $\sigma\gamma \notin \mu(\Sigma \times \Gamma)$.

The case $\gamma \notin \mu(\Pi J \times \Gamma) \upharpoonright J'$ is analogous. □

Fact 9.3.14
(R1) implies $(\mu * 1)$

Proof
Let $\sigma\gamma \in \Sigma \times \Gamma - \mu_I(\Sigma \times \Gamma)$, we show $\sigma\gamma \notin \mu_J(\Sigma) \times \mu_{J'}(\Gamma)$. So $\exists \sigma'\gamma' \in \Sigma \times \Gamma$, $\sigma'\gamma' \prec \sigma\gamma$. Let $K := \{i : (\sigma'\gamma')_i \prec (\sigma\gamma)_i\} \neq \emptyset$. Suppose $K \cap J \neq \emptyset$. Then for any $\sigma\gamma''$, $\sigma'\gamma'' \in \Sigma \times \Gamma$, $\sigma'\gamma'' \prec \sigma\gamma''$, so $\sigma \notin \mu_J(\Sigma)$, and $\sigma\gamma \notin \mu_J(\Sigma) \times \mu_{J'}(\Gamma)$. The case $K \cap J'$ is analogous.

Conversely, let $\sigma\gamma \in \Sigma \times \Gamma$, $\sigma\gamma \notin \mu_J(\Sigma) \times \mu_{J'}(\Gamma)$. Suppose $\sigma \notin \mu_J(\Sigma)$, then there is $\sigma' \in \Sigma$ such that $\sigma' \prec \sigma$, so $\sigma'\gamma \in \Sigma \times \Gamma$ and $\sigma'\gamma \prec \sigma\gamma$, so $\sigma\gamma \notin \mu(\Sigma \times \Gamma)$. The case $\gamma \notin \mu_{J'}(\Gamma)$ is analogous. □

Fact 9.3.15
(12) If (R1) holds, and all individual relations are transitive, then so is the global relation.

(13) In the finite and loop-free, two-value case, if (R1) holds, then smoothness holds, $(\mu * 1)$ holds, and equivalence interpolation holds, too, by Proposition 9.3.25.

Proof
(12) This is trivial by $\rho \prec \sigma \prec \tau \rightarrow \rho_i \prec \sigma_i \prec \tau_i \rightarrow \rho \prec \tau$.

(13) By Fact 9.3.14, (R1) implies $(\mu * 1)$, so by smoothness (which follows from transitivity, finiteness, and the absence of loops), and Proposition 9.3.25, there is a common interpolant for both directions. □

9.3.4 *Some Examples*

Example 9.3.2
Consider the following examples. The relations should always be closed under transitivity. All work with the language $\{p, q\}$, except (I4), which works with $\{p, q, r\}$. We only give the relation between the models.

(I1) $pq \prec \neg pq \prec \neg p\neg q$, $pq \prec p\neg q \prec \neg p\neg q$

(I2) $pq \prec \neg pq \prec \neg p\neg q$, $pq \prec p\neg q \prec \neg p\neg q$, $p\neg q \prec \neg pq$

(I3) $p\neg q \prec \neg pq$

(I4) $p\neg qr \prec \neg pqr$, $p\neg q\neg r \prec \neg pq\neg r$

(I5) $pq \prec \neg p\neg q$

(I6) $pq \prec p\neg q \prec \neg pq$, $p\neg q \prec \neg p\neg q$

(I7) $pq \prec p\neg q$, $\neg p\neg q \prec \neg pq$

See Diagrams 9.3.1, 9.3.2 and 9.3.3. The lower end of the lines indicates the smaller element, the upper end the bigger one. The lower boxes indicate the valid properties, (I) stands for interpolation. A summary of the properties is also found in Table 9.1.

Diagram 9.3.1

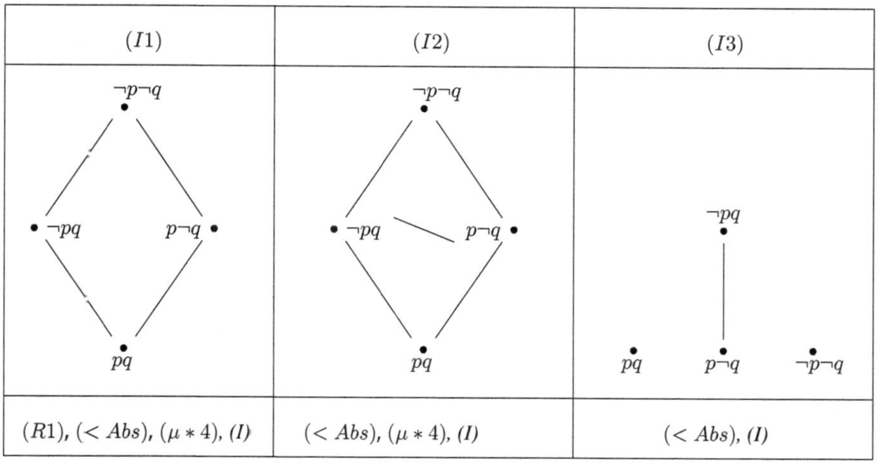

Diagram 9.3.2

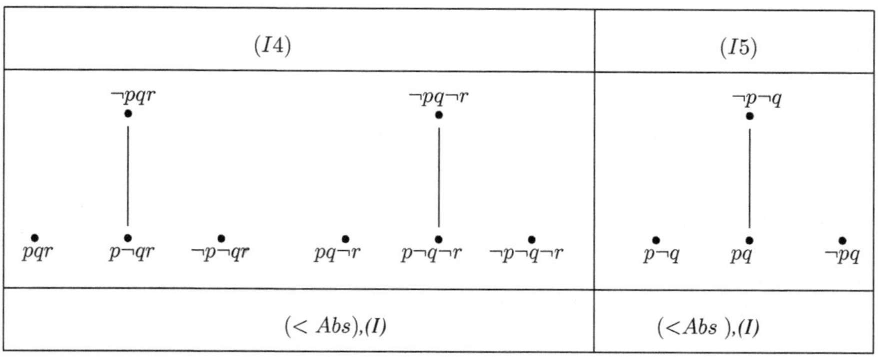

Diagram 9.3.3

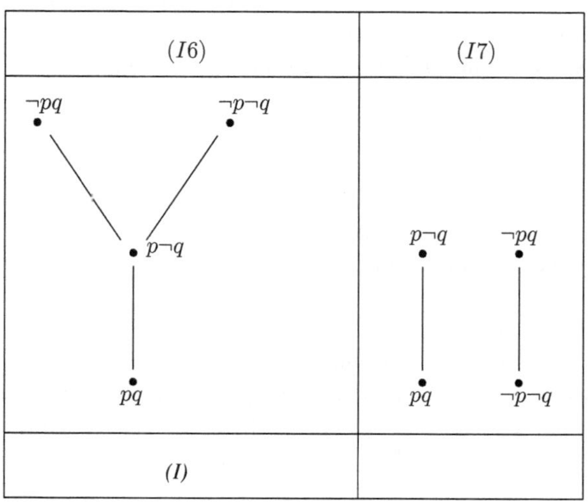

Table 9.1 Properties of (I1)–(I7)

Properties of (I1)–(I7)				
Diagram	$(R1)$	$(< Abs)$	$(\mu * 4)$	Interpolation
$(I1)$	+	+	+	+
$(I2)$	−	+	+	+
$(I3)$	−	+	−	+
$(I4)$	−	+	−	+
$(I5)$	−	+	−	+
$(I6)$	−	−	−	+
$(I7)$	−	−	−	−

Fact 9.3.16
In usual NML, if $L(\phi) \subseteq L(\psi)$ or $L(\psi) \subseteq L(\phi)$, then $\phi \mid\sim \psi$ has an interpolant.

Proof
The simpler formula is an interpolant, by the identity rule $\alpha \mid\sim \alpha$. □

Fact 9.3.17
If $\phi \mid\sim \psi$, $\phi \not\mid\sim \bot$, and $L(\phi) \cap L(\psi) = \emptyset$, the only possible interpolant is TRUE, so it suffices to check if $TRUE \mid\sim \psi$, □

We check now interpolation for $(I1)-(I6)$.

Fact 9.3.18
$(I1)-(I6)$ have interpolation, (I7) does not.

Proof
We use Fact 9.3.16 without mentioning. In particular, any ϕ on the left-hand side which needs all variables, has interpolation. Thus, for all except (I4), the only interesting formulas, where interpolation might fail, are p, $\neg p$, q, $\neg q$.

(I1) We have
$p \mid\sim q$, $\neg p \mid\sim q$ - but $TRUE \mid\sim q$
$q \mid\sim p$, $\neg q \mid\sim p$ - but $TRUE \mid\sim p$
Thus, $p \mid\sim TRUE \mid\sim q$ etc., and (I1) has interpolation.

(I2) has interpolation by the same argument as for (I1).

(I3) and (I5) The only interesting formulas have no consequences beyond classical logic.

(I4) We go by cases.

(a) $TRUE \mid\sim \neg(\neg pq))$ etc. This will an important interpolant.

(b) Formulas with one essential variable: p, $\neg p$, q, $\neg q$: they have only classical consequences, and we can use the classical interpolant by supraclassicality (i.e. $\phi \vdash \psi$ implies $\phi \mid\sim \psi$). r, $\neg r$: This is like TRUE.

(c) Formulas with two essential variables, p, r, like $pr, \neg pr, p\neg r, \neg p\neg r, pr \vee \neg p\neg r$, etc.: they have only classical consequences. Formulas involving q, r: similar. Formulas involving p, q: As they do not involve r, and the structure does not distinguish between r and $\neg r$, they will have the same consequences as TRUE.

(d) Formulas with three essential variables: there is nothing to show by Fact 9.3.16.

(I6) We look again at the interesting formulas

$p \mid\sim q$, $\neg p \mid\sim TRUE$, $q \mid\sim p$, $\neg q \mid\sim p$, but $TRUE \mid\sim pq$

(I7) $p \mid\sim q$, $\neg p \mid\sim \neg q$. The only interpolant for both is TRUE, but it is not the case that $TRUE \mid\sim q, \neg q$. □

Fact 9.3.19
$(\mu * 4)$ holds for $(I1)-(I2)$, and fails for $(I3)-(I7)$.

Proof
(I1) The interesting cases are

(A) $\Sigma = \{p\}$ and $\Gamma = \{q, \neg q\}$,

(B) $\Sigma = \{p, \neg p\}$ and $\Gamma = \{q, \neg q\}$,

and symmetrical ones.

(A): $\mu(\Sigma \times \Gamma) = \{pq\}$. $\mu(\Sigma \times \Pi\{q\}) \upharpoonright \{p\} = \{p\}$, $\mu(\Pi\{p\} \times \Gamma) \upharpoonright \{q\} = \{q\}$, so $(\mu * 4)$ holds. (Note that $\Pi\{p\} = \{p, \neg p\}$, etc., as we look at all functions from $L = \{p\}$ to $\{0, 1\}$.)

(B): Analogous.

(I2) All Σ and Γ would give the same result as for (I1), so $(\mu * 4)$ holds here, too.

(I3) Let $\Sigma = \{p, \neg p\}$, $\Gamma = \{q, \neg q\}$, then $\mu(\Sigma \times \Gamma) = \{pq, p\neg q, \neg p\neg q\}$, but this cannot be a product for cardinality reasons, so $(\mu * 4)$ fails.

(I4) Let $\Sigma = \{p, \neg p\}$, $\Gamma = \{q, \neg q\} \times \{r, \neg r\}$, then $\mu(\Sigma \times \Gamma) = \mu(\Sigma \times \Pi\{q, r\}) = \mu(\Pi\{p\} \times \Gamma) = \{pq, p\neg q, \neg p\neg q\} \times \{r, \neg r\}$, $\mu(\Sigma \times \Pi\{q, r\}) \upharpoonright \{p\} = \{p, \neg p\}$, $\mu(\Pi\{p\} \times \Gamma) \upharpoonright \{q, r\} = \{q, \neg q\} \times \{r, \neg r\}$, so we have 6 versus 8 elements, and $(\mu * 4)$ fails.

But note that we can "split" the full product into $\Sigma = \{p, \neg p\} \times \{q, \neg q\}$, $\Gamma = \{r, \neg r\}$, for which $(\mu * 4)$ applies, so we still have limited $(\mu * 4)$.

(I5) The same argument as for (I3) shows that $(\mu * 4)$ fails.

(I6) Consider $\Sigma = \{\neg p\}$, $\Gamma = \{q, \neg q\}$. $\mu(\Sigma \times \Gamma) = \{\neg pq, \neg p\neg q\}$. $\mu(\Pi\{p\} \times \Gamma) \upharpoonright \{q\} = \{pq\} \upharpoonright \{q\} = \{q\}$, so $(\mu * 4)$ fails.

(I7) Let $\Sigma = \{p\}$, $\Gamma = \{q, \neg q\}$, then $\mu(\Sigma \times \Gamma) = \{pq\}$, but $\mu(\{p, \neg p\} \times \{q, \neg q\}) \upharpoonright \{q\} = \{pq, \neg p\neg q\} \upharpoonright \{q\} = \{q, \neg q\}$, so $(\mu * 4)$ fails. □

Fact 9.3.20
(R1) holds for (I1), and fails for $(I2)-(I7)$.

Proof
(I1) The order can be defined by $p \prec \neg p$, $q \prec \neg q$.

(I2) If $p\neg q \prec \neg pq$ were defined componentwise, then $\neg q \prec q$, contradicting $pq \prec p\neg q$.

(I3) If (R1) were to hold, then by $p\neg q \prec \neg pq$, and as $p \neq \neg p$, $q \neq \neg q$, both $p \prec \neg p$ and $\neg q \prec q$ have to hold, but then, e.g. $pq \prec \neg pq$ has to hold contradiction too.

$(I4)-(I5)$: similar to (I3).

$(I6)-(I7)$ From $pq \prec p\neg q$ conclude $\neg pq \prec \neg p\neg q$. □

Fact 9.3.21
$(< Abs)$ holds for $(I1)-(I5)$, and fails for $(I6)-(I7)$.

Proof
If there is no $\sigma\tau \prec \sigma\tau'$, then $(< Abs)$ holds trivially.

(I3), (I5): $(< Abs)$ holds thus trivially.

(I1) We have $p \prec \neg p$, $q \prec \neg q$, irrespective of $q/\neg q$, $p/\neg p$, so it holds.

(I2) Same as (I1), the condition does not apply to the new pair $p\neg q \prec \neg pq$.

(I4) The only comparisons are $(p\neg q)r \prec (\neg pq)r$, and $(p\neg q)\neg r \prec (\neg pq)\neg r$ with constant parts $\sigma = r$, $\sigma' = \neg r$, $\tau = p\neg q$, $\tau' = \neg pq$. For these, the condition holds, and τ, τ' do not occur with any other σ''.

$(I6)-(I7)$: We have $pq \prec p\neg q$, but not $\neg pq \prec \neg p\neg q$. □

Remark 9.3.22
Note that we have limited (R1) for (I4), when we treat $\{p, q\}$ together as a group. This is important for "real life", where we may treat parts of the language separately.

9.3.5 Interpolation and $(\mu * 1)$

Proposition 9.3.23
$(\mu * 1)$ entails semantical interpolation of the form $\phi \mid\sim \alpha \mid\sim \psi$ in two-valued non-monotonic logic generated by minimal model sets.

As the model sets might not be definable, syntactic interpolation does not follow automatically.

So, more precisely, we show the following:

Let $(\mu * 1)$ hold, and

(1) $I = J \cup J' \cup J''$

(2) $\mu_I(X) \subseteq Y$

(3) $\mu_I(Y) \subseteq X$

(4) $X = X \restriction J \cup J' \times \Pi J''$

(5) $Y = \Pi J \times Y \restriction J' \cup J''$.

Then $Z := \Pi J \times (\mu_I(X) \restriction J') \times \Pi J''$. is an interpolant, i.e. $\mu_I(X) \subseteq Z$, $\mu_I(Z) \subseteq Y$.

This is Proposition 5.3.5 of Gabbay and Schlechta (2011), slightly re-written.

Proof
First, a short comment on the translation of the syntactic to the semantic version: Let the product be defined on $I := J \cup J' \cup J''$ (intuitively I is the set of propositional variables of the language). Let ϕ be defined on $J \cup J'$, ψ on $J' \cup J''$. This translates to $X = X \upharpoonright (J \cup J') \times \Pi J''$, $Y = \Pi J \times Y \upharpoonright (J' \cup J'')$. In our translation, we have $\mu_I(X) \subseteq Y$, and look for Z such that $\mu_I(X) \subseteq Z$ and $\mu_I(Z) \subseteq Y$, with $Z = \Pi J \times Z \upharpoonright J' \times \Pi J''$.

For clarity, even if it clutters up notation, we will be precise about where μ is formed. Thus, we write $\mu_I(X)$ when we take the minimal elements in the full product, $\mu_J(X)$ when we consider only the product on J, etc.

We now prove the result.

$\mu_I(X) \subseteq Z$ is trivial; we turn to $\mu_I(Z) \subseteq Y$.

(1) As $X = X \upharpoonright (J \cup J') \times \Pi J''$, $\mu_I(X) = \mu_{J \cup J'}(X \upharpoonright (J \cup J')) \times \mu_{J''}(\Pi J'')$ by $(\mu * 1)$.

(2) By $(\mu * 1)$ again, $\mu_I(Z) = \mu_I(\Pi J \times (\mu_I(X) \upharpoonright J') \times \Pi J'') = \mu_J(\Pi J) \times \mu_{J'}(\mu_I(X) \upharpoonright J') \times \mu_{J''}(\Pi J'')$.

So it suffices to show $\mu_J(\Pi J) \times \mu_{J'}(\mu_I(X) \upharpoonright J') \times \mu_{J''}(\Pi J'') \subseteq Y$.

Proof Let $\sigma = \sigma_J \sigma_{J'} \sigma_{J''} \in \mu_J(\Pi J) \times \mu_{J'}(\mu_I(X) \upharpoonright J') \times \mu_{J''}(\Pi J'')$.

By $\mu_{J'}(\mu_I(X) \upharpoonright J') \subseteq \mu_I(X) \upharpoonright J'$, there is a $\sigma' = \sigma'_J \sigma'_{J'} \sigma'_{J''} \in \mu_I(X)$ such that $\sigma'_{J'} = \sigma_{J'}$, i.e., $\sigma' = \sigma'_J \sigma_{J'} \sigma'_{J''}$.

By (1) and $\sigma_{J''} \in \mu_{J''}(\Pi J'')$, also $\sigma'_J \sigma_{J'} \sigma_{J''} \in \mu_I(X) \subseteq Y$.

But Y does not depend on J, so also $\sigma = \sigma_J \sigma_{J'} \sigma_{J''} \in Y$. □

Remark 9.3.24
We chose $Z := \Pi J \times (\mu_I(X) \upharpoonright J') \times \Pi J''$, and not $Z := \Pi J \times X \upharpoonright J' \times \Pi J''$ for the following reason.

We used $\mu_{J'}(\mu_I(X) \upharpoonright J') \subseteq \mu_I(X) \upharpoonright J'$, but $\mu_{J'}(X \upharpoonright J') \subseteq \mu_I(X) \upharpoonright J'$ is not necessarily the case, as was shown in Example 9.3.1, (1).

9.3.5.1 Remarks for the Converse: From Interpolation to $(\mu * 1)$

Example 9.3.3
We show here in (1.1) and (1.2) that half of the condition $(\mu * 1)$ is not sufficient for interpolation, and in (2) that interpolation may hold, even if $(\mu * 1)$ fails. When looking closer, the latter is not surprising: μ of subproducts may be defined in a funny way, which has nothing to do with the way μ on the big product is defined.

Consider the language based on p, q, r. We define two orders

(a) $\prec$ on sequences of length 3 by $\neg p \neg q \neg r \prec p \neg q \neg r$, and leave all other sequences of length 3 $\prec$-incomparable.

(b) $<$ on sequences of the same length by $\sigma < \tau$ iff there is a $\neg x$ in σ, an x in τ, but no y in σ, $\neg y$ in τ. E.g., $\neg p < p$, $\neg pq < pq$, $\neg p \neg q < pq$, but $\neg pq \not< p \neg q$.

Work now with $\prec$. Let $\phi = \neg q \wedge \neg r$, $\psi = \neg p \wedge \neg q$, so $\mu(\phi) = \neg p \wedge \neg q \wedge \neg r$, and $\phi \mid\sim \psi$. Suppose there is an α, $\phi \mid\sim \alpha \mid\sim \psi$, α written with q only, so α is equivalent to FALSE, TRUE, q, or $\neg q$. $\phi \not\sim FALSE$, $\phi \not\sim q$. $TRUE \not\sim \psi$, $\neg q \not\sim \psi$. Thus, there is no such α, and $\mid\sim$ has no interpolation.

We show in (1.1) and (1.2) that we can make both directions of $(\mu * 1)$ true separately, so they do not suffice to obtain interpolation.

(1.1) We make $\mu(X \times Y) \subseteq \mu(X) \times \mu(Y)$ true, but not the converse. Take the order $\prec$ on sequences of length 3 as described above. Do not order any sequences of length 2 or 1, i.e. μ is there always identity. Thus, $\mu(X \times Y) \subseteq X \times Y = \mu(X) \times \mu(Y)$ holds trivially.

(1.2) We make $\mu(X \times Y) \supseteq \mu(X) \times \mu(Y)$ true, but not the converse. We order all sequences of length 1 or 2 by $<$, and all sequences of length 3 by $\prec$. Suppose $\sigma \in X \times Y - \mu(X \times Y)$. Case 1: $X \times Y$ consists of sequences of length 2. Then, by definition, $\sigma \notin \mu(X) \times \mu(Y)$. Case 2: $X \times Y$ consists of sequences of length 3. Then $\sigma = p\neg q\neg r$, and $\neg p\neg q\neg r \in X \times Y$. So $\{p, \neg p\} \subseteq X$ or $\{p\neg q, \neg p\neg q\} \subseteq X$, but in both cases $\sigma \upharpoonright X \notin \mu(X)$, so $\sigma \notin \mu(X) \times \mu(Y)$. Finally, note that $\mu(TRUE) \not\subseteq \{\neg p\neg q\neg r\} = \mu(\{p, \neg p\}) \times \mu(\{\langle q, r\rangle, \langle q, \neg r\rangle, \langle \neg q, r\rangle, \langle \neg q, \neg r\rangle\})$, so full $(\mu * 1)$ does not hold.

(2) We make interpolation hold, but $\mu(X) \times \mu(Y) \not\subseteq \mu(X \times Y)$: We order all sequences of length 3 by $<$. Shorter sequences are made incomparable, so for shorter sequences $\mu(X) = X$. Obviously, in general $\mu(X) \times \mu(Y) \not\subseteq \mu(X \times Y)$. But the proof of Proposition 9.3.23 goes through as above, only directly, without the use of factorising and taking μ of the factors.

The reason is that the order on sequences of length 3 behaves in a modular way.

Thus, we can have the result, even if subspaces behave differently, if the completions of the subspace in the whole space behave as they should. □

9.3.6 $(\mu * 1)$ *and Interpolation for Equivalent Formulas*

The question was posed by D. Makinson: If $\phi \mid\sim \psi$ and $\psi \mid\sim \phi$, is there an interpolant α such that $\phi \mid\sim \alpha \mid\sim \psi$ and $\psi \mid\sim \alpha \mid\sim \phi$?

The answer is: yes, but we need for our proof that μ is generated by a smooth relation. We do not know if this additional requirement is necessary. To simplify notation, we work with separate μ's on subspaces, as in Proposition 9.3.23.

Proposition 9.3.25

Let $(\mu * 1)$ *hold, and*

(1) $I = J \cup J' \cup J''$

(2) μ_I *be generated by a smooth relation*

(3) $\mu_I(X) \subseteq Y$

(4) $\mu_I(Y) \subseteq X$

(5) $X = X \upharpoonright J \cup J' \times \Pi J''$

(6) $Y = \Pi J \times Y \upharpoonright J' \cup J''$.

Then

(7) $Z := \Pi J \times (\mu_I(X) \cap \mu_I(Y)) \restriction J' \times \Pi J''$. *is an interpolant in both directions, i.e.* $\mu_I(X) \subseteq Z$, $\mu_I(Z) \subseteq Y$, *and, conversely,* $\mu_I(Y) \subseteq Z$, $\mu_I(Z) \subseteq X$.

Proof
See the proof of Proposition 9.3.23 for a comment on the translation of the syntactic version to the semantic version.

By Fact 9.3.4, $\mu_I(X) = \mu_I(Y) = \mu_I(X) \cap \mu_I(Y)$.

By Proposition 9.3.23 $Z_X := \Pi J \times (\mu_I(X) \restriction J') \times \Pi J''$ is an interpolant in one direction, i.e. $\mu_I(X) \subseteq Z_X$ and $\mu_I(Z_X) \subseteq Y$, and $Z_Y := \Pi J \times (\mu_I(Y) \restriction J') \times \Pi J''$ is an interpolant in the other direction, i.e. $\mu_I(Y) \subseteq Z_Y$ and $\mu_I(Z_Y) \subseteq X$. As $\mu_I(X) = \mu_I(Y)$, they are the same, thus interpolants in both directions. □

9.3.6.1 Discussion of the Constructions for Equivalence

Let $\mu(A) \subseteq B$, $\mu(B) \subseteq A$, and $A = (A \restriction J \cup J') \times \Pi J''$, $B = \Pi J \times (B \restriction J' \cup J'')$. We look for Z such that $\mu(A) \cup \mu(B) \subseteq Z$ and $\mu(Z) \subseteq A \cap B$, with $Z = \Pi J \times (Z \restriction J') \times \Pi J''$.

Diagram 9.3.4

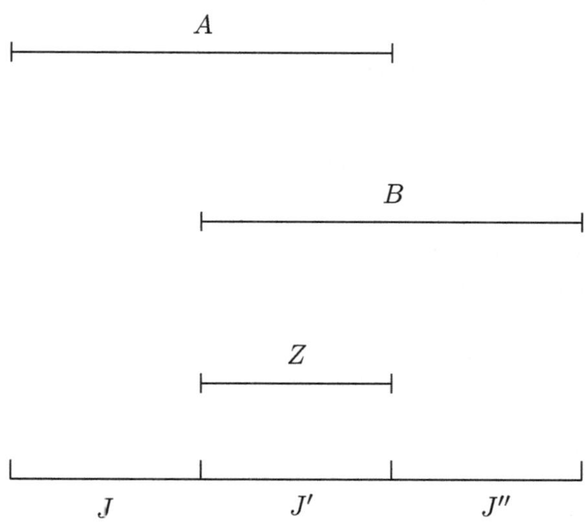

Interpolation for Equivalence

The condition $\mu(A) \cup \mu(B) \subseteq Z$ forces Z (or $Z \restriction J'$) to be relatively big, but making $Z \restriction J'$ too big, may destroy the condition $\mu(Z) \subseteq A \cap B$, as any new element in $Z \restriction J'$ may introduce a multitude of new sequences with all possible combinations over J and J'', whereas A may be restricted on J, and B on J''. We have to make sure that the new minimal sequences are both in A and B.

Smoothness guarantees that $\mu(A) = \mu(B)$, but as shown in Example 9.3.1, (3), this may fail even in the presence of $(\mu * 1)$, and we may use the difference between $\mu(A)$ and $\mu(B)$ to make a construction of the interpolant in both directions difficult. The above example shows there might be some $c \in \mu(A) - \mu(B)$ (or $c \in \mu(B) - \mu(A)$), despite $\mu(A) \subseteq B$ and $\mu(B) \subseteq A$.

The problem of adding new elements is illustrated by the following example:

Example 9.3.4
Let $\mu(A) \approx Z$, more precisely $Z' := Z \upharpoonright J' = \mu(A) \upharpoonright J'$, $Z = \Pi J \times Z' \times \Pi J''$.

Suppose that for each $c' \in \Pi J' - Z'$ there is $bc'b'' \notin B$, as $B = \Pi J \times (B \upharpoonright J' \cup J'')$, b'' must be the problem.

Thus

(1) for $z' \in Z'$: $0z'0 \in B$,

(2) for $c' \in \Pi J' - Z'$: $0c'0 \notin B$ - the problem being on J''.

If $*$ stands for both values, 0/1, then

$B = \{*z'*, *c'1\}$ (or, alternatively, $B = \{*z'0, *c'1\}$)

$A = \{0z'*, 1c'*\}$, $z' \prec c'$, and $0 \prec 1$, so $\mu(A) = \{0z'0\}$.

$Z' = \{z'\}$, $Z = \{*z'*\}$, $\mu(Z) = \mu(A) = 0z'0$, $B = \{*z'0, *c'1\}$, and $0c'0 \notin B$, so adding c' to Z' is impossible. □

9.3.7 *Interpolation and* $(\mu * 4)$

Recall that we examine here the case $\phi \mid\!\sim \alpha \mid\!\sim \psi$.

We use condition $(\mu * 4)$, which is conceptually simpler, as it does not involve a change of the sets where μ is defined. The price to pay is a more complicated notation, which is partly hidden in Fact 9.3.9.

Proposition 9.3.26
$(\mu * 4)$ entails semantical interpolation of the form $\phi \mid\!\sim \alpha \mid\!\sim \psi$ in two-valued non-monotonic logic generated by minimal model sets.

Again, as the model sets might not be definable, syntactic interpolation does not follow automatically.

So, more precisely, we show the following:

Let $(\mu * 4)$ hold, and

(1) $I = J \cup J' \cup J''$

(2) $\mu_I(X) \subseteq Y$

(3) $\mu_I(Y) \subseteq X$

(4) $X = X \upharpoonright J \cup J' \times \Pi J''$

(5) $Y = \Pi J \times Y \upharpoonright J' \cup J''$.

Then

$Z := \Pi J \times (\mu_I(X) \upharpoonright J') \times \Pi J''$.

is an interpolant, i.e. $\mu_I(X) \subseteq Z$, $\mu_I(Z) \subseteq Y$.

Note that we do not change the set where μ is defined, so we can write $Z := \Pi J \times (\mu(X) \upharpoonright J') \times \Pi J''$, as we will do in the sequel.

Proof
Again, see the proof of Proposition 9.3.23 for a comment on the translation of the syntactic version to the semantic version.

We show that $Z := \Pi J \times (\mu(X) \restriction J') \times \Pi J''$ is as desired.

$\mu(X) \subseteq Z$ is trivial; we turn to $\mu(Z) \subseteq Y$.

As $X = \Big(X \restriction (J \cup J')\Big) \times \Big(\Pi J''\Big)$, we have by $(\mu * 4)$:

(1) $\mu(X) = \Big(\mu[(X \restriction (J \cup J')) \times \Pi J''] \restriction (J \cup J')\Big) \times \Big(\mu(\Pi I) \restriction J''\Big)$

By Fact 9.3.9,

$\mu(Z) = \mu[\Pi J \times (\mu(X) \restriction J') \times \Pi J''] = \Big(\mu(\Pi I) \restriction J\Big) \times \Big(\mu[\Pi J \times (\mu(X) \restriction J') \times \Pi J''] \restriction J'\Big) \times \Big(\mu(\Pi I) \restriction J''\Big)$.

Let $\sigma = \sigma_J \sigma_{J'} \sigma_{J''} \in \Big(\mu(\Pi I) \restriction J\Big) \times \Big(\mu[\Pi J \times (\mu(X) \restriction J') \times \Pi J''] \restriction J'\Big) \times \Big(\mu(\Pi I) \restriction J''\Big)$, we show $\sigma \in Y$.

By $\mu[\Pi J \times (\mu(X) \restriction J') \times \Pi J''] \subseteq \Pi J \times (\mu(X) \restriction J') \times \Pi J''$, and Fact 9.3.2, (1), there is a $\sigma' = \sigma'_J \sigma'_{J'} \sigma'_{J''} \in \mu(X)$ such that $\sigma'_{J'} = \sigma_{J'}$, i.e., $\sigma' = \sigma'_J \sigma_{J'} \sigma'_{J''} \in \mu(X)$.

By (1) and $\sigma_{J''} \in \mu(\Pi I) \restriction J''$, also $\sigma'_J \sigma_{J'} \sigma_{J''} \in \mu(X) \subseteq Y$.

But Y does not depend on J, so also $\sigma = \sigma_J \sigma_{J'} \sigma_{J''} \in Y$. □

Chapter 10
Probabilistic and Abstract Independence

10.1 Summary

Independence as an abstract ternary relation $\langle X \mid Y \mid Z \rangle$ for probability and other situations has been examined by W. Spohn, see Spohn (1980), A.P. Dawid, see Dawid (1979), J. Pearl, see, e.g. Pearl (1988), where the situation in graphs is discussed, too.

This chapter is organised as follows:

(1) We first give the basic definitions for probabilistic, set, and general function independence in Sect. 10.2,
(2) We then compare the different concepts using some simple examples in Sect. 10.3,
(3) We turn to a more systematic investigation of (set and) function independence in Sect. 10.4 and Sect. 10.5. Given a set of functions Σ with constant domain I, and $X, Y, Z \subseteq I$ pairwise disjoint, $\langle X \mid Y \mid Z \rangle_\Sigma$ expresses "X and Z are independent, given Y" with the meaning: if $f, g \in \Sigma$ and f and g agree on Y, i.e. $f \restriction Y = g \restriction Y$, then we can "splice together" f and g, i.e. there is $h \in \Sigma$ such that $h \restriction X = f \restriction X, h \restriction Y = f \restriction Y = g \restriction Y, h \restriction Z = g \restriction Z$. This property expresses a "richness" of Σ.
(4) Investigation of ever longer loop conditions lead to the most important result of this chapter, that there is no fixed size finite characterisation of function independence, see Sect. 10.6. The crucial idea is to combine independent layers, see Sect. 10.6.2.
(5) We conclude with ideas for a systematic construction of new rules in Sect. 10.7.

D.M. Gabbay and K. Schlechta, *A New Perspective on Nonmonotonic Logics*,
DOI 10.1007/978-3-319-46817-4_10

10.2 Introduction, Basic Definitions, and Notation

We recall:

Definition 10.2.1
If f is a function, Y a subset of its domain, we write $f \restriction Y$ for the restriction of f to elements of Y.

If F is a set of functions over Y, then $F \restriction Y := \{f \restriction Y : f \in F\}$.

Definition 10.2.2
The constant functions 0_c and 1_c:
$0_c(i) = 0$ for all $i \in I$
$1_c(i) = 1$ for all $i \in I$
Moreover, when we define a fixed function $\sigma : I \to \{0, 1\}$ argument by argument, we abbreviate $\sigma(a) = 0$ by $a = 0$, etc.

Sometimes, we also give (a fragment of) a function just by the sequence of the values, so instead of writing $a = 0$, $b = 1$, $c = 1$, we just write 011 - context will disambiguate.

Notation 10.2.1
Let $I \neq \emptyset$ be an arbitrary (index) set, for $i \in I$ $U_i \neq \emptyset$ arbitrary sets. Let $\mathcal{U} := \Pi\{U_i : i \in I\}$, and $\Sigma \subseteq \mathcal{U}$, an arbitrary set of sequences.

If all $U_i = \{0, 1\}$, then $\mathcal{U} = \mathcal{P}(I)$, and $\Sigma \subseteq \mathcal{P}(I)$.

Recall, that for $X \subseteq I$, $\mathcal{U} \restriction X := \Pi\{U_i : i \in X\}$, etc.

10.2.1 Probabilistic Independence

Definition 10.2.3

(1) Let $P : \mathcal{P}(\mathcal{U}) \to [0, 1]$ be a probability measure. (As we treat the finite case, we may assume that P is defined by its value on singletons.)
(2) Let $\emptyset \neq \Sigma \subseteq \mathcal{U}$ be fixed. Then we define a special probability measure Q, relative to this Σ, on $\mathcal{U}$ by

$$Q(\sigma) := Q_\Sigma(\sigma) := \begin{cases} \frac{1}{card(\Sigma)} & iff\ \sigma \in \Sigma \\ 0 & iff\ \sigma \notin \Sigma \end{cases}$$

We call this Q the counting probability for set Σ.
(3) By abuse of language, for $X \subseteq I$, $x \in U \restriction X$, let $P(x) := P(\{u \in U : u \restriction X = x\})$.
Analogously, for $X, Y \subseteq I$, $X \cap Y = \emptyset$, $x \in U \restriction X$, $y \in U \restriction Y$, let $P(x, y) := P(\{u \in U : u \restriction X = x \text{ and } u \restriction Y = y\})$. We sometimes set $P(xy) := P(x, y)$, when there is no ambiguity.

(4) Finally, for $X, Y, Z \subseteq I$ pairwise disjoint, $x \in U \restriction X$, $y \in U \restriction Y$, $z \in U \restriction Z$, let $P(x \mid y) := \frac{P(x,y)}{P(y)}$, $P(x \mid y, z) := \frac{P(x,y,z)}{P(y,z)}$, etc.
(We have, of course, to pay attention that we do not divide by 0.)

Definition 10.2.4

P as above defines a 3-place relation of independence on pairwise disjoint $X, Y, Z \subseteq I$ $\langle X \mid Y \mid Z\rangle_P$ by

$$\langle X \mid Y \mid Z\rangle_P :\leftrightarrow \begin{cases} \forall x \in U \restriction X, \forall y \in U \restriction Y, \forall z \in U \restriction Z (P(y,z) > 0 \rightarrow P(x \mid y) = P(x \mid y,z)), & if\ Y \neq \emptyset \\ i.e., P(x,y)/P(y) = P(x,y,z)/P(y,z), or & \\ P(x,y,z) * P(y) = P(x,y) * P(y,z) & \\ & \\ \forall x \in U \restriction X, \forall z \in U \restriction Z (P(z) > 0 \rightarrow P(x) = P(x \mid z)), & if\ Y = \emptyset \\ i.e., P(x) = P(x,z)/P(z), or & \\ P(x,z) = P(x) * P(z) & \end{cases}$$

If $Y = \emptyset$, we shall also write $\langle X \mid\mid Z\rangle_P$ for $\langle X \mid Y \mid Z\rangle_P$. When the context is clear, we may omit the "P".

We also write $\neg\langle X \mid\mid Z\rangle_P$, when $\langle X \mid\mid Z\rangle_P$ is not the case.

Here, and for other types of independence, we will sometimes write $\langle X, X' \mid Y \mid Z\rangle$, etc., for $\langle X \cup X' \mid Y \mid Z\rangle$.

When we write $\langle X, X' \mid Y \mid Z\rangle$, etc., we will tacitly assume that all sets X, X', Y, Z are pairwise disjoint.

We call this notion probabilistic independence.

We list now some well-known conditions for probabilistic independence.

Pearl discusses the rules $(Ba) - (Be)$ of Definition 10.2.5 for the relation defined in Definition 10.2.4, see e.g. Pearl (1988).

Definition 10.2.5

The basic rules $(Ba) - (B\emptyset)$:

Assume X, Y, W, Z pairwise disjoint.

(Ba) Symmetry: $\langle X \mid Y \mid Z\rangle \leftrightarrow \langle Z \mid Y \mid X\rangle$
(Bb) Decomposition: $\langle X \mid Y \mid Z \cup W\rangle \rightarrow \langle X \mid Y \mid Z\rangle$
(Bc) Weak Union: $\langle X \mid Y \mid Z \cup W\rangle \rightarrow \langle X \mid Y \cup W \mid Z\rangle$
(Bd) Contraction: $\langle X \mid Y \mid Z\rangle$ and $\langle X \mid Y \cup Z \mid W\rangle \rightarrow \langle X \mid Y \mid Z \cup W\rangle$
(Be) Intersection: $\langle X \mid Y \cup W \mid Z\rangle$ and $\langle X \mid Y \cup Z \mid W\rangle \rightarrow \langle X \mid Y \mid Z \cup W\rangle$
$(B\emptyset)$ Empty outside: $\langle X \mid Y \mid Z\rangle$ if $X = \emptyset$ or $Z = \emptyset$.

Proposition 10.2.1

If P is a probability measure, and $\langle X \mid Y \mid Z\rangle_P$ defined as above, then $(Ba) - (Bd)$ of Definition 10.2.5 hold for $\langle \ldots\rangle = \langle \ldots\rangle_P$, and if P is strictly positive, (Be) will also hold.

The proof is elementary, well known, and will not be repeated here.

10.2.2 Set and Function Independence

Let Σ be a set of subsets of I, and X, Y, Z, etc., pairwise disjoint subsets of I.

Definition 10.2.6
We now define set independence:

(1) We say that $\langle X \mid Y \mid Z\rangle$ holds for Σ, and write $\langle X \mid Y \mid Z\rangle_\Sigma$ when necessary, iff
for all $A, A' \in \Sigma$ such that $A \cap Y = A' \cap Y$ there is $A'' \in \Sigma$ with $A'' \cap X = A \cap X$, $A'' \cap Y = A \cap Y = A' \cap Y$, $A'' \cap Z = A' \cap Z$.
(2) Using characteristic functions, $f : I \to \{0, 1\}$, Σ is a set of such functions, and $\langle X \mid Y \mid Z\rangle$ holds for Σ iff for $f', f'' \in \Sigma$ such that $f' \upharpoonright Y = f'' \upharpoonright Y$ there is $g \in \Sigma$ such that $g \upharpoonright X = f' \upharpoonright X$, $g \upharpoonright Y = f' \upharpoonright Y = f'' \upharpoonright Y$, $g \upharpoonright Z = f'' \upharpoonright Z$.
Thus, we can take two suitable fragments of total functions $f', f'' : I \to \{0, 1\}$ in Σ, put them together, and find a corresponding total function g, again in Σ.

The latter reading of set independence has a straightforward generalisation to arbitrary products: $\Pi\{U_i : i \in I\}$ is the set of functions f from I to $\bigcup\{U_i : i \in I\}$ with $f(i) \in U_i$.

Thus, we define:

Definition 10.2.7

(1) If $\Sigma \subseteq \Pi\{U_i : i \in I\}$, then $\langle X \mid Y \mid Z\rangle$ holds for Σ, $\langle X \mid Y \mid Z\rangle_\Sigma$, iff
for $f', f'' \in \Sigma$ such that $f' \upharpoonright Y = f'' \upharpoonright Y$ there is $g \in \Sigma$ such that $g \upharpoonright X = f' \upharpoonright X$, $g \upharpoonright Y = f' \upharpoonright Y = f'' \upharpoonright Y$, $g \upharpoonright Z = f'' \upharpoonright Z$, and for $Y = \emptyset$, $\langle X \mid\mid Z\rangle$ holds for Σ iff for $f', f'' \in \Sigma$ there is $g \in \Sigma$ such that $g \upharpoonright X = f' \upharpoonright X$, $g \upharpoonright Z = f'' \upharpoonright Z$.
(2) When $Y = \emptyset$, we will also write $\langle X \mid\mid Z\rangle$, $\langle X \mid\mid Z\rangle_\Sigma$, etc.
(3) Finally, we set for $y \in \mathcal{U} \upharpoonright Y$ $\Sigma_y := \{\sigma \in \Sigma : \sigma \upharpoonright Y = y\}$.

As we will always use the second reading of set independence, there is no need for new notation—context will tell whether we consider two-valued functions, or the more general case.

Y may be empty, then the condition $f \upharpoonright Y = g \upharpoonright Y$ is void.

Note that nothing is said about $I - (X \cup Y \cup Z)$, so we look at the projection to $X \cup Y \cup Z$.

Our main result in this chapter (absence of a finite characterisation), see Sect. 10.6, will concern this general case.

We note

Fact 10.2.2

(1) Let $Y = \emptyset$, $X \cup Z \subseteq I$. Then $\langle X \mid\mid Z\rangle_\Sigma$ iff $\Sigma \upharpoonright (X \cup Z) = \Sigma \upharpoonright X \times \Sigma \upharpoonright Z$.
(2) $\langle X \mid Y \mid Z\rangle_\Sigma$ iff $\forall \tau \in \Sigma$
$\{\sigma \upharpoonright X \cup Z : \sigma \in \Sigma, \sigma \upharpoonright Y = \tau \upharpoonright Y\} = \{\sigma \upharpoonright X : \sigma \in \Sigma, \sigma \upharpoonright Y = \tau \upharpoonright Y\} \times \{\rho \upharpoonright Z : \rho \in \Sigma, \rho \upharpoonright Y = \tau \upharpoonright Y\}$.

Proof

(1) "$\Rightarrow$":
"$\subseteq$" is trivial. "$\supseteq$": Let $\sigma \in \Sigma \restriction X$, $\sigma' \in \Sigma \restriction Z$, then there is $\tau \in \Sigma$ such that $\tau \restriction X = \sigma \restriction X$, $\tau \restriction Z = \sigma' \restriction Z$ by prerequisite, so $\tau \restriction X \cup Z \in \Sigma \restriction X \cup Z$, and $\tau \restriction X \cup Z = \sigma\sigma'$.
"$\Leftarrow$":
Let $\sigma \in \Sigma \restriction X$, $\sigma' \in \Sigma \restriction Z$, by prerequisite $\sigma\sigma' \in \Sigma \restriction X \cup Z$, so there is $\tau \in \Sigma$ such that $\tau \restriction X \cup Z = \sigma\sigma'$.

(2) Let $C := \{\sigma \restriction X \cup Z : \sigma \in \Sigma, \sigma \restriction Y = \tau \restriction Y\}$, $A := \{\sigma \restriction X : \sigma \in \Sigma, \sigma \restriction Y = \tau \restriction Y\}$, $B := \{\rho \restriction Z : \rho \in \Sigma, \rho \restriction Y = \tau \restriction Y\}$.
"$\Rightarrow$":
"$\subseteq$": Let $\sigma \in \Sigma$, $\sigma \restriction X \cup Z \in C$, then $\sigma \restriction X \in A$, $\sigma \restriction Z \in B$, so $\sigma \restriction X \cup Z \in A \times B$.
"$\supseteq$": Let $\sigma \restriction X \in A$, $\rho \restriction Z \in B$, then $\sigma \restriction Y = \tau \restriction Y = \rho \restriction Y$. So, by prerequisite, there is $\sigma' \in \Sigma$ such that $\sigma' \restriction X = \sigma \restriction X$, $\sigma' \restriction Y = \sigma \restriction Y = \rho \restriction Y$, $\sigma' \restriction Z = \rho \restriction Z$, so $\sigma' \restriction X \cup Z \in C$.
"$\Leftarrow$":
Let $\sigma, \rho \in \Sigma$ such that $\sigma \restriction Y = \rho \restriction Y$. Set $\tau \restriction Y := \sigma \restriction Y = \rho \restriction Y$. Then $\sigma \restriction X \in A$, $\rho \restriction Z \in B$, thus there is $\sigma' \in \Sigma$, $\sigma' \restriction Y = \tau \restriction Y$, $\sigma' \restriction X \cup Z = (\sigma \restriction X)(\rho \restriction Z)$.

□

10.3 Discussion of Some Simple Examples and Connections

We consider here $X = Y = Z = W = \{0, 1\}$ and their products, etc. To simplify notation, we write $X \times Y \times Z$, etc., where we would normally write $I = X \cup Y \cup Z$, etc., but the meaning will always be obvious.

We will later generalise, but the main ideas stay the same. First, we look at $X \times Z$ (the Cartesian product of X with Z), then at $X \times Z \times W$, at $X \times Y \times Z$, finally at $X \times Y \times Z \times W$. Elements of these products, i.e. sequences, will be written for simplicity 00, 01, 10, etc., context will disambiguate. General sequences will often be written σ, τ, etc. We will also look at subsets of these products, like $\{00, 11\} \subseteq X \times Z$, and various probability measures on these products, and various notions of independence, $\langle \ldots \rangle_P$, $\langle \ldots \rangle_{Q_\Sigma}$, $\langle \ldots \rangle_\Sigma$. As a matter of fact, the main part of this chapter concerns subsets Σ of products $X_1 \times \cdots \times X_n$ and a suitable notion of independence for Σ, roughly, if we can write Σ as $A_1 \times \cdots \times A_m$. This will be made more precise and discussed in progressively more complicated cases in this section.

10.3.1 X × Z

For $\Sigma \subseteq \mathcal{U} := X \times Z$, we will set $P(\Sigma) := \Sigma\{P(\sigma) : \sigma \in \Sigma\}$ - here, $\Sigma\{.\}$ denotes the sum!

If $\Sigma_x := \{\sigma \in \Sigma : \sigma(X) = x\}$, we will write $P(x)$ for $P(\Sigma_x)$, likewise $P(z)$ for $P(\Sigma_z)$, if $\Sigma_z := \{\sigma \in X \times Z : \sigma(Z) = z\}$. When these are ambiguous, we will e.g. write $\Sigma_{X=0}$ for $\{\sigma \in X \times Z : \sigma(X) = 0\}$, etc.

Example 10.3.1
Consider $\Sigma = X \times Z = \{00, 01, 10, 11\}$.

(1) Define $P(00) = P(01) = 1/6$, $P(10) = P(11) = 1/3$.
Then $P(X = 0) = 1/6 + 1/6 = 1/3$, and $P(X = 1) = 2/3$, $P(Z = 0) = 1/6 + 1/3 = 1/2$, and $P(Z = 1) = 1/2$, so $\langle X \parallel Z \rangle_P$.
In more detail:
$P(00) = 1/6$, $P(X = 0) * P(Z = 0) = 1/3 * 1/2 = 1/6$,
$P(01) = 1/6$, $P(X = 0) * P(Z = 1) = 1/3 * 1/2 = 1/6$,
$P(10) = 1/3$, $P(X = 1) * P(Z = 0) = 2/3 * 1/2 = 1/3$,
$P(11) = 1/3$, $P(X = 1) * P(Z = 1) = 2/3 * 1/2 = 1/3$.
(2) Define $P(00) = P(11) = 1/3$, $P(01) = P(10) = 1/6$.
Then $P(X = 0) = P(X = 1) = P(Z = 0) = P(Z = 1) = 1/2$, but $P(00) = 1/3 \neq 1/2 * 1/2 = 1/4$, so $\neg\langle X \parallel Z \rangle_P$.
(3) Consider Q_Σ.
Then $\langle X \parallel Z \rangle_{Q_\Sigma}$ holds, as does $\langle X \parallel Z \rangle_\Sigma$, as $\Sigma = X \times Z$.

Example 10.3.2

(1) Let $\Sigma := \{00, 01\}$,
then $Q_\Sigma(00) = Q_\Sigma(01) = 1/2$, $Q_\Sigma(10) = Q_\Sigma(11) = 0$,
$Q_\Sigma(X = 0) = 1$, $Q_\Sigma(X = 1) = 0$, $Q_\Sigma(Z = 0) = Q_\Sigma(Z = 1) = 1/2$, and we have $\langle X \parallel Z \rangle_{Q_\Sigma}$.
(2) Let $\Sigma := \{00, 11\}$,
then $Q_\Sigma(X = 0) = Q_\Sigma(X = 1) = 1/2$, $Q_\Sigma(Z = 0) = Q_\Sigma(Z = 1) = 1/2$,
but $Q_\Sigma(00) = 1/2 \neq Q_\Sigma(X = 0) * Q_\Sigma(Z = 0) = 1/4$, and we have $\neg\langle X \parallel Z \rangle_{Q_\Sigma}$.
(3) Let $\Sigma := \{00, 01, 11\}$,
then $Q_\Sigma(X = 0) = 2/3$, $Q_\Sigma(X = 1) = 1/3$, $Q_\Sigma(Z = 0) = 1/3$, $Q_\Sigma(Z = 1) = 2/3$,
but $Q_\Sigma(00) = 1/3 \neq Q_\Sigma(X = 0) * Q_\Sigma(Z = 0) = 2/3 * 1/3 = 2/9$, and we have $\neg\langle X \parallel Z \rangle_{Q_\Sigma}$.

Note that in (1) above, $\Sigma = \{0\} \times \{0, 1\}$, but neither in (2), nor in (3), Σ can be written as such a product. This is no coincidence, as we will see in Fact 10.3.1.

10.3.2 $X \times Z \times W$

Here, W will not be mentioned directly.

Example 10.3.3

(1) Define $P(000) = P(001) = P(010) = P(011) = 1/12$, $P(100) = P(101) = P(110) = P(111) = 1/6$, then X and Z are independent for $\Sigma = \mathcal{U}$, $\langle X \parallel Z\rangle_P$ by $P(00) = 1/6$, $P(01) = 1/6$, $P(10) = 1/3$, $P(11) = 1/3$, $P(X = 0) = 1/3$, $P(X = 1) = 2/3$, $P(Z = 0) = 1/6 + 1/3 = 1/2$, $P(Z = 1) = 1/6 + 1/3 = 1/2$.

(2) Define $P(100) = P(101) = P(010) = P(011) = 1/12$, $P(000) = P(001) = P(110) = P(111)=1/6$, then $P(X = 0)=P(X = 1) = P(Z = 0) = P(Z=1) = 1/2$, but $P(X = 0, Z = 0) = 1/3 \neq 1/2 * 1/2 = 1/4$, so $\neg\langle X \parallel Z\rangle_P$ for $\Sigma = \mathcal{U}$.

Consider now Q_Σ for $\emptyset \neq \Sigma \subseteq X \times Z \times W$.

Example 10.3.4

(1) $\Sigma := \{000, 001, 010, 011\}$. Then $Q_\Sigma(X = 0, Z = 0) = Q_\Sigma(X = 0, Z = 1) = 1/2$, $Q_\Sigma(X = 1, Z = 0) = Q_\Sigma(X = 1, Z = 1) = 0$, $Q_\Sigma(X = 0) = 1$, $Q_\Sigma (X = 1) = 0$, $Q_\Sigma(Z = 0) = Q_\Sigma(Z = 1) = 1/2$, so X and Z are independent.

(2) For $\Sigma := \{000, 001, 110, 111\}$, we see that X and Z are not independent for Q_Σ, as $Q_\Sigma(X = 0, Z = 1) = 0$, $Q_\Sigma(X = 0) = Q_\Sigma(Z = 1) = 1/2$.

Considering possible decompositions of Σ into set products, we are not so much interested how many continuations into W we have, but if there are any or none. This is often the case in logic, we are not interested how many models there are, but if there is a model at all.

The equivalence between probabilistic independence, $\langle X \parallel Z\rangle_{Q_\Sigma}$ and set independence, $\langle X \parallel Z\rangle_\Sigma$ is lost now, as the second part of the following example shows:

Example 10.3.5

(1) $\Sigma := \{000, 010, 100, 110\}$ satisfies both forms of independence, $\langle X \parallel Z\rangle_{Q_\Sigma}$ and set independence, $\langle X \parallel Z\rangle_\Sigma$.
First, $Q_Z(X = 0, Z = 0) = Q_Z(X = 0, Z = 1) = Q_Z(X = 1, Z = 0) = Q_Z (X = 1, Z = 1) = 1/4$, and $Q_Z(X = 0) = Q_Z(X = 1) = Q_Z(Z = 0) = Q_Z (Z = 1) = 1/2$.
Second, $\{00, 01, 10, 11\} = \Sigma \upharpoonright X \cup Z = \Sigma \upharpoonright X \times \Sigma \upharpoonright Z = \{0, 1\} \times \{0, 1\}$.

(2) Consider $\Sigma := \{000, 001, 010, 100, 110\}$.
Here, we have $Q_\Sigma(X = 0) = 3/5$, $Q_\Sigma(X = 1) = 2/5$, $Q_\Sigma(Z = 0) = 3/5$, $Q_\Sigma(Z = 1) = 2/5$, but $Q_\Sigma(X = 0, Z = 0) = 2/5 \neq 3/5 * 3/5$.
Consider now $\langle X \parallel Z\rangle_\Sigma$: Take $\sigma, \tau \in \Sigma$, then for all possible values $\sigma(X)$, $\tau(Z)$, there is ρ such that $\rho(X) = \sigma(X)$, $\rho(Z) = \tau(Z)$ - the value $\rho(W)$ is without importance.

10.3.3 $X \times Y \times Z$

We consider now independence of X and Z, given Y.

As we are interested mainly in subsets $\Sigma \subseteq X \times Y \times Z$ and the resulting Q_Σ, and combinations of function fragments, we work immediately with these.

The following example shows that $\langle X \mid Y \mid Z \rangle_\Sigma$ and $\langle X \mid\mid Z \rangle_\Sigma$ are independent from each other:

Example 10.3.6

(1) $\langle X \mid Y \mid Z \rangle_\Sigma$ may hold, but not $\langle X \mid\mid Z \rangle_\Sigma$:
Consider $\Sigma := \{000, 111\}$. $\langle X \mid Y \mid Z \rangle_\Sigma$ is obvious, as only one σ goes through each element in the middle. But there is no 0x1, so $\langle X \mid\mid Z \rangle_\Sigma$ fails.

(2) $\langle X \mid\mid Z \rangle_\Sigma$ may hold, but not $\langle X \mid Y \mid Z \rangle_\Sigma$:
Consider $\Sigma := \{000, 101, 110, 011\}$. Fixing, e.g., 0 in the middle shows that $\langle X \mid Y \mid Z \rangle_\Sigma$ fails (this leaves 000, 101, but there is no 001), but neglecting the middle, we can combine arbitrarily, so $\langle X \mid\mid Z \rangle_\Sigma$ holds.

Example 10.3.7
This example shows that $\langle X \mid Y \mid Z \rangle_\Sigma$ does not mean that Σ is some product $\Sigma_X \times \Sigma_Y \times \Sigma_Z$:

Let $\Sigma := \{000, 111\}$, then clearly $\langle X \mid Y \mid Z \rangle_\Sigma$, but Σ is no such product.

10.3.4 $X \times Y \times Z \times W$

The equivalence between probabilitistic independence, $\langle X \mid Y \mid Z \rangle_{Q_\Sigma}$, and set independence, $\langle X \mid Y \mid Z \rangle_\Sigma$ is lost again, as the following example shows:

Example 10.3.8
$\Sigma := \{0000, 0001, 0010, 1000, 1010\}$.

Here, we have $Q_\Sigma(X=0, Y=0) = 3/5$, $Q_\Sigma(X=1, Y=0) = 2/5$, $Q_\Sigma(Y=0, Z=0) = 3/5$, $Q_\Sigma(Y=0, Z=1) = 2/5$, $Q_\Sigma(Y=0) = 1$, but $Q_\Sigma(X=0, Y=0, Z=0) = 2/5 \neq 3/5 * 3/5$.

Consider now $\langle X \mid Y \mid Z \rangle_\Sigma$: Take $\sigma, \tau \in \Sigma$, such that $\sigma(Y) = \tau(Y)$, then for all possible values $\sigma(X)$, $\tau(Z)$, there is ρ such that $\rho(X) = \sigma(X)$, $\rho(Y) = \sigma(Y) = \tau(Y)$, $\rho(Z) = \tau(Z)$ - the value $\rho(W)$ is without importance.

We summarise the connections in the following Fact:

Fact 10.3.1

(1) Consider above situation $\Sigma \subseteq X \times Z$. Then $\langle X \mid\mid Z \rangle_{Q_\Sigma}$ iff $\langle X \mid\mid Z \rangle_\Sigma$.
(Example 10.3.1 (2) shows that this does not necessarily hold for arbitrary P instead of Q_Σ.)

(2) $\langle X \mid\mid Z \rangle_{Q_\Sigma} \Rightarrow \langle X \mid\mid Z \rangle_\Sigma$ for $\Sigma \subseteq X \times Z \times W$.

(3) Let $\emptyset \neq \Sigma \subseteq X \times Y \times Z$, then $\langle X \mid Y \mid Z\rangle_\Sigma$ and $\langle X \mid Y \mid Z\rangle_{Q_\Sigma}$ are equivalent.
(4) $\langle X \mid Y \mid Z\rangle_{Q_\Sigma} \Rightarrow \langle X \mid Y \mid Z\rangle_\Sigma$ for $\Sigma \subseteq X \times Y \times Z \times W$.

Proof

(1) "$\Rightarrow$":
$\Sigma \subseteq \{\sigma(X) : \sigma \in \Sigma\} \times \{\sigma(Z) : \sigma \in \Sigma\}$ is trivial. Suppose $Q_\Sigma(x,z) = Q_\Sigma(x) * Q_\Sigma(z)$, but there are $\sigma, \tau \in \Sigma$, $\sigma(X)\tau(Z) \notin \Sigma$. Let $x = \sigma(X)$, $z = \tau(Z)$, then $Q_\Sigma(x), Q_\Sigma(z) > 0$, but $Q_\Sigma(x,z) = 0$, a contradiction.
"$\Leftarrow$":
Case 1: $Q_\Sigma(x) = 0$, then $Q_\Sigma(x,z) = 0$, and we are done. Likewise for $Q_\Sigma(z) = 0$.
Case 2: $Q_\Sigma(x), Q_\Sigma(z) > 0$.
By definition and prerequisite,

$$Q_\Sigma(x) = \frac{card\{\sigma\in\Sigma:\sigma(X)=x\}}{card(\Sigma)} = \frac{card\{\sigma(Z):\sigma\in\Sigma\}}{card(\Sigma)}, \text{ the latter as } \Sigma = \Sigma(X) \times \Sigma(Z).$$

$$Q_\Sigma(z) = \frac{card\{\sigma\in\Sigma:\sigma(Z)=z\}}{card(\Sigma)} = \frac{card\{\sigma(X):\sigma\in\Sigma\}}{card(\Sigma)},$$
$$Q_\Sigma(x,z) = \frac{card\{\sigma\in\Sigma:\sigma(X)=x,\sigma(Z)=z\}}{card(\Sigma)} = \frac{1}{card(\Sigma)}.$$

By prerequisite again, $card(\Sigma) = card\{\sigma(X) : \sigma \in \Sigma\} * card\{\sigma(Z) : \sigma \in \Sigma\}$, so
$$\frac{card\{\sigma(Z):\sigma\in\Sigma\}}{card(\Sigma)} * \frac{card\{\sigma(X):\sigma\in\Sigma\}}{card(\Sigma)} = \frac{1}{card(\Sigma)}$$

(2) Let $\sigma, \tau \in \Sigma$, but suppose there is no $\rho \in \Sigma$ such that $\rho(X) = \sigma(X)$ and $\rho(Z) = \tau(Z)$. Then $Q_\Sigma(\sigma(X)), Q_\Sigma(\tau(Z)) > 0$, but $Q_\Sigma(\sigma(X), \tau(Z)) = 0$.
(3) "$\Leftarrow$":
Suppose there are $\sigma, \tau \in \Sigma$ such that $\sigma(Y) = \tau(Y)$, but there is no $\rho \in \Sigma$ such that $\rho(X) = \sigma(X)$, $\rho(Y) = \sigma(Y) = \tau(Y)$, $\rho(Z) = \tau(Z)$. Then $Q_\Sigma(\sigma(X), \sigma(Y))$, $Q_\Sigma(\tau(Y), \tau(Z))$, $Q_\Sigma(\sigma(Y)) > 0$, but $Q_\Sigma(\sigma(X), \sigma(Y), \tau(Z)) = 0$.
"$\Rightarrow$":
Case 1: $Q_\Sigma(x,y)$ or $Q_\Sigma(y,z) = 0$, then $Q_\Sigma(x,y,z) = 0$, and we are done.
Case 2: $Q_\Sigma(x,y), Q_\Sigma(y,z) > 0$. By definition and prerequisite,

$$Q_\Sigma(x,y) = \frac{card\{\sigma\in\Sigma:\sigma(X)=x,\sigma(Y)=y\}}{card(\Sigma)} = \frac{card\{\sigma(Z):\sigma\in\Sigma,\sigma(Y)=y\}}{card(\Sigma)} \text{ and}$$

$$Q_\Sigma(y,z) = \frac{card\{\sigma\in\Sigma:\sigma(Y)=y,\sigma(Z)=Z\}}{card(\Sigma)} = \frac{card\{\sigma(X):\sigma\in\Sigma,\sigma(Y)=y\}}{card(\Sigma)}, \text{ so}$$

$$Q_\Sigma(x,y) * Q_\Sigma(y,z) = \frac{card\{\sigma\in\Sigma:\sigma(Y)=y\}}{card(\Sigma)card(\Sigma)}. \text{ Moreover,}$$
$$Q_\Sigma(y) = \frac{card\{\sigma\in\Sigma:\sigma(Y)=y\}}{card(\Sigma)},\ Q_\Sigma(x,y,z) = \frac{1}{card(\Sigma)}, \text{ so}$$

$$Q_\Sigma(y) * Q_\Sigma(x,y,z) = \frac{card\{\sigma\in\Sigma:\sigma(Y)=y\}}{card(\Sigma)card(\Sigma)} = Q_\Sigma(x,y) * Q_\Sigma(y,z)$$

(4) Let $\sigma, \tau \in \Sigma$ such that $\sigma(Y) = \tau(Y)$, but suppose there is no $\rho \in \Sigma$ such that $\rho(X) = \sigma(X)$, $\rho(Y) = \sigma(Y) = \tau(Y)$, $\rho(Z) = \tau(Z)$. Then $Q_\Sigma(\sigma(X), \sigma(Y))$, $Q_\Sigma(\sigma(Y), \tau(Z))$, $Q_\Sigma(\sigma(Y)) > 0$, but $Q_\Sigma(\sigma(X), \sigma(Y), \tau(Z)) = 0$.

□

10.4 Basic Results for Set and Function Independence

We interpret independence here differently, but in a related way, as introduced in Definition 10.2.6 and Definition 10.2.7. Thus, $\langle \ldots \rangle$ will abbreviate $\langle \ldots \rangle_\Sigma$ throughout this section.

Obviously, we can generalise the equivalence results for probabilistic and set independence for $X \times Z$ and $X \times Y \times Z$ to the general situation with W in Sect. 10.3, as long as we do not consider the full functions σ, but only their restrictions to X, Y, Z, $\sigma \upharpoonright (X \cup Y \cup Z)$. As we will stop the discussion of probablistic independence here, and restrict ourselves to set independence, this is left as an easy exercise to the reader.

In all triples $\langle X \mid Y \mid Z\rangle$, etc., X, Y, Z, etc., will be pairwise disjoint.

Notation 10.4.1
In more complicated cases, we will often write ABC for $\langle A \mid B \mid C\rangle$, and $\neg ABC$ if $\langle A \mid B \mid C\rangle$ does not hold. Moreover, we will often just write $f(A)$ for $f \upharpoonright A$, etc.

For $\langle A \cup A' \mid B \mid C\rangle$, we will then write $(AA')BC$,, etc.

If only singletons are involved, we will sometimes write abc instead of ABC, etc.

When we speak about fragments of functions, we will often write just $A : \sigma$ for $\sigma \upharpoonright A$, $B : \sigma = \tau$ for $\sigma \upharpoonright B = \tau \upharpoonright B$, etc.

Remark 10.4.1
This remark gives an intuitive justification of (some of) above rules of Definition 10.2.5 in our context.

Rule (Ba) is trivial.

It is easiest to set $Y := \emptyset$ in the following to see the intuitive meaning.

Rule (Bb) is a trivial consequence. If we can combine longer sequences, then we can combine shorter ones, too.

Rule (Bc) is again a trivial consequence. If we can combine arbitrary sequences, then we can also combine those which agree already on some part.

Rule (Bd) is the most interesting one, it says when we may combine *longer* sequences. Having just $\langle X \parallel Z\rangle$ and $\langle X \parallel W\rangle$ as prerequisite does not suffice, as we might lose when applying $\langle X \parallel W\rangle$ what we had already by $\langle X \parallel Z\rangle$. The condition $\langle X \mid Z \mid W\rangle$ guarantees that we do not lose this.

In our context, it means the following - recall that $Y = \emptyset$:

We want to combine $\sigma \upharpoonright X$ with $\tau \upharpoonright Z \cup W$. By $\langle X \parallel Z\rangle$, we can combine $\sigma \upharpoonright X$ with $\tau \upharpoonright Z$. Fix ρ such that $\rho \upharpoonright X = \sigma \upharpoonright X$, $\rho \upharpoonright Z = \tau \upharpoonright Z$. As $\rho \upharpoonright Z = \tau \upharpoonright Z$, by $\langle X \mid Z \mid W\rangle$, we can combine $\rho \upharpoonright X \cup Z$ with $\tau \upharpoonright W$, and have the result.

Note that we change the functions here, too: we start with σ, τ, then continue with ρ, τ.

We can use what we constructed already as a sort of scaffolding for constructing the rest.

We show now that above Rules $(Ba) - (Bd)$ hold in our context, but (Be) does not hold.

Fact 10.4.2
In our interpretation,
(1) rule (Be) does not hold,
(2) all $\langle X \mid Y \mid \emptyset \rangle$ (and thus also all $\langle \emptyset \mid Y \mid Z \rangle$) hold.
(3) rules (Ba) to (Bd) hold, even when one or both of the outside elements of the triples is the empty set.

Proof
(1) (Be) does not hold:

Consider $I := \{x, y, z, w\}$ and $\Sigma := \{1111, 0100\}$. Then $x(yw)z$ and $x(yz)w$, as for all $\sigma \upharpoonright yw$ there is just one τ this σ can be. The same holds for $x(yz)w$. But for $y = 1$, there are two different paths through $y = 1$, which cannot be combined.

(2) Trivial.

(3) Rules (Ba), (Bb), (Bc) are trivial, by definition, also for $X, Z = \emptyset$. In (Bc), if $W = \emptyset$, there is nothing to show.

Rule (Bd): The cases for $X, W, Z = \emptyset$ are trivial. Assume σ, τ such that $\sigma \upharpoonright Y = \tau \upharpoonright Y$, we want to combine $\sigma \upharpoonright X$ with $\tau \upharpoonright Z \cup W$. By $\langle X \mid Y \mid Z \rangle$, there is ρ such that $\rho \upharpoonright X = \sigma \upharpoonright X$, $\rho \upharpoonright Y = \sigma \upharpoonright Y = \tau \upharpoonright Y$, $\alpha \upharpoonright X = \rho \upharpoonright Z = \tau \upharpoonright Z$. Thus ρ and τ satisfy the prerequisite of $\langle X \mid Y \cup Z \mid W \rangle$, and there is α such that $\alpha \upharpoonright X = \rho \upharpoonright X = \sigma \upharpoonright X$, $\alpha \upharpoonright X = \rho \upharpoonright Y = \sigma \upharpoonright Y = \tau \upharpoonright Y$, $\alpha \upharpoonright W = \tau \upharpoonright W$. □

We saw already in Example 10.3.6 that $\langle X \mid Y \mid Z \rangle$ and $\langle X \parallel Z \rangle$ are independent, and give a more complicated example now:

Example 10.4.1
Consider $I := \{x, a, b, c, d, z\}$.

Let $\Sigma := \{111111, 011110, 011101, 111100, 110111, 010000\}$.

Then $\neg x(abcd)z$, $x(abc)z$, $\neg x(ab)z$.

For $\neg x(abcd)z$, fix $abcd = 1111$, then $111111, 011110 \in \Sigma$, but, e.g., $011111 \notin \Sigma$.

For $x(abc)z$, the following combinations of abc exist: $111, 101, 100$. The result is trivial for 101 and 100. For 111, all combinations for x and z with 0 and 1 exist.

For $\neg x(ab)z$, fix $ab = 10$, then $110111, 010000 \in \Sigma$, but there is, e.g. no $110xy0 \notin \Sigma$.

See Diagram 10.4.1. □

Diagram 10.4.1

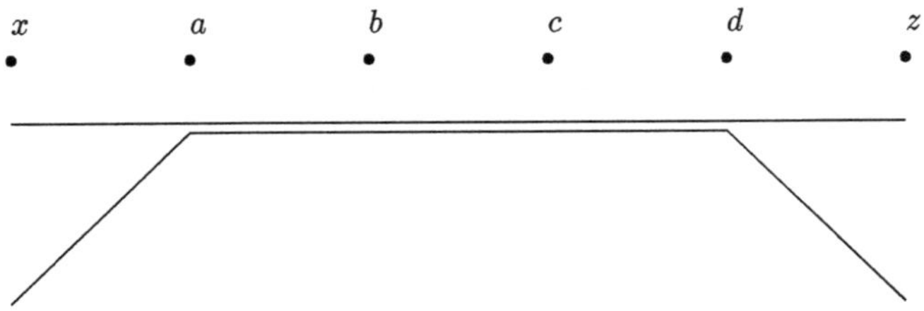

(1) $\neg\langle x \mid abcd \mid z\rangle$

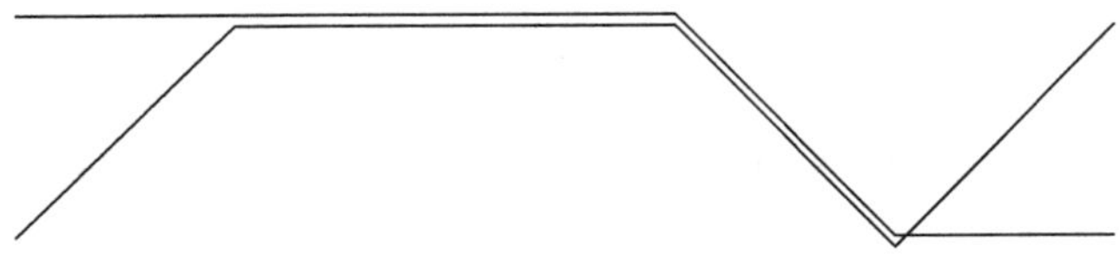

(2) $\langle x \mid abc \mid z\rangle$

add paths equal on abc, different on d, to compensate lacking paths in (1)

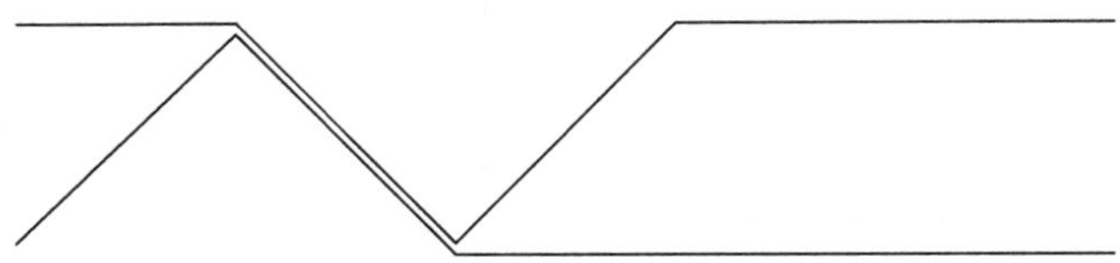

(3) $\neg\langle x \mid ab \mid z\rangle$

add paths different on ab, singletons on c, so they don't disturb on abc: seen on abc, the added paths are singletons, so they respect automatically $\langle x \mid abc \mid z\rangle$

10.5 New Rules, Examples and Discussion for Function Independence

Again, $\langle \ldots \rangle$ will abbreviate $\langle \ldots \rangle_\Sigma$, see Definition 10.2.7.

All X, Y, Z, X', A, B, B_i, etc., are assumed to be pairwise disjoint.

10.5.1 *Example of a Rule Derived from the Basic Rules*

We will use the following definition.

Definition 10.5.1

Given Σ as above, set

$\Sigma_\mu := \{\langle X, Y, Z\rangle$: X, Y, Z are pairwise disjoint subsets of I, $\langle X \mid Y \mid Z\rangle \notin \Sigma$, but for all $X' \subset X$ and all $Z' \subset Z$ $\langle X' \mid Y \mid Z\rangle \in \Sigma$ and $\langle X \mid Y \mid Z'\rangle \in \Sigma\}$.

Remark 10.5.1

(1) Σ_μ contain thus the minimal X and Z for fixed Y, such that $\langle X \mid Y \mid Z\rangle \notin \Sigma$.
(2) By rule (Bb), for all $\langle X \mid Y \mid Z\rangle \in \Sigma$, there is $\langle X', Y, Z'\rangle \in \Sigma_\mu$ $X \subseteq X'$, $Z \subseteq Z'$, unless all σ, τ such that $\sigma \restriction Y = \tau \restriction Y$ can be combined.

As the cases can become a bit complicated, it is important to develop a good intuition and representation of the problem. We do this now in the proof of the following fact, where we use the result we want to prove to guide our intuition.

Fact 10.5.2

Let Σ be closed under rules $(Ba) - (Bd)$. Then, if $\langle X, X', X'' \mid Y \mid Z, Z', Z''\rangle \in \Sigma_\mu$, then $\langle X, Z' \mid X', Y, Z'' \mid X'', Z\rangle \notin \Sigma$.

Proof

Diagram 10.5.1

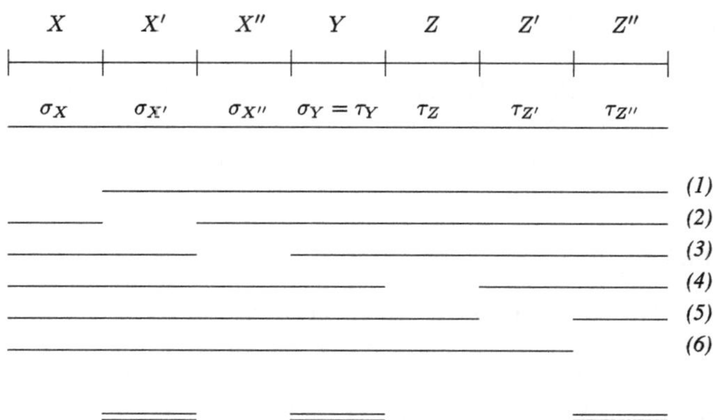

Prerequisite: $\sigma_{X'} = \tau_{X'}$, $\sigma_Y = \tau_Y$, $\sigma_{Z''} = \tau_{Z''}$

The upper line is the final aim. Line (1) expresses that we can combine all parts except σ_X, by $\langle X', X'' \mid Y \mid Z, Z', Z''\rangle$, which holds by $\langle X, X', X'' \mid Y \mid Z, Z', Z''\rangle \in \Sigma_\mu$, by similar arguments, we can combine as indicated in lines (2)–(6). We now assume $\langle X, Z' \mid X', Y, Z'' \mid X'', Z\rangle \in \Sigma$. So we have to look at fragments, which agree on X', Y, Z''. This is, for instance, true for (1) and (3).

We turn this argument now into a formal proof:
Assume
(A) $\langle X, Z' \mid X', Y, Z'' \mid X'', Z\rangle \in \Sigma$, and
(B) $\langle X, X', X'' \mid Y \mid Z, Z', Z''\rangle \in \Sigma_\mu$.
(C) $\langle X, X' \mid Y \mid Z, Z', Z''\rangle$ by (B), minimality, see line (3)
(D) $\langle X \mid X', Y, Z', Z'' \mid X'', Z\rangle$ by (A) and rule (Bc)
(E) $\langle X \mid X', Y \mid Z, Z', Z''\rangle$ by (C) and rule (Bc)
(F) $\langle X \mid X', Y \mid Z', Z''\rangle$ by (E) and (Bb)
(G) $\langle X \mid X', Y \mid X'', Z, Z', Z''\rangle$ by (D) and (F) and (Bd)
(K) $\langle X \mid X', X'', Y \mid Z, Z', Z''\rangle$ by (G) and (Bc)
(L) $\langle X', X'' \mid Y \mid Z, Z', Z''\rangle$ by (B), minimality, see line (1)
(M) $\langle Z, Z', Z'' \mid X', X'', Y \mid X\rangle$ by (K) and (Ba)
(N) $\langle Z, Z', Z'' \mid Y \mid X', X''\rangle$ by (L) and (Ba)
(O) $\langle Z, Z', Z'' \mid Y \mid X, X', X''\rangle$ by (M) and (N) and (Bd)
(P) $\langle X, X', X'' \mid Y \mid Z, Z', Z''\rangle$ by (O) and (Ba).

So we conclude $\langle X, X', X'' \mid Y \mid Z, Z', Z''\rangle \in \Sigma$, a contradiction.

Comment:

We first move Z', Z'' to the right, and then X', X'' to the left.

Moving Z', Z'':

We use X'' (or Z) on the right, which is not changed, therefore we can use line (3), resulting in

(C) $\langle X, X' \mid Y \mid Z, Z', Z''\rangle$, or, directly
(C') $\langle X, X' \mid Y \mid Z', Z''\rangle$, again by Σ_μ,
which is modified to
(F) $\langle X \mid X', Y \mid Z', Z''\rangle$, so we have on the right Z', Z'' which we want to move.
We put Z' in the middle (Z'' is there already) of (A), resulting in
(D) $\langle X \mid X', Y, Z', Z'' \mid X'', Z\rangle$.
Now we can apply (Bd) to (D) and (F), and have moved Z', Z'' to the right:
(G) $\langle X \mid X', Y \mid X'', Z, Z', Z''\rangle$.

We still have to move X' and X'' to the left of (G), and do this in an analogous way. □

Corollary 10.5.3

Let Σ be closed under rules $(Ba)-(Bd)$. Then, if $\langle X, X', X'' \mid Y, Y', Y'' \mid Z, Z', Z''\rangle \in \Sigma_\mu$, then $\langle X, Y', Z' \mid X', Y, Z'' \mid X'', Y'', Z\rangle \notin \Sigma$.

Thus, if, for given $Y \cup Y' \cup Y''$, $\langle X, X', X'' \mid Y, Y', Y'' \mid Z, Z', Z''\rangle \in \Sigma_\mu$, then for no distribution of $X \cup X' \cup X'' \cup Y \cup Y' \cup Y'' \cup Z \cup Z' \cup Z''$ such that the outward elements are non-empty, $\langle X, Y', Z' \mid X', Y, Z'' \mid X'', Y'', Z\rangle \in \Sigma$.

Proof

Suppose $\langle X, Y', Z' \mid X', Y, Z'' \mid X'', Y'', Z\rangle \in \Sigma$. Then by rule (Bc) $\langle X, Z' \mid X', Y, Y', Y'', Z'' \mid X'', Z\rangle \in \Sigma$. Set $Y_1 := Y \cup Y' \cup Y''$. Then $\langle X, Z' \mid X', Y_1, Z'' \mid X'', Z\rangle \in \Sigma$, and $\langle X, X', X'' \mid Y_1 \mid Z, Z', Z''\rangle \in \Sigma_\mu$, contradicting Fact 10.5.2. □

10.5.2 More New Rules

Above rules $(Ba) - (Bd)$, and the one in Fact 10.5.2, are not the only ones to hold, and we introduce now more complicated ones, and show that they hold in our situation. Of the possibly infinitary rules, only (Loop1) is given in full generality, (Loop2) is only given to illustrate that even the infinitary rule (Loop1) is not all there is.

For warming up, we consider the following short version of (Loop1):

Example 10.5.1
$ABC, ACD, ADE, AEB \Rightarrow ABE.$

We show that this rule holds in all Σ.

Suppose $A : \sigma,\ B : \sigma = \tau,\ C : \tau$, so by ABC, there is ρ_1 such that $A : \rho_1 = \sigma,\ B : \rho_1 = \sigma = \tau,\ C : \rho_1 = \tau$. So by ACD, there is ρ_2 such that $A : \rho_2 = \sigma,\ C : \rho_2 = \rho_1 = \tau,\ D : \rho_2 = \tau$. So by ADE, there is ρ_3 such that $A : \rho_3 = \sigma,\ D : \rho_3 = \rho_2 = \tau,\ E : \rho_3 = \tau$. So by AEB, there is ρ_4 such that $A : \rho_4 = \sigma,\ E : \rho_4 = \rho_3 = \tau,\ B : \rho_4 = \tau = \sigma.$

So ABE.

We abbreviate this reasoning by:

(1) ABC: $A : \sigma,\ B : \sigma = \tau,\ C : \tau$
(2) ACD: $(1) + \tau$
(3) ADE: $(2) + \tau$
(4) AEB: $(3) + \tau$
So ABE.

It is helpful to draw a little diagram as given in the table in Sect. 10.5.2.

Validity of $ABC, ACD, ADE, AEB \Rightarrow ABE$						
	A	B	C	D	E	
	σ	$\sigma = \tau$			τ	ABE?
(1) ρ_1	σ	$\sigma = \tau$	τ			ABC
(2) ρ_2	σ		$\rho_1 = \tau$	τ		ACD
(3) ρ_3	σ			$\rho_2 = \tau$	τ	ADE
(4) ρ_4	σ	$\sigma = \tau$			$\rho_3 = \tau$	AEB

We introduce some more new rules.

Definition 10.5.2
As always, we assume that all sets occurring in a rule are pairwise disjoint.

- (Bin1) $XYZ, XY'Z, Y(XZ)Y' \Rightarrow X(YY')Z$
  Note that this rule is not trivial. The prerequisites $X : \sigma,\ Y, Y' : \sigma = \tau,\ Z : \tau$ are stronger than those for XYZ: $X : \sigma,\ Y : \sigma = \tau,\ Z : \tau$, but by XYZ alone we do not know what happens on Y', it might be any $\rho \neq \sigma = \tau$.
- (Bin2) $XYZ, XZY', Y(XZ)Y' \Rightarrow X(YY')Z$

- (Loop1) $AB_1B_2, AB_2B_3, \ldots, AB_{i-1}B_i, AB_iB_{i+1}, AB_{i+1}B_{i+2}, \ldots, AB_{n-1}B_n, AB_nB_1 \Rightarrow AB_1B_n$
  so we turn AB_nB_1 around to AB_1B_n.
  When we have to be more precise, we will denote this condition $(Loop1_n)$ to fix the length.
  Note that $(Loop1_1)$ and $(Loop1_2)$ will always hold trivially, as the conclusion is one of the prerequisites.
- (Loop2) $ABC, ACD, DAE, DEF, FDG, FGH, HFB \Rightarrow HBF$

The complicated structure of these rules suggests already that the ternary relations are not the right level of abstraction to speak about construction of functions from fragments. This is made formal by our main result below, which shows that there is no finite characterisation by such relations. In other words, the main things happen behind the screen.

Fact 10.5.4
The new rules in Definition 10.5.2 are valid in our situation.

Proof

- (Bin1)
  Let $X:\sigma,\ Y:\sigma=\tau,\ Y':\sigma=\tau,\ Z:\tau$. We construct ρ_3 such that $X:\rho_3=\sigma$, $Y:\rho_3=\sigma=\tau,\ Y':\rho_3=\sigma=\tau,\ Z:\rho_3=\tau$.
  Let, by XYZ ρ_1 be such that $X:\rho_1=\sigma,\ Y:\rho_1=\sigma=\tau,\ Z:\rho_1=\tau$, and by $XY'Z$ ρ_2 be such that $X:\rho_2=\sigma,\ Y':\rho_2=\sigma=\tau,\ Z:\rho_2=\tau$, combine by $Y(XZ)Y'$ ρ_1 with ρ_2 to ρ_3 such that $Y:\rho_3=\rho_1=\sigma=\tau,\ X:\rho_3=\rho_2=\rho_1=\sigma$, $Z:\rho_3=\rho_2=\rho_1=\tau,\ Y':\rho_3=\rho_2=\sigma=\tau$.
  We abbreviate this as follows:
  (1) XYZ: $X:\sigma,\ Y:\sigma=\tau,\ Z:\tau$
  (2) $XY'Z$: $X:\sigma,\ Y':\sigma=\tau,\ Z:\tau$
  (3) $Y(XZ)Y'$: (1) + (2)
  So $X(YY')Z$.
- (Bin2)
  Let again $X:\sigma,\ Y:\sigma=\tau,\ Y':\sigma=\tau,\ Z:\tau$. We construct ρ_3 such that $X:\rho_3=\sigma,\ Y:\rho_3=\sigma=\tau,\ Y':\rho_3=\sigma=\tau,\ Z:\rho_3=\tau$.
  We present the abbreviation only.
  Let $X:\sigma,\ Y:\sigma=\tau,\ Y':\sigma=\tau,\ Z:\tau$
  (1) XYZ: $X:\sigma,\ Y:\sigma=\tau,\ Z:\tau$
  (2) XZY': (1) $+\ \tau$
  (3) $Y(XZ)Y'$: (1) + (2)
  So $X(YY')Z$.
- (Loop1)
  Let $A:\sigma,\ B_1:\sigma=\tau,\ B_n:\tau$, we construct ρ_n such that $A:\rho_n=\sigma,\ B_1:\rho_n=\sigma=\tau,\ B_n:\rho_n=\tau$.
  We start with ρ_1 combining σ with τ by AB_1B_2 such that $A:\rho_1=\sigma,\ B_1:\rho_1=\sigma=\tau,\ B_2:\rho_1=\tau$, then there is ρ_2, combining ρ_1 with τ by AB_2B_3 such that

$A : \rho_2 = \rho_1 = \sigma,\ B_2 : \rho_2 = \rho_1 = \tau,\ B_3 : \rho_2 = \tau$, etc. up to ρ_{n-1} combining ρ_{n-2} with τ by $AB_{n-1}B_n$ such that $A : \rho_{n-1} = \rho_{n-2} = \sigma,\ B_{n-1} : \rho_{n-1} = \rho_{n-2} = \tau,\ B_n : \rho_{n-1} = \tau$, and finally ρ_n combining ρ_{n-1} with τ by AB_nB_1 such that $A : \rho_n = \rho_{n-1} = \sigma,\ B_n : \rho_n = \rho_{n-1} = \tau,\ B_1 : \rho_n = \tau$.
This is abbreviated by:
(1) AB_1B_2: $A : \sigma,\ B_1 : \sigma = \tau,\ B_2 : \tau$
(2) AB_2B_3: $(1) + \tau$
....
(i-1) $AB_{i-1}B_i$: $(i-2) + \tau$
(i) AB_iB_{i+1}: $(i-1) + \tau$
$(i+1)$ $AB_{i+1}B_{i+2}$: $(i) + \tau$
....
$(n-1)$ $AB_{n-1}B_n$: $(n-2) + \tau$
(n) AB_nB_1: $(n-1) + \tau$
So AB_1B_n.

- (Loop2)
  Let $H : \tau,\ B : \sigma = \tau,\ F : \sigma$. We construct ρ_7 such that $H : \rho_7 = \tau,\ B : \rho_7 = \sigma = \tau,\ F : \rho_7 = \sigma$.
  Let
  (1) ABC: $A : \sigma,\ B : \sigma = \tau,\ C : \tau$
  (2) $ACD : 1 + \tau$
  (3) $DAE : 2 + \sigma$
  (4) $DEF : 3 + \sigma$
  (5) $FDG : 4 + \tau$
  (6) $FGH : 5 + \tau$
  (7) $HFB : 6 + \sigma$
  So HBF by $B : \sigma = \tau$.

Note that we use here $B : \sigma = \tau,\ E : \sigma = \tau,\ H : \sigma = \tau$, whereas the other triples are used for other functions. □

Next we show that the full (Loop1) cannot be derived from the basic rules (Ba) – (Bd) and (Bin1), and shorter versions of (Loop1). (This is also a consequence of the sequel, but we want to point it out right away.)

Fact 10.5.5
Let $n \geq 1$, then $(Loop1_n)$ does not follow from the rules $(Ba) - (Bd)$, $(B\emptyset)$, (Bin1), and the shorter versions of (Loop1)

Proof
Consider the following set of triples $L \cup L'$ over $I := \{a, b_1, \ldots, b_n\}$:
$L := \{ab_1b_2,\ \ldots,\ ab_ib_{i+1},\ \ldots,\ ab_{n-1}b_n,\ ab_nb_1\}$,
$L' := \{\emptyset AB : A \cap B = \emptyset,\ A \cup B \subseteq I\}$,
and close $L \cup L'$ under symmetry (rule (Ba)).
L is essentially the set of $ab_ib_{i+_{mod(n)}1}$.
Call the resulting set $\mathcal{A}$.

Note that, if the inside set of a triple is $\emptyset$, then one of the outside sets of the triple must also be $\emptyset$, and if both outside sets are $\neq \emptyset$, then all three sets are singletons.

When we look at L, and define a relation $<$ by $x < y$ iff $axy \in L$, we see that the shortest $<$-loop is $b_1 < b_2 < \ldots < b_{n-1} < b_n < b_1$, of length n.

We show first that $\mathcal{A}$ is closed under rules $(Ba) - (Bd)$, $(B\emptyset)$ (see Definition 10.2.5), and (Bin1).

(Ba) and $(B\emptyset)$ are trivial.

(Bb) If $X = \emptyset$ or $Z = \emptyset$, or $W = Z$, this is trivial. So assume none of these is the case, so we are in L, $Z \cup W$ is a singleton, and $W = \emptyset$, so this is trivial, too.

(Bc) Suppose $\langle X \mid Y \mid Z \cup W \rangle$, but not $< X \mid Y \cup W \mid Z >$. Then $X \neq \emptyset$ and $Z \neq \emptyset$, otherwise $< X \mid Y \cup W \mid Z > \in L'$ (or $\langle Z \mid Y \cup W \mid X \rangle \in L'$). As $\langle X \mid Y \mid Z \cup W \rangle$ and $X \neq \emptyset$, $Z \neq \emptyset$, $X, Y, Z \cup W$ are all singletons. As $Z \neq \emptyset$, Z must be a singleton and $Z = W$ or $W = \emptyset$. If $W = \emptyset$, then the conclusion is the same as the prerequisite. But $Z \cap W = \emptyset$, so $Z = W$ is impossible.

(Bd) If $X = \emptyset$ or $Z \cup W = \emptyset$, this is trivial, so assume $X \neq \emptyset$, $Z \cup W \neq \emptyset$.

Case 1: $Z \neq \emptyset$, so X, Y, Z are singletons, so by $Z \cup W \neq \emptyset$, $Z \cup W$ is a singleton, so $W = \emptyset$ or $Z = W$, and the conclusion is the first prerequisite. (In addition, $Z = W$ is excluded by disjointness.)

Case 2: $W \neq \emptyset$, so $X, W, Y \cup Z$ are singletons. Case 2.1: $Y = Z$ is excluded, as $Y \cap Z \neq \emptyset$ otherwise. Case 2.2: $Y = \emptyset$, so Z is a singleton, and Y is a singleton, contradiction. Case 2.3: $Z = \emptyset$, then the conclusion is the second prerequisite.

(Bin1) The cases $X = \emptyset$ and $Z = \emptyset$ are trivial. So assume $X \neq \emptyset$, $Z \neq \emptyset$. Then X, Z, Y, Y', XZ are all singletons. Thus $X = Z$, but this is impossible, as $X \cap Z = \emptyset$.

Obviously, $(Loop1_n)$ does not hold.

We show now that all $(Loop1_k)$, $1 \leq k < n$ hold.

Consider the case $2 < k < n$.

This has the form $AB_1B_2, AB_2B_3, \ldots, AB_{k-2}B_{k-1}, AB_{k-1}B_k, AB_k B_1 \Rightarrow AB_1B_k$. Suppose it fails.

If $A = \emptyset$ or $B_k = \emptyset$, the condition holds, so assume $A \neq \emptyset$, $B_k \neq \emptyset$.

By $AB_{k-1}B_k$, $B_{k-1} \neq \emptyset$, so by $AB_{k-2}B_{k-1}$, $B_{k-2} \neq \emptyset$, etc., and we see that all $B_i \neq \emptyset$, $1 \leq i \leq k$. So we have $AB_1B_2, \ldots, AB_{k-1}B_k, AB_kB_1 \in L$, but $AB_1B_k \notin L$. Moreover, A has to be a, which is the only element occuring repeatedly on the outside. But $AB_1B_2, \ldots, AB_{k-1}B_k, AB_kB_1$ form now a cycle shorter than n, a contradiction. So the prerequisites cannot all be in $\mathcal{A}$. □

10.6 There Is No Finite Characterisation for Function Independence

We turn to our main result.

Again, $\langle \ldots \rangle$ will abbreviate $\langle \ldots \rangle_\Sigma$, see Definition 10.2.7.

10.6.1 Discussion

Consider the following simple, short, loop for illustration:

$ABC, ACD, ADE, AEF, AFG, AGB \Rightarrow ABG$—so we can turn AGB around to ABG.

Of course, this construction may be arbitrarily long.

The idea is now to make ABG false, and, to make it coherent, to make one of the interior conditions false, too, say ADE. We describe this situation fully, i.e. enumerate all conditions which hold in such a situation. If we make now ADE true again, we know this is not valid, so any (finite) characterisation must say "NO" to this. But as it is finite, it cannot describe all the interior triples of the type ADE in a sufficiently long loop, so we just change one of them which it does not "see" to FALSE, and it must give the same answer NO, so this fails.

Basically, we cannot describe parts of the loop, as the <||>-language is not rich enough to express it, we see only the final outcome. Bundling singletons into bigger sets will not help, either, as we will see below (cases with size > 3).

The problem is to fully describe the situation.

10.6.2 Composition of Layers

A very helpful fact is the following:

Definition 10.6.1

Let Σ_j be function sets over I into some set K, $j \in J$, $f_j : I \to K$ for $f_j \in \Sigma_j$.

Let $\Sigma := \{ f : I \to K^J : f(i) = \{\langle f_j(i), j\rangle : j \in J, f_j \in \Sigma_j\} \}$.

So any $f \in \Sigma$ has the form $f(i) = \langle f_1(i), f_2(i), \ldots, f_n(i)\rangle$, $f_m \in \Sigma_m$, $m \in J$ (we may assume J to be finite).

Thus, given $f \in \Sigma$, $f_m \in \Sigma_m$ is defined for $m \in J$.

Fact 10.6.1

For the above Σ $\langle A \mid B \mid C\rangle$ holds iff it holds for all Σ_j.

Thus, we can destroy the $\langle A \mid B \mid C\rangle$ independently, and collect the results.

Proof

The proof is trivial, and a direct consequence of the fact that $f = f'$ iff for all components $f_j = f'_j$.

Suppose for some Σ_k, $k \in J$, $\neg\langle A \mid B \mid C\rangle$.

So for this k there are $f_k, f'_k \in \Sigma_k$ such that $f_k(B) = f'_k(B)$, but there is no $f''_k \in \Sigma_k$ such that $f''_k(A) = f_k(A)$, $f''_k(B) = f_k(B) = f'_k(B)$, $f''_k(C) = f'_k(C)$ (or conversely). Consider now some $h \in \Sigma$ such that $h_k = f_k$, and h' is like h, but $h'_k = f'_k$, so also $h' \in \Sigma$. Then $h(B) = h'(B)$, but there is no $h'' \in \Sigma$ such that $h''(A) = h(A)$, $h''(B) = h(B) = h'(B)$, $h''(C) = h'(C)$.

Conversely, suppose $\langle A \mid B \mid C\rangle$ for all Σ_j. Let $h, h' \in \Sigma$ such that $h(B) = h'(B)$, so for all $j \in J$ $h_j(B) = h'_j(B)$, where $h_j \in \Sigma_j$, $h'_j \in \Sigma_j$, so there are $h''_j \in$

Σ_j with $h''_j(A) = h_j(A)$, $h''_j(B) = h_j(B) = h'_j(B)$, $h''_j(C) = h'_j(C)$ for all $j \in J$. Thus, h'' composed of the h''_j is in Σ, and $h''(A) = h(A)$, $h''(B) = h(B) = h'(B)$, $h''(C) = h'(C)$. □

10.6.3 Systematic Construction

Recall the general form of (Loop1) for singletons:
$ab_1b_2, \ldots, ab_{i-1}b_i, ab_ib_{i+1}, ab_{i+1}b_{i+2}, \ldots, ab_{n-1}b_n, ab_nb_1 \Rightarrow ab_1b_n$
We will fully describe a model of above triples, with the exception of ab_1b_n and ab_ib_{i+1} which will be made to fail, and all other $\langle X \mid Y \mid Z\rangle$ which are not in above list of triples to preserve, will fail, too (except for $X = \emptyset$ or $Z = \emptyset$).

Thus, the triples to preserve are:
$P := \{ab_1b_2, \ldots, ab_{i-1}b_i$, (BUT NOT ab_ib_{i+1}) , $ab_{i+1}b_{i+2}, \ldots, ab_{n-1}b_n, ab_nb_1\}$

We use the following fact:

Fact 10.6.2
Let $X, Y, Z \subseteq I$ be pairwise disjoint. Let $M \subseteq I$, $card(M) > 1$, be fixed, consider $\Sigma := \Sigma_M := \{\, \sigma : I \to \{0, 1\}$: $card\{m \in M : \sigma(m) = 0\}$ is even $\}$.

Then $\neg\langle XYZ\rangle_{\Sigma_M}$ iff ($X \cap M \neq \emptyset$ and $Z \cap M \neq \emptyset$ and $M \subseteq X \cup Y \cup Z$).

Proof
"$\Leftarrow$":

Suppose $X \cap M \neq \emptyset$, $Z \cap M \neq \emptyset$, $M \subseteq X \cup Y \cup Z$.

Take σ such that $card\{m \in M : \sigma(m) = 0\}$ is odd, then $\sigma \notin \Sigma_M$. As $M \not\subseteq X \cup Y$, there is $\tau \in \Sigma_M$ such that $\sigma \restriction X \cup Y = \tau \restriction X \cup Y$. As $M \not\subseteq Y \cup Z$, there is $\rho \in \Sigma_M$ such that $\rho \restriction Y \cup Z = \sigma \restriction Y \cup Z$. Thus, $\tau \restriction Y = \rho \restriction Y = \sigma \restriction Y$. If there were $\alpha \in \Sigma_M$ such that $\alpha \restriction X = \tau \restriction X = \sigma \restriction X$, $\alpha \restriction Y = \tau \restriction Y = \rho \restriction Y = \sigma \restriction Y$, $\alpha \restriction Z = \rho \restriction Z = \sigma \restriction Z$, then $\alpha \restriction X \cup Y \cup Z = \sigma \restriction X \cup Y \cup Z$, contradiction
"$\Rightarrow$":

Suppose $X \cap M = \emptyset$ or $Z \cap M = \emptyset$, or $M \not\subseteq X \cup Y \cup Z$. We show $\langle XYZ\rangle_{\Sigma_M}$.
Case 1: $Z \cap M = \emptyset$. Let $\sigma, \tau \in \Sigma_M$ such that $\sigma \restriction Y = \tau \restriction Y$. As $Z \cap M = \emptyset$, we can continue $\sigma \restriction X \cup Y$ on Z as we like, it will not change $card\{m \in M : \sigma(m) = 0\}$, so any continuation will be in Σ_M.
Case 2: $X \cap M = \emptyset$, analogous.
Case 3: $M \not\subseteq X \cup Y \cup Z$. But then there is no restriction in $X \cup Y \cup Z$. (Any σ such that $\sigma \restriction X \cup Y \cup Z$ is odd, can always be "corrected" outside $X \cup Y \cup Z$, and any σ such that $\sigma \restriction X \cup Y \cup Z$ is even, can always be continued as 1 outside $X \cup Y \cup Z$.)
□

We will have to make ab_1b_n false, but ab_nb_1 true. On the other hand, we will make ab_1b_3 false, but ab_3b_1 need not be preserved.

This leads to the following definition, which helps to put order into the cases.

Definition 10.6.2
The definition is unnecessarily complicated, considering the cases where $ayx \in P$, and $ayx \notin P$ would be enough, still, it seems good for the intuition to measure the distance to P. For this reason, we use the more complex approach.

Suppose we have to destroy axy. Then

$dmin(axy) := min\{d(\{a, x, y\}, \{a, u, v\}) : auv$ has to be preserved $\}$ - d the counting Hamming distance (as set, not as sequence).

Thus, $dmin(ab_1b_n) = 0$ (as ab_nb_1 has to be preserved), $dmin(ab_1b_3) = 1$ (because ab_1b_2 has to be preserved, but not ab_3b_1).

We introduce the following order defined from the loop prerequisites to be preserved.

Definition 10.6.3
Order the elements by following the string of sequences to be preserved as follows:
$b_{i+1} \prec b_{i+2} \prec \cdots \prec b_{n-1} \prec b_n \prec b_1 \prec b_2 \prec \cdots \prec b_{k-1} \prec b_k \cdots \prec b_{i-1} \prec b_i$

Note that the interruption at ab_ib_{i+1} is crucial here—otherwise, there would be a cycle.

As usual, $\preceq$ will stand for $\prec$ or $=$.

10.6.4 The Cases to Consider

The elements to consider are: $a, b_1, \ldots, b_n$.

Recall that the triples to preserve are:

$P := \{ab_1b_2, \ldots, ab_{i-1}b_i,$ (BUT NOT ab_ib_{i+1}) $, ab_{i+1}b_{i+2}, \ldots, ab_{n-1}b_n, ab_nb_1\}$

The $\langle X \mid Y \mid Z\rangle$ to destroy are (except when $X = \emptyset$ or $Z = \emptyset$) :

(1) all $\langle X \mid\mid Z\rangle$
(2) all $\langle X \mid Y \mid Z\rangle$ such that $X \cup Y \cup Z$ has > 3 elements
(3) all triples which do not have a on the outside, e.g. $b_1b_2b_3$
(4) and the following triples:
(the (0) will be explained below - for the moment, just ignore it)
$ab_1b_3, \ldots, ab_1b_{n-1}, ab_1b_n$ (0)
ab_2b_1 (0), $ab_2b_4, \ldots, ab_2b_n$
ab_3b_1, ab_3b_2 (0), $ab_3b_5, \ldots, ab_3b_n$
....
$ab_ib_1, ab_ib_2, , \ldots,$ ALSO $ab_ib_{i+1}, \ldots, ab_ib_n$
....
$ab_{n-2}b_1, \ldots, ab_{n-2}b_{n-3}$ (0), $ab_{n-2}b_n$
$ab_{n-1}b_1, \ldots, ab_{n-1}b_{n-2}$ (0),
ab_nb_1 $, \ldots, ab_nb_{n-1}$ (0)

(Note that triples of the kind ab_1b_1 are excluded by disjointness.)

10.6.5 Solution of the Cases

We show how to destroy all triples $\langle X \mid Y \mid Z\rangle$ mentioned above by suitable Σ's, while preserving all triples in P. (When helpful, we will mark the first part (destruction) with "(a)", the second (preservation) with "(b)".)

0_c and 1_c will stand for the constant functions 0 and 1.

Assume $X \neq \emptyset$, $Z \neq \emptyset$.

(1) all $\langle X \mid Y \mid Z\rangle$ where $X \cup Y \cup Z$ has > 3 elements:
Set $M := X \cup Y \cup Z$, consider $\Sigma_{\langle X|Y|Z\rangle} := \Sigma_M$ as in Fact 10.6.2, then
(a) $\neg\langle X \mid Y \mid Z\rangle$, but
(b) all triples in P are preserved, as $M \not\subseteq \{a, b_i, b_j\}$, by the same Fact.

(2) all $\langle X \mid Y \mid Z\rangle$ with 1 element: There is nothing to do, as $X = \emptyset$ or $Z = \emptyset$.

(3) all $\langle X \mid\mid Z\rangle$:
This can be done by considering $\Sigma_{\langle X||Z\rangle} := \{0_c, 1_c\}$.
(a) Then, say for a, c, we have to examine the fragments 00 and 11, but there is no 10 or 01.
(b) For $\langle a \mid b \mid c\rangle$ this is no problem, as we have only the two 000, 111, which do not agree on b.

(4) all $\langle X \mid Y \mid Z\rangle$ with 2 elements: eliminated by $\langle X \mid\mid Z\rangle$, or $X = \emptyset$, or $Z = \emptyset$.

(5) all $\langle X \mid Y \mid Z\rangle$ with 3 elements:

(5.1) a is not on the outside

(5.1.1) a is in the middle, we need $\neg xay$: Consider Σ with 2 functions, 0_c, and the second defined by $a = 0$, and $u = 1$ for all $u \neq a$.
(a) Obviously, $\neg xay$.
(b) Recall that all triples to be preserved have a on the outside, and some other element x in the middle. Then the two functions are different on x, so for these axy is trivially true.

(5.1.2) a is not in xyz, we need $\neg xyz$: Consider Σ with 2 functions, 0_c, and the second defined by $a = y = 0$, $u = 1$ for all other u.
(a) As a is neither x nor z, $\neg xyz$.
(b) If some uvw has a on the outside, say $u = a$, then both functions are 000 or 0vw on this triple, so uvw holds.

(5.2) a is on the outside, say ayz, but $ayz \notin P$, we destroy ayz :

(5.2.1) Case $dmin(ayz) > 0$:
Take as Σ the set of all functions with values in $\{0, 1\}$, but without those with $a = y = z = 0$.
(a) Then $\neg\langle ayz\rangle$ (we have on ayz 100, 001, but not 000).
(b) But for all auv with $d(\{a, y, z\}, \{a, u, v\}) > 0$ $\langle auv\rangle$ holds, as auv has all possible combinations in Σ. (Suppose e.g. $y \notin \{a, u, v\}$, then all σ such that $\sigma(y) = 1$ are in Σ.)

(5.2.2) Case $dmin(ayz) = 0$.
The elements ayz with $dmin(ayz) = 0$ to be destroyed are:
ab_1b_n, ab_2b_1, ..., ab_ib_{i-1}, NOT $ab_{i+1}b_i$, $ab_{i+2}b_{i+1}$, ..., $ab_{n-1}b_{n-2}$,

ab_nb_{n-1}, they were marked with (0) above.
Σ will again have 2 functions, the first is always 0_c.
The second function: Always set $a = 1$.
We see that the triples with $dmin = 0$ to be destroyed have the form ayz, where z is the immediate $\prec$-predecessor of y in above order - see Definition 10.6.3. Conversely, those to be preserved (in P) have the form azy, where again z is the immediate $\prec$-predecessor of y.
We set $z' = 1$ for all $z' \preceq z$, and $y' = 0$ for all $y' \succeq y$. Recall that $z \prec y$, so we have the picture $b_{i+1} = 1, \ldots, z = 1, y = 0, \ldots, b_i = 0$.
See Diagram 10.6.1.
(a) Then $\neg ayz$, as we have the fragments {000, 101}.
(b) But azy, as we have the fragments {000, 110}. Moreover, considering the successors of the sequence, we give the values 11, or 10, or 00, recall $a = 1$. This results in the function fragments for auv as 111, or 110, or 100. But the resulting fragment sets (together with 0_c) are then: {000, 111}, {000, 110}, {000, 100}. They all make auv true. Thus, all triples in P are preserved.

Diagram 10.6.1

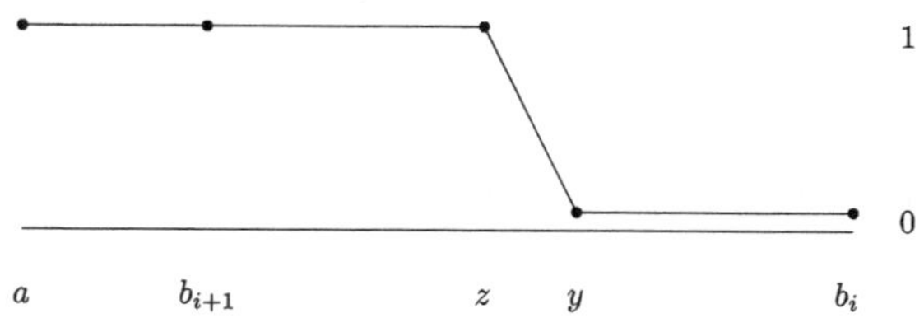

10.6.6 Final Argument

Suppose there is a fixed size finite characterisation, which we can summarise in one conjunction with finitely many variables. Let $\phi = \forall x_1, \ldots, x_k \phi'(x_1, \ldots, x_k)$ be a purely universal expression, with ϕ' quantifier free, characterising $\langle \ldots \mid \ldots \mid \ldots \rangle$. So, given any ternary relation $R(.,.,.)$, it will tell whether R corresponds to some $\langle . \mid . \mid . \rangle$ or not. Take a loop which is sufficiently big ($> k$). Consider a wrong loop example, as described above, e.g. we add ab_ib_{i+1} to P, so this relation $R(X, Y, Z)$ cannot correspond to any $\langle X \mid Y \mid Z \rangle$.

Then for this relation R, ϕ' fails for some constants, i.e. $R \models \neg\phi'[c_1, \ldots, c_k]$. Now change outside the $c_1, \ldots, c_k$ a triple in P, e.g. making it wrong, or adding ab_nb_1, so the new loop is possible for some $\langle X \mid Y \mid Z \rangle$, everything else stays the same, but $\phi'[c_1, \ldots, c_k]$ still fails, whereas it must now be true, a contradiction.

The important point of the construction is that we can change some single axy, without affecting any other XYZ, in particular none with bigger cardinality. So we cannot summarize several small axy in a big XYZ, as they are independent in the construction.

10.6.6.1 Comment

Note that the construction and result apply to bigger products only, as constructed by Fact 10.6.1. We do not know if a similar result holds for binary sets.

10.7 Systematic Construction of New Rules for Function Independence

We give here a general way how to construct new rules of the type ABC, DEF, ... $\Rightarrow$ XYZ which are valid in our situation. It is an outline, not a formal proof.

Again, $\langle \ldots \rangle$ will abbreviate $\langle \ldots \rangle_\Sigma$, see Definition 10.2.7.

10.7.1 Consequences of a Single Triple

Let $(XX'X'')Y(ZZ'Z'')$ be a triple, then all consequences of this single triple have the form $X(X'YZ')Z$ (up to symmetry).

Obviously, such $X(X'YZ')Z$ are consequences, using rules (Bb) and (Bc).

We now give counter examples to other forms to show that they are not consequences in our setting. We always assume that the outside is not $\emptyset$. We consider $A = B = C = \{0, 1\}$, and subsets of $A \times B \times C$.

(1) Y decreases:
Consider $\{000, 111\}$, then ABC, but not $A\emptyset C$.
(2) Z increases:
Consider $\{000, 101\}$, then $A\emptyset B$, but not $A\emptyset(BC)$.
(3) X goes from left to right:
Consider $\{000, 110\}$, then $(AB)\emptyset C$, but not $A\emptyset(BC)$
(4) Y increases by some arbitrary W :
Consider $\{000, 101, 110, 011\}$, then $A\emptyset C$, but not ABC.

10.7.2 Construction of Function Trees

We can construct new functions from two old functions using triples ABC, so, in a more general way, we have a binary function construction tree, where the old functions are the leaves, and the new function is the root. The form of such a tree is obvious, the triples used are either directly given, or consequences of such triples. In example 10.7.3, for instance, in the construction of ρ_2, we used ACD, but we could also have used e.g. $AC(DD')$, for some D'.

Note that, if we want to piece two functions, σ, τ, together, we may have nothing more than those two functions, and all other functions we may construct from them. So we have to start with just σ and τ, and nothing else. Not all such function construction trees are proof trees for a rule $T_1, \ldots, T_n \Rightarrow T$, where the T_i and T are triples.

We have to look at the logical structure of the triples to see what we need. In order to show $T = ABC$, we assume given two arbitrary functions σ and τ, which agree on B, and construct ρ such that on A $\rho = \sigma$, on B $\rho = \sigma = \tau$ (the latter, $\sigma = \tau$ by prerequisite), and on C $\rho = \tau$. We will write this as $A : \rho = \sigma,\ B : \rho = \sigma = \tau,\ C : \rho = \tau$.

Thus, we have no functions at the beginning, except σ and τ, so all leaves in a proof tree for $T_1, \ldots, T_n \Rightarrow T$ have to be σ or τ. Moreoever, all we know about σ and τ is that they agree on B. Thus, we can only use some $T_i' = A'B'C'$ on σ and τ if $B' \subseteq B$. Likewise, in the interior of the tree, we can only use $\sigma \upharpoonright B = \tau \upharpoonright B$, and, of course, all equalities which hold be construction. E.g., in example 10.7.3, in the construction of ρ_2, by construction of ρ_1, $C : \rho_1 = \tau$, so we can use ACD to construct ρ_2 from ρ_1 and τ. At the root, we must have a function ρ of the form $A : \rho = \sigma,\ B : \rho = \sigma = \tau,\ C : \rho = \tau$. In example 10.7.3, ρ_4, at the root, was constructed using AEB from ρ_3 and τ. But we do not interpret ρ_4 as AEB, but as ABE, which is possible, as $A : \rho_4 = \sigma,\ B : \rho_4 = \sigma = \tau,\ E : \rho_4 = \tau$. Intermediate nodes can be read as an intermediate result $A'B'C'$ by the same criteria: They must be functions ρ' such that $A' : \rho' = \sigma,\ B' : \rho' = \sigma = \tau,\ C' : \rho' = \tau$ and all B'' such that $B'' : \sigma = \tau$ used up to this node must be subsets of B', as $B' : \sigma = \tau$ is the only hypothesis we then have.

The following is a universal proof for XYZ: It is a binary tree, whose leaves are all f or g. It uses as prerequisite only $Y : f = g$ (and equalities constructed on the way). It makes f, g, and all other functions as different as possible. For instance, in example 10.7.3, where we show that $ABC, ACD, ADE, AEB \Rightarrow ABE$, let us assume all sets A, etc., are singletons, we then set: $\sigma = 00000,\ \tau = 10111,\ \rho_1 = 00122,\ \rho_2 = 03113,\ \rho_3 = 04411,\ \rho_4 = 00551$. So each new node has a new default value (2, 3, 4, 5 here). Then we have no chance equalities, but only those we constructed. In particular, if we write the equalities with σ, τ for every ρ_i thus constructed, we can read off the derived equalities. There are no others. The root of the tree must be a function h, which agrees on X, Y with f, and on Y, Z with g. This is a universal proof tree, as it works for any other pair f, g, and any other internally constructed ρ_i, too.

10.7.3 Examples

Explanation:

By "prerequisite" of ρ_i we mean the set X we used in the construction, where $X : \sigma = \tau$. For instance, in the construction of ρ_2 in example 10.7.1, we used only that $B \cup C : \rho_1 = \tau$ by the construction of ρ_1, no additional use of $\sigma = \tau$ on some set X other than B was made.

By "common part" of ρ_i we mean the set X such that $X : \rho_i = \sigma = \tau$.

In example 10.7.5, α and β are parts of ρ_1 or ρ_2, whose actual value is unimportant.

Example 10.7.1
(Contraction), ABC, $A(BC)D \rightarrow AB(CD)$:

(See Diagram 10.7.1 upper part.)

- $\rho_1 : A : \sigma,\ B : \sigma = \tau,\ C : \tau$
  generated by ABC from σ, τ
  prerequisite B,
  common part: B
  ρ_1 can be interpreted as the (trivial) derived triple ABC
- $\rho_2 : A : \rho_1 = \sigma,\ B : \rho_1 = \sigma = \tau,\ C : \rho_1 = \tau,\ D : \tau$
  generated by $A(BC)D$ from ρ_1, τ
  prerequisite -,
  common part: B.
  ρ_2 can be interpreted as a derived triple by $AB(CD)$.
  ρ_2 can also be interpreted as a derived triple by $A(BC)D$ or $A(BD)C$. Note that these possibilities can be derived from $AB(CD)$ by rule (Bc), Weak Union (Table 10.1).

Diagram 10.7.1

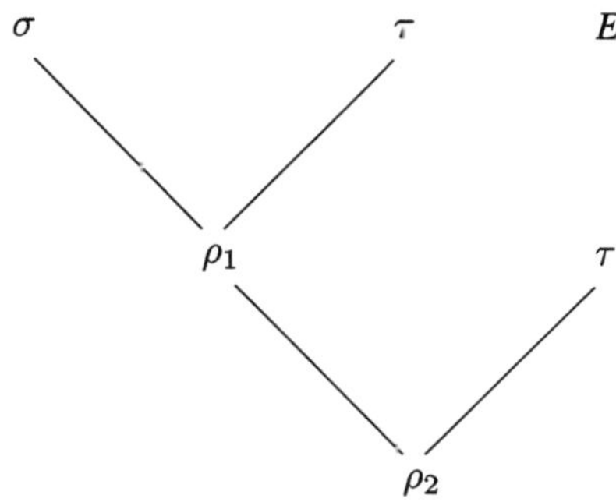

Example 10.7.1

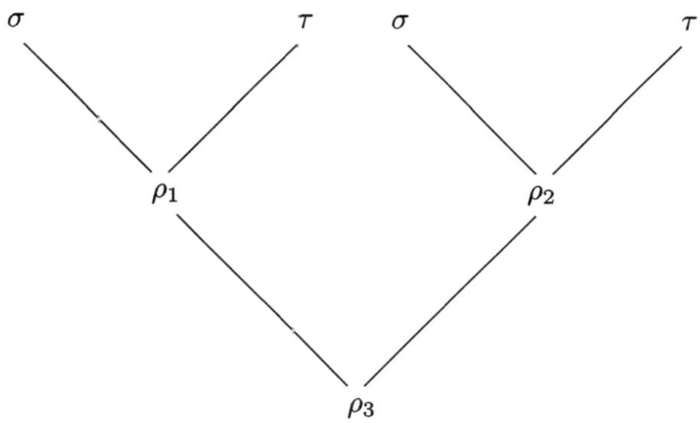

Examples 10.7.2 and 10.7.4

Table 10.1 Table for Example 10.7.1

Fct.	A	B	C	D	Law	from	Prereq.
ρ_1	σ	$\sigma = \tau$	τ		ABC	σ, τ	B
ρ_2	$\rho_1 = \sigma$	$\rho_1 = \sigma = \tau$	$\rho_1 = \tau$	τ	$A(BC)D$	ρ_1, τ	
Result: AB(CD)							

Example 10.7.2
(Bin1), XYZ, $XY'Z$, $Y(XZ)Y' \Rightarrow X(YY')Z$:
(See Diagram 10.7.1 lower part.)

- $\rho_1 : X : \sigma,\ Y : \sigma = \tau,\ Z : \tau$
  generated by XYZ from σ, τ
  prerequisite Y
  common part: Y
- $\rho_2 : X : \sigma,\ Y' : \sigma = \tau,\ Z : \tau$
  generated by $XY'Z$ from σ, τ
  prerequisite Y'
  common part: Y'
- $\rho_3 : Y : \rho_1 = \sigma = \tau,\ X : \rho_1 = \rho_2 = \sigma,\ Z : \rho_1 = \rho_2 = \tau,\ Y' : \rho_2 = \sigma = \tau$
  generated by $Y(XZ)Y'$ from ρ_1, ρ_2
  prerequisites -
  common part: YY'
  ρ_3 can be interpreted as a derived triple by $X(YY')Z$.

Diagram 10.7.2

Example 10.7.3

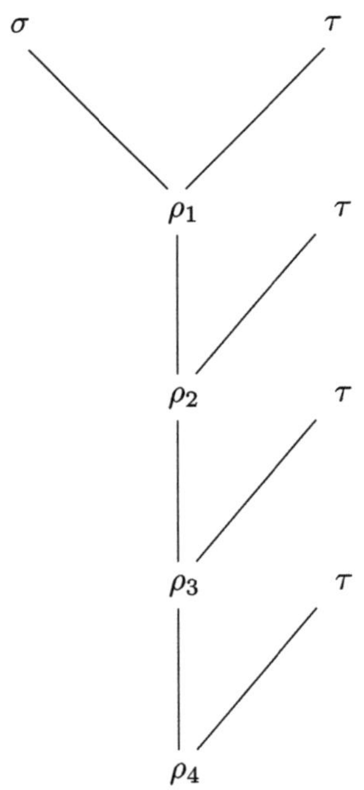

ρ_1, *using ABC* - $A : \sigma, B : \sigma = \tau, C : \tau$

ρ_2, *using ACD* - $A : \rho_1 = \sigma, C : \rho_1 = \tau, D : \tau$

ρ_3, *using ADE* - $A : \rho_2 = \sigma, D : \rho_2 = \tau, E : \tau$

ρ_4, *using AEB* - $A : \rho_3 = \sigma, E : \rho_3 = \tau, B : \sigma = \tau$

Interpretation: ABE*, common part* $B : \sigma = \tau$

Example 10.7.3
(Loop1) ABC, ACD, ADE, AEB $\Rightarrow$ ABE:
(See Diagram 10.7.2.)

- $\rho_1 : A : \sigma,\ B : \sigma = \tau,\ C : \tau$
generated by ABC from $\sigma,\ \tau$
prerequisite B
common part B
- $\rho_2 : A : \rho_1 = \sigma,\ C : \rho_1 = \tau,\ D : \tau$
generated by ACD from $\rho_1,\ \tau$
prerequisite -
common part -
ρ_2 cannot be interpreted as a derived triple, as there was a prerequisite used in its derivation (B), but the common part in ρ_2 is $\emptyset$.
- ρ_3 similar to ρ_2 :
$\rho_3 : A : \rho_2 = \sigma,\ D : \rho_2 = \tau,\ E : \tau$
generated by ADE from $\rho_2,\ \tau$
prerequisite -
common part -
ρ_3 cannot be interpreted as a derived triple, as there was a prerequisite used in its derivation (B), but the common part in ρ_3 is $\emptyset$.
- $\rho_4 : A : \rho_3 = \sigma,\ E : \rho_3 = \tau,\ B : \sigma = \tau$
generated by AEB from $\rho_3,\ \tau$
prerequisites -
common part B
ρ_4 can be interpreted as the common part B contains all prerequisites used in its derivation. ABE is the only non-trivial derived triple.
Note that we could, e.g. also have replaced ACD by $AC'(DC'')$, where $C = C' \cup C''$, using rule (Bc), Weak Union.

Example 10.7.4
$BA(CD),\ DF(CE),\ (AB)(CD)(EF) \Rightarrow B(ADF)(CE)$:
(See Diagram 10.7.1 lower part.)

This example shows that we may need an assumption in the interior of the tree (in the construction of ρ_3, we use $D : \sigma = \tau$; Tables 10.2 and 10.3).

- $\rho_1 : A : \sigma = \tau,\ B : \sigma,\ C : \tau,\ D : \tau$
generated by $BA(CD)$ from $\sigma,\ \tau$
prerequisites A
common part A
- $\rho_2 : C : \tau,\ D : \sigma,\ E : \tau,\ F : \sigma = \tau$
generated by $DF(CE)$ from $\sigma,\ \tau$
prerequisite F
common part F

Table 10.2 Table for Example 10.7.3

Fct.	A	B	C	D	E	Law	from	Prereq.
ρ_1	σ	$\sigma = \tau$	τ			ABC	σ, τ	B
ρ_2	$\rho_1 = \sigma$		$\rho_1 = \tau$	τ		ACD	ρ_1, τ	
ρ_3	$\rho_1 = \rho_2 = \sigma$			$\rho_2 = \tau$	τ	ADE	ρ_2, τ	
ρ_4	$\rho_1 = \rho_2 = \rho_3 = \sigma$	$\sigma = \tau$			$\rho_3 = \tau$	AEB	ρ_3, τ	

Result: ABE

Table 10.3 Table for Example 10.7.4

Fct.	A	B	C	D	E	F	Law	from	Prereq.
ρ_1	$\sigma = \tau$	σ	τ	τ			$BA(CD)$	σ, τ	A
ρ_2			τ	σ	τ	$\sigma = \tau$	$DF(CE)$	σ, τ	F
ρ_3	$\sigma = \tau$	σ	τ	$\sigma = \tau$	τ	$\sigma = \tau$	$(AB)(CD)(EF)$	ρ_1, ρ_2	D

Result: B(ADF)(CE)

- ρ_3 : A: $\rho_1 = \sigma = \tau$, $B : \rho_3 = \sigma$, $C : \rho_1 = \rho_2 = \tau$, $D : \rho_1 = \rho_2 = \sigma = \tau$, E : $\rho_2 = \tau$, $F : \rho_2 = \sigma = \tau$
  generated by $(AB)(CD)(EF)$ from ρ_1, ρ_2
  prerequisite D
  common part ADF
  So ρ_3 can be seen as the derived triple $B(ADF)(CE)$ (but NOT as $(AB)(DF)(CE)$, etc., as DF does not contain ADF.

Example 10.7.5
$(AA')BC$, $AD(CD')$, $(AB')C(C'D)$, $(A'B')C(C'D')$, $(AD)(B'CC')(A'D')$, $BC(ADD') \Rightarrow A(BD)(CD')$:

(See Diagram 10.7.3.)

This example shows that we may need an equality (here α and β in the construction of ρ_5) which is not related to σ and τ. Of course, we cannot use it as an assumption, but we know the equality by construction.

α and β will not be known, they are fixed, unknown fragments of ρ_1 and ρ_2 respectively.

- $\rho_1 : A : \sigma$, $A' : \sigma$, $B : \sigma = \tau$, $B' : \alpha$, $C : \tau$
  generated by $(AA')BC$ from σ, τ
  prerequisites B
  common part B
  α is the (arbitrary) value of ρ_1 for B'.

- $\rho_2 : A : \sigma,\ C : \tau,\ C' : \beta,\ D : \sigma = \tau,\ D' : \tau$
  generated by $AD(CD')$ from σ and τ
  prerequisite D
  common part D
  β is the (arbitrary) value of ρ_2 for C'.
- $\rho_3 : A : \sigma,\ B' : \alpha,\ C : \tau,\ C' : \beta,\ D : \sigma = \tau$
  generated by $(AB')C(C'D)$ from ρ_1 and ρ_2
  prerequisite -
  common part of σ, τ mentioned: D

- $\rho_4 : A' : \sigma,\ B' : \alpha,\ C : \tau,\ C' : \beta,\ D' : \tau$
  Generated by $(A'B')C(C'D')$ from ρ_1 and ρ_2
  prerequisites -
  common part -
- $\rho_5 : A : \sigma,\ A' : \sigma,\ B' : \alpha,\ C : \tau,\ C' : \beta,\ D : \tau,\ D' : \tau$
  generated by $(AD)(B'CC')(A'D')$ from ρ_3 and ρ_4
  prerequisites -
  (note that equality on B' and C' is by construction of ρ_3 and ρ_4, and not by a prerequisite on σ and τ)
  common part of σ, τ mentioned: D
- $\rho_6 : A : \sigma,\ B : \sigma = \tau,\ C : \tau,\ D : \sigma = \tau,\ D' : \tau$
  generated by $BC(ADD')$ from ρ_1 and ρ_5
  prerequisites -
  common part of σ, τ mentioned: BD
  Thus, ρ_6 may be seen as derived triple $A(BD)(CD')$ (Table 10.4)

Diagram 10.7.3

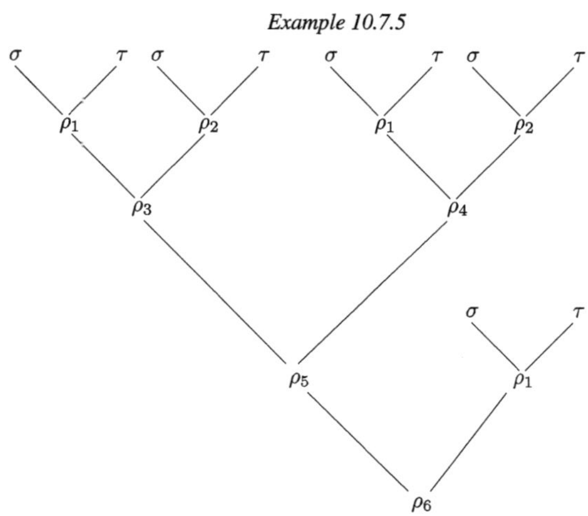

Table 10.4 Table for Example 10.7.5

Fct.	A	A'	B	B'	C	C'	D	D'	Law	from	Prereq.
ρ_1	σ	σ	$\sigma=\tau$	α	τ				$(AA')BC$	σ, τ	B
ρ_2	σ				τ	β	$\sigma=\tau$	τ	$AD(CD')$	σ, τ	D
ρ_3	σ			α	τ	β	$\sigma=\tau$		$(AB')C(C'D)$	ρ_1, ρ_2	
ρ_4		σ		α	τ	β		τ	$(A'B')C(C'D')$	ρ_1, ρ_2	
ρ_5	σ	σ		α	τ	β	$\sigma=\tau$	τ	$(AD)(B'CC')(A'D')$	ρ_3, ρ_4	
ρ_6	σ		$\sigma=\tau$		τ		$\sigma=\tau$	τ	$BC(ADD')$	ρ_1, ρ_5	

Result: A(BD)(CD')

Chapter 11
Formal Construction

11.1 Summary

We construct here a special preferential structure, based on our analysis of various desirable properties of a logic. The basic ideas are more important than the actual realisation of the construction, bringing several ideas together. Different aims will result in different constructions. We present the building blocks, and the construction in Sect. 11.4.1.

- We give an abstract description of the construction and its advantages in this section.
- We discuss several interesting properties with our final construction in mind in Sect. 11.2:
  - the notions of strategy, heuristics, and expectation/surprise, and their epistemic meaning,
  - natural revision,
  - use of (postulated) homogenousness of concepts,
  - modularity, and others.
- We detail desirable properties of the logic to be constructed in Sect. 11.3, in particular:
  - a clear and intuitive semantics
  - inference pressure, the pressure to make as many instances as possible of a default true,
  - graceful degradation,
  - inheritance,
  - a modular and easily adaptable construction,
  - expression of some epistemic considerations,
  - and, above all, taking homogenousness (or rarity of influence change) into account.

D.M. Gabbay and K. Schlechta, *A New Perspective on Nonmonotonic Logics*,
DOI 10.1007/978-3-319-46817-4_11

- We turn to the formal construction in Sect. 11.4.
  The construction is basically a six-step process, which we describe by a toy example now. We have birds and penguins. Birds normally fly, birds normally do not dive. Penguins do both.
  Some ducks dive and fly—but we will not discuss the class of ducks as an exceptional subset.

(1) Attached defaults:
We have the sets birds, penguins, and penguins-birds (set difference), say $B, P, B - P$. The defaults attached to B are "fly", "don't dive". The defaults attached to P are "don't fly", "dive".
See Definition 11.3.1, here $\mathcal{A} = \{B, P\}$, the set of sets to which defaults are attached.

(2) Valid defaults:
The defaults valid at B and $B - P$ will be "fly", "don't dive", the defaults valid at P will be "don't fly", "dive".
More precisely, the defaults "fly", "don't dive" are inherited downward to P, but are contradicted and overruled by the more specific defaults attached to P. The choice of the valid defaults is made by inheritance, conflicts are solved by (some) theory revision strategy.
See Definition 11.4.1.

(3) All elements in $B - P$ and P should satisfy their respective defaults: elements in $B - P$ those valid in B, elements in P those valid in P. In our toy example, $\mathcal{R}_f$ as defined in Definition 11.4.2 will be $\{B - P, P\}$.

(4) Exceptions *not* explained by subclasses and their defaults:
There are birds which do not fly, but are not penguins (ostriches etc.), and birds which dive, but are not penguins (some ducks etc.). For the sake of argument, suppose there are some birds which do not fly, dive, but are no penguins.
We order the class $B - P$ now, e.g. as follows:
The best elements, the set denoted $\mu(B - P)$ as usual, are those which satisfy all defaults valid at $B - P$, the set of the others will be denoted $\omega(B - P)$. The second best are those which satisfy all defaults except 1, the third best are those which satisfy all defaults except 2, etc. This takes care of graceful degradation and other important properties. We will denote this order by $\sqsubset$ see Definition 11.4.3. This is the intra-group relation.
Those elements in $B - P$, which do not satisfy all defaults, are surprises, as we have no explanation for their abnormal behaviour—they are not in P, in our toy example. So, we will put them above the normal P elements, they are even more exceptional than the explained exceptions. This takes care of the epistemic aspect.
(Similarly, we will order elements of P, e.g. penguins which fly, contrary to expectation, etc.)

(5) We order $\{B - P, P\}$, in the general case $\mathcal{R}_f$, as follows:
We put P above $B - P$, as P is the more specific set than B, and penguins are more exceptional than birds, more precisely, as P is in more sets in $\mathcal{A}$ (namely,

in P and B), than $B - P$ is (which is only in $B \in \mathcal{A}$). This order is denoted $\lhd$, see Definition 11.4.4. This is the inter-group relation.

(6) We are now ready to construct the final order, written $\prec$, see Construction 11.4.1.

  (6.1) The best elements $\mu(B - P)$ are totally normal birds, they fly and do not dive.
  (6.2) The second best elements are totally normal penguins, $\mu(P)$, they are exceptional birds, but we know why they behave the way they do. We follow the relation $\lhd$.
  (6.3) Birds which do not fly (or dive), $\omega(B - P)$, will be put above normal penguins, as they are surprises. Birds which do not fly and do dive, will be put even above the latter birds, as they are doubly surprising, they satisfy even less defaults valid for birds. We follow here the relation $\sqsubset$.
  (6.4) (Roughly) Above those, we put surprising penguins, $\omega(P)$, etc.

- The construction is discussed in Sect. 11.5, and, in particular, compared to the list of desirable properties, as detailed in Sect. 11.3.
- What we achieve

(1) A clear and well-understood semantics with nice logical properties in the form of a special preferential structure.

  – In contrast to general preferential structures, our construction is not purely algebraic, but guided by language, inheritance, and conflict resolution through theory revision.
  General preferential structures can be compared to the bizarre example 1.2.1 which has little normal properties, as both are arbitrary compared to our usual concepts of thinking.

(2) Throughout, the construction is guided by our basic principle that changes are rare. This is reflected in:

  (2.1) downward inheritance of defaults, see Definition 11.4.1,
  (2.2) the treatment of subideal cases, by the relation $\sqsubset$, see Definition 11.4.3,
  (2.3) resulting general graceful degradation.

(3) A notion of consistency: conflicts are seen and resolved in a principled and intuitive way, thus eliminating cyclic behaviour as possible with Reiter defaults.
(4) Epistemic considerations are expressed by putting the “unexplained” exceptions higher up.
(5) The construction is modular and flexible, and can be adapted and extended easily, for instance to

  (5.1) negated defaults,
  (5.2) the first order situation, etc.

(6) We combine the expressive power and natural properties of defaults and inheritance systems with the good logical properties of preferential structures.

11.2 Discussion of Various Properties

11.2.1 Inference Pressure

11.2.1.1 Introduction

Minimal preferential logic is static, the relation is given. More dynamic systems like Reiter defaults try to make as many instances true as possible. This is a fundamental difference. In particular, there must be something to stop them, if the rule is not equivalent to a universal quantifier. Let us call the rule which tries to make the instances true the positive agent, and the stopping "something" the opposing or negative agent. Of course, there may be several positive agents, and several negative agents. The negative agent may also be reality, when the instance just simply is not true. The agents may also be real, even human, agents in an argumentation system.

The positive agent or agents may launch a question whether there is any opposition, or trust the whole system (e.g. the brain) that someone might come up with opposition by itself, or the positive agent may believe that he knows already all the exceptions.

If we can assume that, in most cases, the positive agent is right, we can build a semantics using a "dynamic filter" which gets better by the superset relation. If not, we can still have a dynamic system without the filter property. Moreover, in many cases, this relation of "better" is not purely algebraic, but defined by the language, and the defaults, as done in our main construction (see Construction 11.4.1). In addition, if ϕ holds, e.g. in a set of cases X, and x is close to X (by whatever criterion), we will be more courageous to extend to x, than if X were distant from x. Thus, a better treatment of the concept requires more information.

Reality should certainly win, but a conflict between several agents may be decided by the (best possible) comparison in a partial order, we leave this open; in many cases, specificity will decide. We also leave open whether the property to maximise is on the object or the meta-level. We try to be as general as possible, and the reader will probably come up with cases we did not consider here.

11.2.1.2 A More Systematic Approach

There seem to be at least four questions:

(1) What should be maximised?
 In particular, object or meta level information?
(2) Which agents try to maximise?
(3) Which agents are opponents?
(4) How are conflicts resolved?
 Is there a superior instance which solves conflicts?

We discuss some examples:

(1) Inheritance systems

(1.1) The information attached to birds tries to maximise e.g. the property "flying".
The information attached to penguins tries to maximise "not flying".
Specificity resolves the conflict.

(1.2) The information attached to Republicans tries to maximise "hawkish behaviour".
The information attached to Quaker tries to maximise "pacifist behaviour".
They are of equal strength, no resolution on this level is possible.

(2) Reiter defaults

(2.1) $\{: \phi, : \neg\phi\}$ has no solution.

(2.2) We can make specificity decide.

(3) Autoepistemic logic

(3.1) This seems to be a special case of a reasoning strategy, see below.

(3.2) Counter examples will come up "automatically" (e.g. in a breadth first search, or we "ask" special reasoning agents for the counter examples first), and may be easier to detect than it is to execute the positive search.

(4) Negation as failure

(4.1) This can be seen as a strategy or heuristics, but it might also model a real world property, where, for instance, it is easy to achieve positive properties, like a disturbance, but difficult to achieve the opposite, a disturbance free situation.

(5) Homogenousness or rarity of influence change
We mention this separately because of its importance in this book.

(5.1) A meta-property is maximised.

(5.2) The background assumption is that our concepts correspond to the world, and that the world is orderly, we try to push this as far as possible, but know that this is not always true.

(5.3) Reality might confront us with "local" failure, but we will still assume that the assumption is mostly true.

(5.4) An expert might also point out a local failure, without questioning the basic assumption.

(6) Expectations

(6.1) We trust our expectations sufficiently well to be surprised by their failure. In this sense, they are a heuristics, which is successful very often.

(6.2) Failure happens in a tiny amount of cases, where "tiny" means, e.g. "small of small".

(6.3) The surprise might result in a re-orientation of the brain, and is thus beyond a logical approach like theory revision.

(7) Reasoning strategies

(7.1) We see a reasoning strategy as a reasonable way to proceed. "Reasonable" might, e.g. mean risk avoidance. This is a question of utility, we assume that failure is detected sufficiently well, or is not so damaging.
(7.2) A strategy might be contradicted by reality, or an other reasoning agent, an expert, special knowledge of the agent itself, etc.
(7.3) There is no general way to resolve conflicts.

(8) Heuristics

(8.1) We see heuristics as similar to strategies, with the additional property that, in most cases, a heuristics is successful. This quantitative aspect was absent for strategies.

(9) Liar paradox

(9.1) We can see the liar paradox as trying to extend the statement "I lie" as much as possible, without a paradox.
(9.2) The limit will then be the meta-level (Table 11.1).

(10) Argumentation

(10.1) Argumentation can be about anything, object or meta-level, as such, it is very general.
(10.2) The agents may be equals, experts in their fields, there might be arbitrars or not, history of past arguments and their quality, etc.
(10.3) This is perhaps the most general approach with, e.g. a partial order between the arguments.

(11) "Conversation of experts"

(11.1) This can be seen as a special case of argumentation.
(11.2) The working of the higher brain has also often been described as a conversation of different areas of the brain, without any superior instance to regulate conflicts.
(11.3) Usually, there is no "meta-expert".

11.2.1.3 Inference Pressure as Operator "As Much As Possible"

We can introduce a new operator, say $\spadesuit\phi$, with the meaning: "Make as often as possible ϕ true". This would be similar to a Reiter default $: \phi/\phi$, but have some properties of the "most" operator ∇, too. Of course, there is then a relativised version,

Table 11.1 Table Inference pressure

Inference pressure			
Object level (o), meta level (m), Reality (R)			
	Type	Opponent Other agents (O)	Conflicts Specificity (Sp) Supervisor (Su) automatic (a)
Inheritance	o	o	Sp
Defaults	o	o	Sp
Autoepistemic	o	o	a
Neg. as failure	o	o	a
Homogenousness	m	R, O	Sp, R
Expectations	o, m	R, O	Sp, R
Strategies	o, m	R, O	a, R
Heuristics	o, m	R, O	Sp, R
Liar paradox	o, m	m	m
Argumentation	o, m	R, O	Sp, Su
Expert conversation	o, m	R, O	Sp, Su

too, like other relativised quantifiers. Seen on the background of above discussion and examples, it would be underspecified. For instance: Who "wants" ♠ϕ? Who determines if it is possible? Reality? Other agents, and if so, which agents? Who or what solves conflicts? Specificity, a superior agent? Do those other agents have to interfere by themselves, or does the "positive" agent have to ask them? Consider a group of experts. Does every expert voice his/her opinion, or do experts consult each other actively? Semantically, it could be expressed by a (meta) preference between preferential structures, preferring those structures with more arrows (in the right direction). Again, this would be underspecified without indicating limitations, etc. More abstractly, we might have a dynamic filter, where bigger sets are preferred, with all the above problems of underspecification.

The term "as much as possible" itself does not seem to pose problems, rather the additional information required to make it precise. A procedural approach does not seem to make these questions easier, only to transfer them from a static (results) to a dynamic problem (how the results are obtained).

11.2.2 Remarks on Various Systems

11.2.2.1 Strategy

It might be a useful strategy to treat some cases differently from others. This might have several reasons, e.g.:

(1) there is an expert (or an efficient algorithm) which allows to treat some cases in a fast way,
(2) the negative or positive consequences of one case are so important that it makes the case and its negation asymmetrical, so erring in one direction is much better than erring in the other direction—this is a question of utility.

As the knowledge of an expert (or algorithm) can be arbitrarily crooked, we should not always expect good logical properties. But, if it is to be integrated in general human reasoning, at least the boundaries between the different strategies should correspond to usual concepts—and thus to the world, by our basic assumption of this fit. There is the problem of deciding which reasoner is called upon. In autoepistemic logic, it is written down in the database. In the brain, it might be decided by which one of the parallel processes comes up first with a result. Experts may decide among themselves in a decentralised way. When a logic is only a strategy, partially determined by factors outside the object world, it seems doubtful whether, e.g. inheritance, or specificity, are a valid criterion for subsets and conflict resolution. Behaviour for subsets can then be very different from the original behaviour.

11.2.2.2 Heuristics

Heuristics work most of the time, but we do not really understand why, otherwise it would be a law. They indicate that our concepts fit well with the world. Thus, they are more than a strategy, as they are correct most of the time, and not only an adaptation to our reasoning processes. They are a good approximation to reality, thus useful. Heuristics can collide with reality. But also another agent can tell us, that in a certain area, they will not work. Until we find a contradiction, we will suppose them to be true. As a heuristics says something about the world, a logic which is a heuristics can be wrong, like any other theory about the world. A dynamic filter will be a good semantics for a heuristics, and determine the laws governing a heuristics. A heuristics has "inference pressure", but perhaps not as strongly as discussed above, and is perhaps more tightly bound to reality than some inference mechanisms "with pressure". A good heuristics may take the cost of errors into account, and thus be based (among other things) on a notion of utility. Generalising from some samples (data from the real world, or "deep" theoretical analysis) to a larger population is also a heuristics, but beyond the subject of this book.

11.2.2.3 Expectation and Surprise

Expectations are somewhat different from heuristics. If they fail, we are surprised.

First, the cases where they fail should be very rare: a small subset of a small subset might formalise this. Note that this corresponds well to our treatment in the main definition (see Sect. 11.4, Construction 11.4.1). Sets of surprises should not be closed under arbitrary finite unions, otherwise, it would be surprising that someone wins the lottery. Only unions which are themselves still small of small should be surprises. We know that someone wins the lottery, but would be surprised if *we* win the lottery. So rarity alone is probably not sufficient, but the surprising event should have some important impact. In addition, we might know exceptions, like penguins, so being an exception, even a rare one, is not necessarily a surprise. But discovering a new species of eagle which does not fly, would be a surprise. So surprise is related to knowledge (and expectations). This (with all aspects) is perhaps best modelled with possible worlds, describing developments of the actual world, and of our knowledge of it.

Traditional logical approaches to the incorporation of contradicting information like theory revision are probably inadequate here, the human reaction seems to go beyond logic. It seems that the brain undergoes some modification (synaptic changes). This is illustrated by the reaction to the news of Kennedy's assassination, see Sect. 8.3.4.3, and also the anecdote about Einstein's boyhood, when he was first given a compass, which seems to have had a profound effect on him, revealing hidden connections "behind the scene". (The second author may add from his personal experience: When he understood a general, important concept, e.g. in mathematics, he often felt "unfamiliar" in everyday life afterwards, everything had a touch of "new".)

11.2.2.4 Argumentation

Argumentations are perhaps the most general of the logics discussed here. They can be about the object or the metalevel, agents may be more or less reliable, counterarguments may attack results or ways the results were achieved, even other agents themselves (e.g. as unreliable), unrestricted argumentation may lead to circular behaviour, like unrestricted Reiter defaults, and thus have similar problems and semantics.

Questions like:

(1) field of expertise of the agents,
(2) trust into agents,
(3) temporal order of arguments,
(4) significance of the argument

are important here.

11.2.2.5 Confirmation

We should note that, instead of opposing agents, we might also have (even only) supporting agents. In particular, positive cases in the neighbourhood of an open case may lend credibility to decide the latter positively, too. This is, of course, different from inductive logic, as we already have the hypothesis, and only try to evaluate it.

We do not pursue this, as it is beyond the main direction of this book.

11.2.3 Further Properties

In the Sect. 11.2.1, we looked at different types of logics which express inference pressure.

We now enumerate some other logical properties, but in a less systematic way.

(1) Does a logic have its own natural revision? Some logics have a natural revision "built in".

  (1.1) For instance, when the set of "best models" in a preferential structure is too small, we can try the second best (below them are only best models), etc.
  (1.2) $\forall$ can be weakened to ∇.
  (1.3) Graded deontic logic (contrary to duty obligations), see Chaps. 6 and 13, is exactly about its own revision.
  (1.4) In inheritance systems, adding a contradictory direct link will negate a property in a strong sense, i.e. adding the opposite. If we just want to contradict, we have to insert a path of length at least 2, as in the Nixon diamond.
  (1.5) Logics which are basically heuristics would not necessarily have a logically nice revision, as we would rather revise the heuristics.
  (1.6) Note that "surprises" may not have any logical revision at all, as our brain seems to work "a-logically" when surprised—see above remark.

(2) Dynamic logics (Reiter defaults, etc.) vs. static static logics (minimal preferential structures).
(3) Logics driven only by the objects (preferential structures), or by the inference mechanism, too (autoepistemic logic).
(4) Do the logics allow a change of the base set? Restricted, like preferential structures, or liberal, like Reiter defaults?
(5) Does the logic have an underlying concept of homogenousness (like Reiter defaults, but not preferential structures)?
(6) Can the logic be broken down into independent parts (like inheritance, Reiter defaults), resulting in graceful degradation?
(7) Does the logic have a degree of varying confidence in the conclusion (like in Bayesian reasoning)?
(8) Are the results closed under logical operators, like AND, OR, etc.?
(9) Can the logic be translated naturally into an object language operator like ∇ for preferential structures, with accompanying laws, like those for size?
(10) Is the logic strongly language dependent (like Reiter defaults), or basically algebraic (as preferential structures)?
(11) Are object and meta level clearly separated? (Natural argumentation does not always do this.)

11.3 Desiderata

11.3.1 Overall Aim

We want to construct a logic which combines the advantages of several existing logics, and we also want to add some new features. In particular, we try to achieve:

(1) a clear and intuitive semantics,
(2) a modular and flexible construction, which is easy to modify for different purposes,
(3) a clear and intuitive handling of contradictions,
(4) a notion of (in)consistency—e.g. a default "if ϕ, then normally ψ" should only hold if there are "sufficiently" many ϕ-cases where ψ holds,
(5) combine the expressive power of Reiter defaults and inheritance systems with the nice logical properties of preferential structures,
(6) express some epistemic concepts.

It should

(1) be based on a notion of rarity of influence changes,
(2) allow graceful degradation for changes of the base sets, i.e. allow treatment of subideal cases,
(3) be adaptable to the first-order case,
(4) be able to express negated defaults—e.g., "it is not the case that normally, if ϕ holds, then so does ψ" without committing to $\neg\psi$.

We will achieve this by using essentially a defeasible inheritance system as scaffolding to build a preferential structure.

11.3.2 Situation

We describe here the situations we will treat in an overview, and will consider "rare influence change" in more detail in Sect. 11.3.3.

(1) Negated and "attached" defaults
We will express negated defaults by formulas like $\phi \not\mid\!\sim \psi$. For instance, in a preferential structure, in some minimal ϕ-models ψ will hold, in others $\neg\psi$ will hold. Negated defaults arise naturally in inheritance systems, e.g. in the Nixon diagram. Some people are Quakers and Republicans, Quakers are normally pacifists, Republicans not, so we have contradictory information, and may assume that some of those people will be pacifists, and some not. Neither the default pacifist nor the default not-pacifist will hold for this set. We allow to introduce negated defaults explicitly. Negated defaults allow to "break" the inheritance mechanism.

So, the general picture is as follows. We have default information $\alpha \mid\sim \phi$, $\alpha' \mid\sim \phi'$, $\alpha'' \not\mid\sim \phi''$, etc., and classical relations like $\alpha \vdash \alpha'$, $\phi' \vdash \alpha'' \wedge \phi''$, etc. Some of those may be expressed by a defeasible inheritance network, some not.

Definition 11.3.1
If $\alpha \mid\sim \phi$ or $\alpha \not\mid\sim \phi$, we say that the default ϕ is attached to α.

Given any fixed default theory, let $\mathcal{A}$ be the set of α (or $M(\alpha)$), to which some default is attached.

(2) Information transfer
Properties of logics are often properties of information transfer. We know that elements of set X have a certain property ϕ, and want to conclude that elements of set X' have the same or a different property ϕ'.
In classical logic, we have the monotony rule. In preferential logic, we have properties like cautious monotony, rational monotony, etc. Reiter defaults and, implicitly, defeasible inheritance systems allow much more information transfer, but this is defeasible itself. We are allowed to make information transfer until we obtain a contradiction. This is, essentially, an encoding of rarity of influence: things stay the same, until there is a reason they must change, a new influence. This happens, but not very often.
Note that we argue here from supersets to subsets, so this is a downward transfer. Analogical reasoning is a "sideways" transfer, whereas induction goes upwards. But we do not develop a general theory of information transfer here.

(3) We see three different possible epistemic phenomena:

   (3.1) We know that possible exceptions are expressed in small language fragments, ("penguins") and the normal case ("birds") in a small language fragment, too. So we can treat the normal case and the exceptions separately, each in a simple language. But we may not have examined whether we really have an exceptional case. We just know where to look, e.g. the beginning ornithologist will know which book to consult, see Sect. 11.6.
   (3.2) We know the exceptions, not only where to look for them, but we actually are familiar with them. This is the case for birds which do not fly, they are penguins, emus, dead, in a cage, etc. This is the situation of an expert, who knows his field.
   (3.3) There are surprises, e.g., birds which do not fly, but they fall in none of the known categories of exceptions. We may consider them very exceptional cases. They might also be difficult to detect, we see them more or less by chance reasoning or discovery.

(4) Rare influence changes
Our underlying assumption is that influence changes are rare, when we go from one set to another, and that our language is chosen to make this true. Our language serves to bring order into chaos—see example 1.2.1 for illustration, and below for a more detailed discussion.

(5) Nice logical properties, like OR
The advantage of constructing a special preferential structure is that the final result has nice and well-understood logical properties (like OR etc.), valid for all preferential logics, whereas we are quite free in the construction itself.

11.3.3 Rare Influence Changes and Its Consequences

As said, we construct a preferential structure based on inheritance of properties to subsets, which we base, in turn, on rarity of influence changes—assuming our concepts are useful.

(1) Rare influence changes:
Our main hypothesis is that influence changes are rare, absolutely, and relative to our knowledge.
If we know that birds fly, and consider European birds, or ravens, or birds of a certain colour, we assume that all those additional properties have no influence on flying. And if they do, we will probably know this. We know that penguins do not fly, neither do dead birds. But we do not exclude that some birds do not fly for unknown reasons, but such cases are surprises. Perhaps we can even deduce this fact from available information, but we cannot be certain that we make this deduction.
(This models the chance element in the functioning of the brain, as well as the cognitive routines, and learning. The chance element provides flexibility for (surprise) learning.)
In addition, if one property changes, we do not assume that other properties do, too (unless they are related). Penguins do not fly, but they probably still lay eggs, etc. The more properties change, the more reasons we expect to be for the changes. ("Reason" may also be a special subclass, here of the class of birds. We assume that our concepts are well chosen!)
Reality does not make a distinction between object level and meta level rules (like default transitivity), logic has to reflect its domain, and not our habits of doing logic.
Note that there are also spurious cases: inhabitants of a certain house in Stockholm might all be small and dark-haired, without any particular reason—just as it happens to win the lottery.

(2) Subideal cases, graceful degradation, continuity, approximation:
General preferential structures are very rigid. They allow to change the set considered only as far as the size notion permits. If $\alpha \mid\sim \phi$, then we cannot change α much. This is expressed by the blond Swedes problem. Swedes are tall and blond. But we would expect even not blond Swedes to be tall. The purely size-defined laws of general preferential structures do not allow this. Not blond Swedes are a subideal case, still, we want to preserve as much as possible, rare influence changes provide a justification: We have to give up one property, but need not give

up others. This can be described as graceful degradation, approximation, or continuity, where abrupt changes are unlikely. (Of course, there are non-continuous situations, too. Adding "dead" changes many properties (and for many (classes of) objects)).
We summarise this:

(2.1) If $\alpha' \vdash \alpha$, and $\alpha \mid\sim \phi$, then we expect $\alpha' \mid\sim \phi$ still to hold, unless we have information to the contrary.
We try to preserve everything which is consistently possible. This will be achieved by a suitable order between models.

(2.2) If $\alpha' \vdash \alpha$, and $\alpha \mid\sim \phi$, $\alpha \mid\sim \phi'$, then we expect $\alpha' \mid\sim \phi$ still to hold, even if $\alpha' \not\mid\sim \phi'$, unless we have information to the contrary.
We try to preserve as much as consistently possible. This will again be achieved by a suitable order between models (better models preserve more properties).

(2.3) In the same spirit, we can approximate in the FOL case $\forall$ by ∇. We see $\forall$ as $\bigwedge$, and try to save as many cases as possible. We have to code the meaning of $\forall$ into its treatment by the logic, just as it is done in classical FOL—we cannot have something for nothing.
If $\alpha \mid\sim \phi \wedge \phi'$, we can try to save ϕ', even if we have to give up ϕ, making the defaults syntax independent, i.e. $\alpha \mid\sim \phi \wedge \phi'$ is equivalent to $\{\alpha \mid\sim \phi, \alpha \mid\sim \phi'\}$.
In the finite case, we can re-write ϕ as the finest conjunction of ϕ_i, where the ϕ_i each eliminate one model of $\neg\phi$.
The justification is, again, rarity of influence change.

(3) Inheritance

(3.1) We want to have a sound core for the defaults.
Thus, if $\alpha \mid\sim \phi_1, \ldots, \alpha \mid\sim \phi_n$, we postulate that $\{\alpha, \phi_1, \ldots, \phi_n\}$ is (classically) consistent.

(3.2) Conflicts, and resolution by reliability (specificity).
When we inherit defaults $\alpha \mid\sim \phi$ downwards to subsets (i.e. α' with $\alpha' \vdash \alpha$), conflicts are possible.
For instance inheriting the default "fly" from birds to penguins leads to a conflict. Or, inheriting "pacifist" from Quaker and "non-pacifist" from Republican leads to a conflict for Quakers who are Republicans.
This is all well known. We need a principled strategy to resolve conflicts, which may also involve more than two formulas, as we work in full logic. Specificity (also based on rarity of influence change) allows to resolve some conflicts. There might be other criteria, we leave this open, and assume that we have a strategy. Note that we have a new type of information, $\alpha \not\mid\sim \phi$, our strategy has to apply to this case, too. Possible conflicts and their solutions by a theory revision approach will be discussed in detail in Sect. 11.4.

(4) Model comparison, different possibilities
An important aspect of our construction will be the way we compare models. Basically, we will prefer those models which satisfy more defaults. This seems the natural way, but it is underspecified. We may

(4.1) compare by subset relation, m is better than m' iff the set of defaults satisfied by m' is a subset of those satisfied by m,
(4.2) compare by cardinality of the defaults satisfied,
(4.3) adopt a more complicated relation, e.g. "lexicographic", giving more priority to defaults considered more important.
This is useful for contrary-to-duty obligations, with primary, secondary etc. obligations.

11.3.4 Varia

(1) Singletons
Applying a negated default to a singleton presents a problem: Either the strong negation of the default is true, or the default itself is true. We can either stipulate that it must be consistently possible, or, that we have at least two copies, Nixon and his twin brother, where one is pacifist, the other is not. (More precisely, this is a possible world approach.)
(2) The set "penguins" is itself abnormal
As $birds \mid\!\sim fly$, but $penguins \mid\!\sim \neg fly$, being a penguin is itself abnormal. This can also be justified by the assumption that influence changes are rare, so this is one of the rare cases.

11.4 The Solution

We work here in propositional logic, and on the semantic level.

We assume a classical background theory B, and a set of classical formulas $\alpha_1, \ldots, \alpha_n$ to which defaults are attached, $\alpha_i \mid\!\sim \phi_{i,1}, \ldots, \alpha_i \mid\!\sim \phi_{i,j_i}$, where some or all of the $\mid\!\sim$ may also be $\not\mid\!\sim$ (without being necessarily $\mid\!\sim \neg$).

Condition 11.4.1
We assume the following consistency conditions:

(1) B is classically consistent
(2) For each α_i the defaults attached to α_i together with B are jointly consistent.
In particular, the theory of defaults attached to one of the α_i must be consistent. For instance, $\alpha_i \vdash \alpha_j$ and $\alpha_i \mid\!\sim \neg\alpha_j$ together are inconsistent. We thus rule out default theories like $\{: \phi/\phi, : \neg\phi/\neg\phi\}$.
A negated default like $\alpha \not\mid\!\sim \phi$ needs a model of $\alpha \wedge \neg\phi$ to be consistent, so we replace $\alpha \not\mid\!\sim \phi$ by $\alpha \wedge \neg\phi$ for the consistency check.

The $\vdash$-relation between the α_i's gives a specificity relation by strict inclusion. We use this for the inheritance relation, and to solve (some) conflicts.

Definition 11.4.1
This definition describes how to obtain a consistent default theory at every point in the universe, using a theory revision approach, with specificity solving some conflicts. Of course, modifications are possible, but the general idea seems sound.

We define the set of valid defaults at some point. This influences the relation $\sqsubset$ as defined in Definition 11.4.3, but not the relation $\lhd$ as defined in Definition 11.4.4, as the latter depends only on the sets to which defaults are attached, and not which defaults are attached.

We consider now some classical formula β—it may be one of the α_i's to which defaults are attached, or not.

(1) Visible defaults at β
The defaults visible at β : All defaults attached to some α_i such that $\beta \vdash \alpha_i$ are considered visible at β.

(2) Valid defaults at β

(2.1) The visible defaults are ordered by the $\vdash$ relation between the α's to which they are attached. The more specific ones are considered stronger defaults (for this β).
Of course, we can plug in here any partial relation, if it seems more suitable, e.g. $\mid\!\sim$ itself, as is often done in defeasible inheritance networks.

(2.2) Consider now the set of visible defaults, together with the classical information available at β. (If $\beta' \vdash \beta$, but not conversely, we need not consider β', etc.) If this set is inconsistent:

(2.2.1) Consider first in parallel all minimal inconsistent subsets involving classical information. They must contain at least one default, as the classical information was supposed to be consistent. Eliminate simultaneously from each such set the weakest (by the $\vdash$ relation) defaults. (If there are several weakest ones, eliminate all the weakest ones.)
As contradictions involving classical, i.e. strongest information, seem to be more serious, we do these sets first.

(2.2.2) Consider now all remaining minimal inconsistent subsets of default information only. (Note that some might already have been eliminated in the previous step, but for other reasons.)
Proceed as in the previous step, i.e. eliminate the weakest information.

(2.2.3) We call the remaining defaults, visible at β, the defaults valid at β.
We will now work with the valid defaults only.

We now define "relevant" sets, sets where some default may change. We work in some fixed default theory, recall from Definition 11.3.1 that $\mathcal{A}$ is the set of α or $M(\alpha)$ to which at least one default is attached.

Definition 11.4.2
Consider the model variant of $\mathcal{A}$.
Let $S(\mathcal{A}) := \{\bigcap \mathcal{A}' : \mathcal{A}' \subseteq \mathcal{A}\}$ and $U(\mathcal{A}) := \{\bigcup \mathcal{A}' : \mathcal{A}' \subseteq \mathcal{A}\}$,
Let $\mathcal{R} := \{X - Y : X - Y \neq \emptyset,\ X \in S(\mathcal{A}),\ Y \in U(\mathcal{A})\}$,
where $X = U$ (the universe), and $Y = \emptyset$ are possible.
$R \in \mathcal{R}$ is called a relevant set (or formula).
Let $\mathcal{R}_f$ be the $\subseteq$-minimal elements of $\mathcal{R}$—"f" for finest.

Obviously, the elements of $\mathcal{R}_f$ are pairwise disjoint. The motivation for $\mathcal{R}_f$ is the following. Let $X - Y \in \mathcal{R}_f$, then all elements of $X - Y$ should satisfy all defaults valid at X, but no other, as they are in no other sets to which defaults are attached.

Example 11.4.1
Let $A, A' \subseteq U$, $A'' \subseteq A'$, let defaults be attached to A, A', A'', let all resulting intersections and set-differences be non-empty, if possible, e.g. $A'' - A \neq \emptyset$.
See Diagram 11.4.1 for illustration.
So $\mathcal{A} = \{A, A', A''\}$,
$\mathcal{R} = \{U, U - A, U - A', U - A'', U - A - A',$
$A, A - A', A - A'',$
$A', A' - A, A' - A'', A' - A - A'',$
$A'', A'' - A,$
$A \cap A', A \cap A' - A'',$
$A \cap A''\}$
and
$\mathcal{R}_f = \{U - A - A',$
$A - A',$
$A' - A - A'',$
$A'' - A,$
$A \cap A' - A'',$
$A \cap A''\}$
It is useful to code the sets A, A', A'' by 3 bits, e.g. the left bit codes A, the middle one A', the right one A''. We then have
$\mathcal{R}_f = \{U - A - A' = 000,\ A - A' = 100,\ A \cap A' - A'' = 110,\ A \cap A'' = 111,$
$A'' - A = 011,\ A' - A - A'' = 010\}$,
where the codes 001, 101 do not exist, as $A'' \subseteq A'$.

We will order the models by groups, and within groups by their "quality" satisfying defaults. We first define the latter relation, to be denoted $\sqsubset$. It will be used to order the models within the sets $\omega(A)$, see Construction 11.4.1.

Definition 11.4.3
Suppose we are at some relevant β, with the valid defaults $\phi_1, \ldots, \phi_n$. We order the models of β according to satisfaction of the ϕ_i. There are different possibilities to define $\sqsubset$:

(1) By subsets: if the set of ϕ_i satisfied by m is a subset of those satisfied by m', then m' is better than m, $m' \sqsubset m$.

(2) By cardinality: if the set of ϕ_i satisfied by m is a smaller than the set of those satisfied by m', then m' is better than m, $m' \sqsubset m$.
(3) Some more complicated order, preferring certain defaults over others. (This might be interesting for contrary-to-duty obligations. Note that we solve here the case of additional information in real-life situations, too.)
(4) In particular, we may order the valid defaults by specificity of the α to which they are attached, and satisfy the most specific ones first, then the next, etc., resulting in a lexicographic order.

Note that the construction is more adapted to sets with rare exceptions, than to a classification, like vertebrates into fish, amphibians, reptiles, birds, and mammals, where all are exceptions—and those outside these subsets are the real exceptions.

Diagram 11.4.1

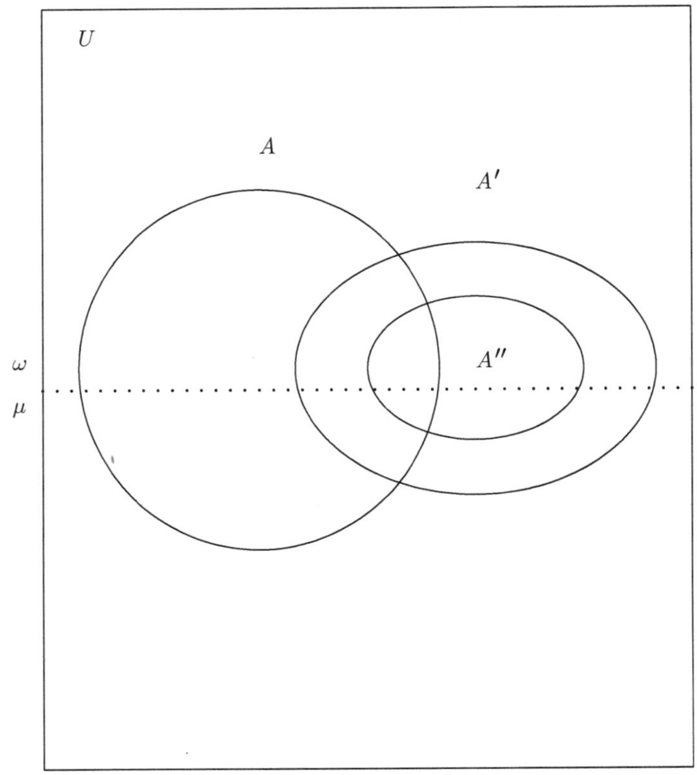

The dotted line separates $\omega(X)$ from $\mu(X)$

Example, the sets

11.4.1 The Construction

Note that, again, the following definition is independent of the actual defaults, and uses only the fact that defaults are attached to the $A \in \mathcal{A}$. The order to be constructed, $\lhd$, will be used as scaffolding in the construction of $\prec$ in Construction 11.4.1, which orders packets of models.

(See example 11.4.2 and Diagrams 11.4.1, 11.4.2.)

Recall Definition 11.4.2.

Definition 11.4.4

We order $\mathcal{R}_f$ as follows:

$X \lhd Y$ iff $\{A_i : X \subseteq A_i\} \subset \{A_i : Y \subseteq A_i\}$ for $X, Y \in \mathcal{R}_f$,

i.e. by the subset relation, of the A_i they are in.

This order expresses exceptionality of the $X \in \mathcal{R}_f$. If $X \lhd Y$, then Y is a subset of more (by the subset relation) A_i than X is, defaults were attached to the A_i, so the A_i are sets of exceptions. For instance, A_0 might be the set of birds, A_1 the set of penguins, being a penguin (an element of A_0 and A_1) is more exceptional than being a bird (an element of A_0 only).

Recall the coding of the $X \in \mathcal{R}_f$ in example 11.4.1. This example is continued in example 11.4.2, and $\lhd$ is then (the transitive closure of) $U - A - A' - A''$ (000) $\lhd$ $A - A'$ (100) $\lhd$ $A \cap A' - A''$ (110) $\lhd$ $A \cap A''$ (111), $U - A - A' - A''$ (000) $\lhd$ $A' - A - A''$ (010) $\lhd$ $A'' - A$ (011) $\lhd$ $A \cap A''$ (111), $A' - A - A''$ (010) $\lhd$ $A \cap A' - A''$ (110).

See Diagram 11.4.2.

We are now ready to construct the preferential relation $\prec$ between model sets, it is implicitly extended to their elements.

Take, e.g. in example 11.4.2 the set $A \cap A' - A''$. It will "see" the defaults valid for $A \cap A'$, but not necessarily those for A''. Every element of $A \cap A' - A''$ can satisfy all those defaults, or only a part of them (or none). The best elements are those which satisfy all, the worst those which satisfy none. The precise relation is described by $\sqsubset$, see Definition 11.4.3. Let us call $\mu(X)$ the set of those elements which satisfy all defaults, the set of the others $\omega(X)$, for given X. Elements which do not satisfy all defaults are doubly exceptional, as they are not in any subclass where this failure is "explained". They are the "unexcused" exceptions, they have no reason to be exceptional, they are surprises, and doubly exceptional. So they should sit higher up in the hierarchy.

We do not think that there is a unique reasonable solution. Two ideas come to mind:

(1) Put all sets $\omega(X)$ above all other elements, in their own hierarchy, defined by $\lhd$, and ordered internally by $\sqsubset$. This is the radical approach.
(2) A less radical idea is to put them above the immediate $\lhd$-successors of X. Example $\omega(A' - A - A'')$ will sit above $A \cap A' - A''$ and $A' \cap A'' - A$—or, rather, above $\mu(A \cap A' - A'')$ and $\mu(A' \cap A'' - A)$. This is what we will do here.

Construction 11.4.1
(See example 11.4.2 and Diagram 11.4.3.)
We define a "packetwise" order $\prec$ between model sets, it will then be extended to the elements.

The relation $\prec$ has two parts:

(1) $\mu(X) \prec \mu(Y)$ iff $X \lhd Y$
for $X, Y \in \mathcal{R}_f$
(2) $\mu(X) \prec \omega(Y)$ iff $X \lhd Y$ or $X = Y$ or X is a direct $\lhd$-successor of Y
for $X, Y \in \mathcal{R}_f$.

Note:

(1) We do not continue the order above the $m \in \omega(A)$, they are maximal elements—except for the interior order $\sqsubset$ among themselves.
(2) The $\mu(X)$ are "flat", all are best possible, There is no interior order in this set $\mu(X)$.
(3) The inside structure of the $\omega(X)$ is given by $\sqsubset$ according to Definition 11.4.3 for this X.
(4) If any of the sets is empty, we just omit it and close under transitivity.

More formally, for $X, X' \in \mathcal{R}_f$, and element wise:

$$m \prec m' := \begin{cases} \text{there is } X \text{ and } m, m' \in \omega(X) \text{ and } m \sqsubset m' \\ or \\ \text{there are } X, X' \text{ and } m \in \mu(X) \text{ and } m' \in \mu(X') \text{ and } X \lhd X' \\ or \\ \text{there are } X, X' \text{ and } m \in \mu(X) \text{ and } m' \in \omega(X') \text{ and} \\ (X \lhd X' \text{ or } X = X' \text{ or } X \text{ is a direct} \lhd -\text{successor of } X') \end{cases}$$

The final order between models is thus basically lexicographic with two parts, first by $\lhd$, then by $\sqsubset$.

Example 11.4.2
We continue example 11.4.1, see also Diagram 11.4.1.

Recall that $\mathcal{A} = \{A, A', A''\}$, and $\mathcal{R}_f = \{U - A - A' - A'' = 000,\ A - A' = 100,\ A \cap A' - A'' = 110,\ A \cap A'' = 111,\ A'' - A = 011,\ A' - A - A'' = 010\}$.

(1) We define $\lhd$ by Definition 11.4.4: $U - A - A' - A'' = 000 \lhd A - A' = 100 \lhd A \cap A' - A'' = 110 \lhd A \cap A'' = 111, U - A - A' - A'' = 000 \lhd A' - A - A'' = 010 \lhd A'' - A = 011 \lhd A \cap A'' = 111, A' - A - A'' = 010 \lhd A \cap A' - A'' = 110$, closed by transitivity.
See Diagram 11.4.2.

(2) We construct the relation between models according to Construction 11.4.1. as follows (considering only the model sets, and neglecting the internal structure of the $\omega(X)$). Constructing the full relation (and closing under transitivity) is then trivial.
$\mu(000) \prec \mu(100) \prec \mu(110) \prec \mu(111)$, $\mu(000) \prec \mu(010) \prec \mu(011) \prec \mu(111)$, $\mu(010) \prec \mu(110)$,
$\mu(100) \prec \omega(000)$, $\mu(010) \prec \omega(000)$, $\mu(110) \prec \omega(100)$, $\mu(110) \prec \omega(010)$, $\mu(011) \prec \omega(010)$, $\mu(111) \prec \omega(110)$, $\mu(111) \prec \omega(011)$.
Finally, we close under transitivity.
See Diagram 11.4.3.

Diagram 11.4.2

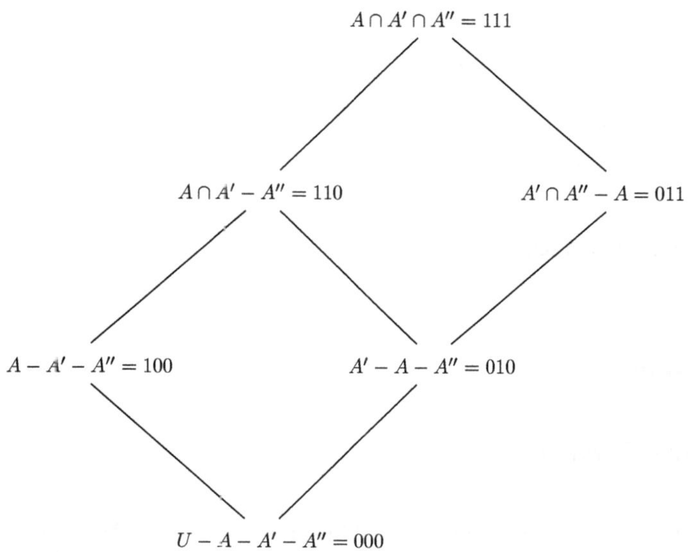

Relation between sets

Diagram 11.4.3

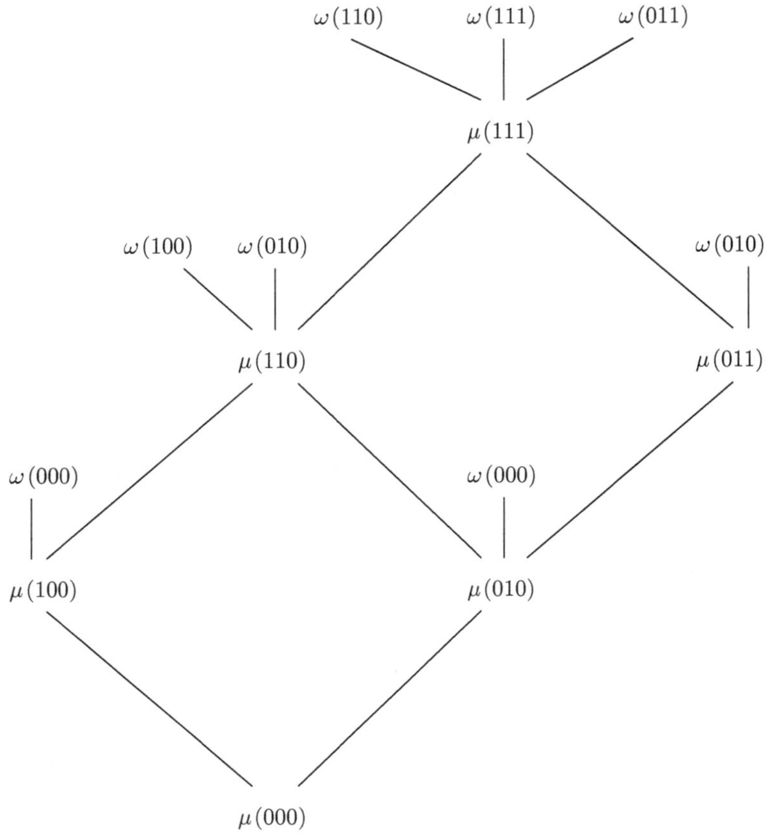

Relation between model sets

11.5 Discussion

11.5.1 General Remarks

(1) It is very important to see that our construction applies only to "reasonable" concepts, and these concepts are constructions of our brain—see example 1.2.1 for a counter example, where we probably lose many good properties, like homogenousness, etc.

(2) Logic as heuristics
Classical logic is basically the theory of its operators. Nonmonotonic logics are useful, if the world and our knowledge have certain properties, and our

knowledge, in particular our language, is adapted to the world, they contain a hypothesis about the world and our relation to it. Heuristics allow to find an efficient way of dealing with a large number of cases. In this sense, NML, in combination with a useful language, are heuristics.

(3) General preferential logics are purely algebraic structures and do not consider the language. $a \prec b$ is independent of any logical properties of a and b. For this reason, the properties of general preferential structures are purely algebraic, too, and best expressed by abstract size—which, again, is, of course, language independent.
Our construction is very language dependent, first by the construction of the scaffolding, and then by the special way we order models (by the number or sets of defaults satisfied). For this reason, we have powerful new logical properties. Our construction has an object and a meta-part. The object part is the preferential relation, and the meta-part is the scaffolding and the resulting order between "packages" of models. The relation is—despite the relatively complex definition—rather simple. We have a "primary" relation between (model) sets, and a "secondary" relation inside the $\omega(A)$. Constructing finally a preferential structure has the advantage of offering a well understood semantics, with nice logical properties, like OR, etc.

(4) Sometimes, there is no ideal case, but a set of prototypes $P_1, \ldots, P_n$ with property sets $\Phi_1, \ldots \Phi_n$. In this case, the ideal cases (prototypes) have the property $\bigwedge \Phi_1 \vee \cdots \vee \bigwedge \Phi_n$, and we can use the same construction.

(5) The adequate abstract semantics for general preferential structures is a notion of size. This will not work here any more, as the construction is too complicated. There are two candidates: inheritance systems, and, still more abstractly, rarity of influences. The latter seems more interesting, as, in principle, it allows to deduce inheritance systems, but, at the time of writing, we have no formal deduction of our approach from rarity of influences (and, of course, adequacy of our language).

11.5.2 Rarity and Its Coding by Inheritance

11.5.2.1 The Meaning of Rarity of Influence Changes

There are at least two different meanings of such rarity:

(1) in most cases (probabilistically, or a qualitative "most"), there is no influence change, or
(2) we assume there is no influence change, unless we are convinced there is one.

In the first case, rarity would have to scale, i.e. irrespective of the set of origin, big or small, such changes are (relatively) rare. In the second case, we would also expect there is no change, irrespective of the origin set. Thus, in both interpretations, rarity should be constant in above sense over the universe.

This assumes that the set of origin, and the subsets, are chosen "reasonably". We could, for instance, choose one element for each influence class, and collect them in a set, then we have changes with every modification of the set. We see again the importance of a "reasonable" choice of the sets. (See also example 1.2.1.)

We think that the second interpretation is intuitively more adequate. "Rarity" seems like a rule of thumb, a heuristic which is reasonable to adopt, without committing to a more quantitative interpretation—but see also below in Sect. 11.5.2.2. It is thus close to the autoepistemic interpretation of Reiter defaults, but with a clear separation between object level (there is a modification) and meta-level (normally, there are no modifications).

What are the consequences of this interpretation?

(1) First, such changes are usually known. They are attached to subsets. This was encoded directly in Definition 11.4.4.
(2) Second, if we do not go into a subset with more defaults attached then things stay as they are in the set of origin—except that we lose some models. Thus, the normality expressed in the fine relation between models does not change.
(3) Third, if one thing changes, the other things usually stay as they are. But we have to re-calculate normality as the new information might be stronger. The small change is now expressed by the minimal change of theory revision, see Definition 11.4.1.
(4) Fourth, we assume that we have a reasonable level of knowledge, and that our sets are well chosen. Thus, surprise exceptions should be rare, most exceptions should be attached to sets we are aware of. Thus, surprise exceptions (the $\omega(A)$) should be even higher up than "normal" exceptions.
(5) The second to fourth point can be seen as "continuity" of the resulting logic, we change as little as possible.
(6) Finally, we can turn the argument around: if our strategy does not work well, then something is wrong with the choice of sets (predicates). It might, e.g. indicate a spurious property.

11.5.2.2 Modifications

Our method can be modified easily.

We assume that our knowledge is well organised. We (more or less) know the exceptions, they are collected in a few, relatively small, classes. Penguins do not fly, dead birds do not fly, birds in cages do not fly (or not very far), etc.

So, we have a second-order default, again related to rare influences (blackbirds fly, black birds fly, white ones do, those in Germany do, etc.). This is already partly expressed in our relation, as the order is changed only for explicitly known counter examples (penguins...). In analogy to the requirements for a valid default (there are enough elements where it holds), we can require that "enough" properties b, c, etc. leave the default valid. This can be expressed again by a (second order) filter on the properties. In most cases, the filter will be relative to the default, so it changes for different defaults. In this way, we may combine both interpretations of "rarity of influence".

Again, we do not commit.

11.5.2.3 The Place of $\omega(X)$

There are arguments for and against our strategy. A flying penguin should not be a normal bird, but doubly abnormal. This is an example which justifies our approach. On the other hand, imagine A to be the set of vertebrates, with subclasses {fish, amphibians, reptiles, birds, mammals}, and one element left, a perfectly fine vertebrate animal, but it has no place in the usual order of vertebrates. We would think this to be a very exceptional case. This example is a counterargument to our approach. There is probably no uniform strategy, it has to be adapted. Of course, there is an ad hoc fix: Vertebrates are normally fish or amphibians or...

We based the construction on the sets to which defaults are attached. We could also have used the set of defaults which change, the more defaults change, the more important and exceptional the set to which they are attached. For instance, the predicate "dead" would be very important, as it changes many defaults (for many objects, plants and animals.)

Note that a set A' with $A \not\subseteq A'$ to which new defaults are attached, even if they only add to the valid defaults, and do not take any away, is still an exception. By prerequisite, we will have normal elements, satisfying all valid defaults, and not being in any exceptional subset.

11.5.2.4 Varia

(1) Implicitly, we also have a length of path argument. The longer the path through sets to which defaults are attached is for a model, the higher up the model will be.
(2) We adapt inheritance, i.e. from supersets to subsets, for information transfer, justified by rarity. We could use soft inclusion, too. We could use any other method of information transfer, too—and work with some similarity condition. Of course, specificity would have to be modified, too.
(3) We can see our construction also as a semantics for a logic of conjectures. Our main conjecture is that things do not change. Only forced by better information, we concede that our conjecture was wrong in this case, but we uphold it as principle.

11.5.3 Modularity

(1) Our approach is very compartmentalised. For example, in Example 11.4.2, we cannot compare elements in $A - A'$ with those in $A' - A$. This might be desirable, or not. There is probably no general best solution. Sometimes, we want to compare apples with pears, and sometimes not.
(2) The modularity of our construction is not to be confused with modular properties of the finally constructed logic.

11.5.4 Graceful Degradation and Coherence

(1) We indicated several methods of comparing individual models. Our approach allows graceful degradation, as we do not only compare the best with other models, but have implicitly a finer grading, this is expressed by $\sqsubset$, see Definition 11.4.3.
(2) We can see our treatment of subideal cases as a case of theory revision: We take the ideal case as start, and revise with the set of (subideal) cases, using the Hamming distance of valid defaults. The closest elements are those which differ the least from the ideal case considering the defaults valid at this point in the hierarchy.
(3) Working without copies (see Sect. 5.2.1.2) allows to apply the information to arbitrary subsets—by "meta-default", i.e. unless other defaults become valid—resulting in graceful degradation or continuity: If the ideal is no longer possible, we are still as good as possible, subideal cases are handled properly.
(4) Coherence
"Coherence" seems another word for "reasonable behaviour" (of a logic).
We have the coherence properties of general smooth preferential structures, we have a simple and "packet-wise" relation which takes care of epistemic ideas—and is thus in principle "reasonable". We have no unnecessary changes (by inheritance), and when necessary, we change as little as possible (graceful degradation). Finally, we avoid the problem of inconsistencies and the EFQ rule by excluding them.
Some properties like "dead" change many things, but this is unavoidable—and to be expected.
We do not have any analogical reasoning, nor any inductive component, but both were not to be expected.
(5) Upward and downward properties
Our construction requires a default theory to be valid, in the sense that there is a filter where all defaults valid at this point are valid together (for at least one model). The OR rule ensures that we have reasonable upward rules: if it holds for subsets (or elements) covering the set, then it holds for the set itself. The

construction guarantees good downward behaviour: if it holds for a set, then it will hold for elements and subsets, as far as is possible.

11.5.5 Core and Extensions

(1) We require the set of defaults valid at one point to be consistent. (We show later how to attenuate this position.) This guarantees that there is "core" of cases, where all defaults are satisfied. Theories like $\{: \phi/\phi, : \neg\phi/\neg\phi\}$ are excluded as nonsensical, just as we exclude classical inconsistent theories.

(2) The extension of the core and treatment of individual elements is automatic, if we have enough models. In Reiter defaults, this is implicit, as we do not specify the full language. Here, we need enough "copies", where the defaults hold or not. Provided we have those copies, "inference greediness" is encoded by the procedure to compare models, see Definition 11.4.3. This is a rule, like Reiter defaults, but it is applied to individual models, not sets, and contradictions are eliminated in an earlier step.

11.5.6 Contradictions

We have a strategy of handling contradictions of inherited information, it uses specificity and minimal inconsistent sets. Other strategies are possible, again, instead of strict specificity, we could use "soft" specificity.

Specificity can be justified by rare influence: There is (at least) one way of transferring information from A via $A \supseteq A' \supseteq A''$ to A'' which has to go through more influences than the way from A' to A''. Thus, it might be less reliable.

In addition, there is an argument about size and probability: Suppose 95 % of A and of A' are normal, and B is 10 % of A, but 50 % of A', then the probability that B falls into the abnormal part of A' is smaller than that it falls in the abnormal part of A. We do not *know*, as about normal cases, we conjecture.

11.5.7 Philosophy of Science

The exception sets (like penguins in the bird case) are now sets of results which do not fit the theory, but for which we have (potentially) an explanation—a disturbing influence (at least, we hope so). We might have several disturbing influences, strong wind, high temperature, strong magnetic field—several at the same time are more exceptional, and probably more difficult to handle. But, in principal, we are optimistic that we can explain the reason why the theory seems to fail. Again, an exception to the exception (which fits again the theory) is not a regular case. Therefore, it sits higher up, and needs a double explanation—why the first exception produces irregular results, and why the "inner" exception reverses this. Exceptions without explanation (the cases in the ω-sets) are more "worrying" than those with a potential explanation. They should incite to look for reasons, or even to doubt the validity of the theory. In the context of philosophy of science, we should put them perhaps even higher up.

So the basic ideas of the NML case carry over to the philosophy of science situation, perhaps with small modifications.

11.5.8 The Different Aspects of Our Construction

Nonmonotonic logics can be distinguished as follows: Some use explicitly or implicity a notion of size ("most") to express normality; this is the case for preferential structures. Some use conflicting information on the same level to express normality, things are normal, unless we have contradictory information; this is the case for autoepistemic logic and the autoepistemic interpretation of Reiter defaults, as well as much of argumentation theory. Some use normality on a meta-level, to be corrected by exceptions on the object level; this is the case for circumscription.

We use all three. Size is used for the final preferential structure. Same level contradictions are used to find a consistent subset of the defaults by a theory revision approach. This avoids the sometimes delicate equilibria between conflicting information, as seen in extensions of, e.g. Reiter defaults. Finally, on the meta-level, we have the hypothesis of rare influence change, which is contradicted on the object level by actual changes. We have a second meta-level assumption: that we know the exceptions, and we can be contradicted on the object level by unexpected cases, elements in the $\omega(A)$'s.

11.5.9 Modifications

(1) If the presence of the ideal case in the valid defaults seems too strict a requirement, we can either soften it by imposing a minimum which should be valid, or, postulating, e.g. $(d \wedge d') \vee (d \wedge d'') \vee (d' \wedge d'')$, instead of requiring that at least 2 of the 3 defaults d, d', d'' should be valid. The lottery paradox falls into this category, too.
In the same spirit, we can attenuate $\forall$ to ∇ in the first order case, as a fixed requirement, or by a preference relation among FOL models. (Read $\forall$ as $\bigwedge$.)

(2) The straightforward construction using the given defaults is syntax dependent. This may be desirable, or not.
The default $\alpha \mid\sim \phi \wedge \phi'$ behaves differently from the set of defaults $\{\alpha \mid\sim \phi, \alpha \mid\sim \phi'\}$. If this is not desirable, in the finite case, we can write any default

ϕ with model set $M(\phi)$ as the (finest possible) conjunction of all ϕ_m, with $M(\phi_m) = M - \{m\}$, $m \not\models \phi$, as done in the following example.
It shows that our inheritance problem is under-specified in the following sense: we can describe the same result by two different sets of rules, which, when we try to use them for approximation, give different results.

Example 11.5.1
Consider the propositional language with three variables, p, q, r, let M be the set of all models. For simplicity, we will replace $\wedge$ by juxtaposition when we describe models.

Suppose we have $TRUE \mid\sim pqr$, but no other information.

Consider now the model set $X := \{\neg pqr, \neg pq\neg r\}$.

What should we inherit by homogenousness from $TRUE \mid\sim pqr$? The natural answer seems to be $\neg pqr \prec \neg pq\neg r$, so $\mu(X) = \{\neg pqr\}$, as we "feel" that p is better than $\neg p$, q better than $\neg q$, r better than $\neg r$. The rules can be described as: make p true, if possible, make q true, if possible, make r true, if possible. Clearly, this gives the described result for X.

But pqr can also be described as a conjunction of 7 different formulas. Consider any $m \in M$, described by ϕ_m (a conjunction of literals). $M - \{m\}$ is described by $\neg\phi_m$, which is the disjunction of negations of those literals. Now, $\{pqr\}$ is the intersection of all 7 $M(\neg\phi_m)$, $m \neq pqr$. We have the 7 rules to make all those $\neg\phi_m$ true. They are just as good as the above 3 rules. We re-consider $X := \{\neg pqr, \neg pq\neg r\}$. Each $m \in X$ violates exactly one of the seven rules—so we cannot decide which model is better.

Consequently, the problem of inheriting properties to subsets is underspecified. The problem lies *not* in the description of X, which is very simple: $\neg pq$. The problem lies in the *description* of the order resulting in $\mu(X)$.

(3) As mentioned already, the approach has to be slightly modified to treat single elements, where a possible worlds strategy seems adequate.
(4) We may consider the last idea in Definition 11.4.3 in a more radical way: Consider a given set A, and all defaults attached to A or its supersets. We order them by specificity, and then satisfy them according to specificity, letting the existence of models treat the consistency question.

11.5.10 Aspects Not Considered

(1) For the problem of "repairing" mechanisms of living beings or crucial human inventions, like aircraft, see the next section.
(2) We did not consider clusters of properties, as "dead" influences many properties, and also of many living beings—see the next section for an outline.
(3) We did not consider changing contexts.
(4) For the problem of language fragments where exceptions are "hidden", see the next section.

11.6 Extensions

It might be a good sign that our construction allows important extensions in an easy and natural way.

11.6.1 "Repairing" Properties

If a bird cannot fly, it will try to walk, if one aircraft engine fails, it will fly with three engines instead of four, etc. Living beings have "desire" to function as well as possible, critical human inventions are constructed with redundancies.

We can put this into the "fine" order between models. "fly" is better than "walk", "walk" is better than "crawl", etc.

11.6.2 Theory Revision

11.6.2.1 Distance Based Approach

There is a natural way to define a distance between two models. We first define a distance between two sets A, A'. The idea is as follows: find a $\subseteq$-minimal B such that $A, A' \subseteq B$. Then count the steps upward from A to B, and then downward to A', through the relevant sets. There are two problems: First, there might be several paths from A to B (and then to A'), of different length. We could chose the shortest or the longest; we have no intuition about this. Second, there might be several such B, with different lengths of paths. Again, we could chose the shortest overall, longest overall, etc.

Within the $\omega(A)$, it seems natural to take the counting variety of the Hamming distance. We then have a pair, (distance between sets, ranks of the models within $\omega(A)$ and $\omega(A')$)—provided, the models are in the ω's. If they are in the μ's, then the second value is 0. Thus, elements in the exception sets (or failing defaults for "unfamiliar" reasons) will be put (progressively) into the interior. This seems natural. We do not want to see strange elements first, but normal ones.

Given a distance, we can do counterfactuals and theory revision in a standard way.

11.6.2.2 Minimal Inconsistent Set Approach

Our technique to obtain consistent (default) theories (see Definition 11.4.1), is a variant of safe contraction, see Sect. 7.3. If defaults attached to a set are inconsistent, we make them consistent by a suitable theory revision technique. This makes sure that we have at least one "perfect" model. We now determine the valid defaults by the same strategy as before, preferring one theory over the other, if so desired.

11.6.3 Properties with Many Influences

For example, the predicate "dead" changes many other properties, and of many things.

First, we can code this into a default attached to "all living things", but should then give it a more robust status for determining consistent subsets of defaults. Even if it is general, it should be stronger than many specific ones. If there are no exceptions, then we can code it classically. Second, we can make the construction parallel for the things concerned and also for the properties concerned. This does not change the logic, but the construction of the logic.

11.6.4 Language Fragments

This concerns the situation where we know that exceptions might be hidden in some language fragments (penguins, etc.), but do not yet know, where. We can reason without considering these fragments. The full models are elements of $X \times F$, where F is the fragment. We have created a relation on X as described above, and turn now to F. We can easily add $\prec'$s to the new, full models, without erasing already existing $\prec'$s. We can use the idea of higher order preferential structures (reactive arrows) to destroy the existing $\prec'$s. These ideas are described in Gabbay and Schlechta (2009), Chap. 6. Neglecting the critical sublanguages is bold reasoning. We know that we might be wrong, but we accept this, as benefits (fast decisions) seem to outweigh costs (wrong decisions).

NML in general can be seen as an abstraction—we forget the exceptional cases. Considering language fragments is then a fortiori an abstraction. We can consider a progressive treatment of the information in three steps:

(1) treat only the basic information, assuming there are no exceptions,
(2) treat only the "packets" (the A's in our example) without looking inside,
(3) treat the full information.

Note that the language does not only consist of propositional variables in the non-monotonic context, but also of relations etc., so a simplification is even more important in our context. Ideally, a sublanguage is like opening a book, which explains details, but leaves the main points unchanged.—A detailed elaboration is left to the reader.

11.7 Formal Properties

(1) We have the following:

(1.1) If $B \subseteq \mu(A) \cup \omega(A)$, and $B \cap \mu(A) \neq \emptyset$, then $\mu(B) = \mu(A) \cap B$, as $\mu(A) \prec \omega(A)$.

(1.2) If satisfying default ϕ in A improves a model, so ϕ is valid in A, and if ϕ is still valid in B, then satisfying ϕ in B improves a model, too. The order is preserved as much as possible. This is coded directly into the inheritance mechanism, see Definitions 11.4.1, 11.4.4, and Construction 11.4.1.

(1.3) Consequently, for inheritance and subideal cases, there is nothing else to do.
Swedes are tall and blond. So the best models through S have t and b. The second best have $S, \neg t, b$, or $S, t, \neg b$. So the best through S and $\neg t$ still have b, etc. So not tall Swedes are blond, and dark haired Swedes are still tall.
It is the universality of the relation which settles the subideal cases, and the relation (sacrificing as little as possible) makes the degradation graceful.

(2) The relation $\prec$ is free of loops.
It follows from the fact that the relations $\sqsubset$ and $\lhd$ are free of loops, more precisely:
Every m is in exactly one $\mu(A)$ or $\omega(A)$.
Now,

(2.1) There are no loops within one $\mu(A)$—obvious.

(2.2) There are no loops within one $\omega(A)$—by the interior ordering.

(2.3) So any loop must involve elements from different $\mu(A)$ and $\omega(A')$. As there are no elements above any $\omega(A')$, the loop must contain at most one $\omega(A)$, and perhaps different $\mu(A)$'s. But the $\mu(A)$'s are ordered by $\lhd$, which has no loops.

(3) **Fact 11.7.1**

Normally, $\alpha \mid\sim \beta \mid\sim \gamma$ implies $\alpha \mid\sim \gamma$, default transitivity.

Proof
By $\alpha \mid\sim \beta$, the best α-models go through β. As $\beta \mid\sim \gamma$, not only the best β-models go through γ, but in the order, valid at β, γ-models are preferred over $\neg\gamma$-models. This order is inherited downward to subsets of β, so also to the set of α-models through β—unless there is a (rare) change of influences, which, in addition, has to contradict the inherited order concerning $\gamma/\neg\gamma$. Thus, also in the set of α-models through β, those through γ are preferred, normally.

(4) Discussion of $\mu * \mu$
For the role of $\mu(X \times Y) = \mu(X) \times \mu(Y)$ for interpolation, see Sect. 9.3.

We cannot expect interpolation, as, e.g., $bird \mid\sim fly$ shows. But we have (sometimes) above rule for μ. Let X, Y be two disjoint parts of the language, let some defaults be formulated in and attached to X, the others similarly to Y. We then have an order constructed as above for the X-part, and one for the Y-part. If they do not contradict each other, the joint-order relation is the same as the separate order relations, put together. (Contradictions are possible in the consequences of the defaults, not the prerequisites.)

By the same argument, we see that we can add properties, as long as they do not contradict other properties, and have the product rule (applied to the formulas to which the defaults are attached to: we separate these formulas in two groups, and examine the orderings separately and jointly). Moreover, the lexicographic variant in Definition 11.4.3 allows to consider groups of same strength defaults separately, and to put the results together—again a "$\mu \times \mu$-rule".

(5) Pre-processing:
We can determine the defaults valid at certain A before looking into A, the latter will not affect the former. As the number of A's will, in most cases, be much smaller than the total number of models, this is a considerable advantage. This is possible, as the order is not arbitrary, but follows well-defined principles, guided by the language and the defaults.

(6) The models are all in some "packages" ($\mu(A)$, $\omega(A)$ for some A), and the packages are ordered themselves. This order will often eliminate already some models in a set X from being minimal, and the others will often be divided into small groups (of incomparable packages). In this case, we have $\mu(X' \cup X'') = \mu(X') \cup \mu(X'')$. Thus, we will often have to compare only a few models, to find the best ones, i.e. $\mu(X)$ This is an informal argument about the average complexity of finding $\mu(X)$ for a given set X of models.

(7) Arbitrary subsets (without change of valid defaults) are treated as the superset is, by the same order, only some models are left out, but the order principle does not change.

11.7.1 A General Remark

A basic problem of logic is—apart from arriving at conclusions—to find the necessary information to arrive at conclusions. In particular, we look for stability of conclusions, and small sets of information, which allow to derive a conclusion, if this is at all possible.

In classical logic, monotony allows to restrict search for arguments. Once a conclusion is derived, adding information will not alter this.

In nonmonotonic logic, cautious and rational monotony are rules of the same kind. Due to the possibly "illogical" definition of the preference relation (and to the possibility of copies), we do not have much more for general preferential structures.

We have shown the connection between factorisation of μ and interpolation in Gabbay and Schlechta (2011), Chap. 10, and discussed factorisation above, see Chap. 9.

The simple construction principles of our relation allowed further simplifications to find minimal model sets, as we saw above.

Chapter 12
The Talmudic Kal Vachomer Rule

12.1 Summary

Abraham, Gabbay and Schild (AGS) translated the Talmudic Kal Vachomer rule in several steps into the problem of representing an abstract partial order by a maximally simple multiset inclusion relation. (The approach is a bit more complicated, but this is the main and most difficult part.)

(1) The problem to be solved by the rule is first translated into a matrix whose elements are 0 or 1, with one exception, whose value is unknown.
(2) Replacing this value once with 0 and then with 1 results in two matrices, and the "simpler" one is chosen, as follows.
(3) To determine simplicity, AGS consider the columns of the two matrices, and their elementwise order. The matrix with the simpler column order is considered the simpler matrix. Thus, in abstract terms, we have to measure the simplicity of two partial orders.
(4) AGS then translate the abstract partial orders into set inclusion relations, more precisely, into multiset inclusion relations, where at most one element is allowed to appear several times.
(5) Finally, the minimal number of elements needed to represent the inclusion relation is a measure of complexity. (For instance, a total order just needs one element, say a, which occurs multiple times, etc.)
(6) AGS give an algorithm to determine this minimal number.

Our contributions here are the following:

- we show that there is a problem with the AGS algorithm,
- we show that there is no straightforward inductive algorithm at all, in a bottom-up approach, as we may have to backtrack,
- we replace the multiset approach by an extremely simple approach, just counting the related pairs of the partial order: the more pairs are related, the simpler the order (note that a total order has the maximal number of related pairs), and

D.M. Gabbay and K. Schlechta, *A New Perspective on Nonmonotonic Logics*,
DOI 10.1007/978-3-319-46817-4_12

- compare our counting approach to the AGS approach. Somewhat surprisingly, in many cases the two very different approaches give the same result—but not always.

As none of the two approaches seems clearly superior in results, the simpler one might be a serious candidate.

12.2 Introduction

12.2.1 The Problem

Our discussion is based on work by A. Abraham, D. Gabbay, U. Schild, see Abraham et al. (2009), we will call their approach the AGS approach, see also Schlechta (2016), Bessos and Gabbay (2016), Laufer and Gabbay (2016).

The Kal Vachomer rule is a rule for principled guessing of an unknown value. It might be best to begin with an example. In the table given in Sect. 12.2.1, we have information about four different computer screens, concerning, e.g. price, size, etc. The information is binary, 0 stands for "not so good", 1 stands for "better", for all qualities. One information is lacking, we do not know if Screen 3 has stereophonic speakers, the entry there is "?". The idea is to try both possible values, 0 and 1, and see which results in the "nicer" partial order between the columns.

Ordering the (vertical) vectors in the table given in Sect. 12.2.1 componentwise, (with $1 < 0$, expressing that best is smallest) we have for "?" replaced by 1: $S < D, C, R$; $C < P$; $R < I$ - closed under transitivity. If "?" is replaced by 0, $S < D$ does not hold any more, and D is now an isolated point. We then consider the first order with 1 in the place of "?" as "nicer" than the second order, and assume that the first guess is the better one.

Thus, leaving all details aside for the moment, the Kal Vachomer rule is based on an assumption that the world is rather orderly, and disorder is the exception. We can also consider a more orderly world as simpler, and thus have an example of Occam's Razor, which considers simpler descriptions of the world as more likely to be true. Thus, the Kal Vachomer rule and its discussion is fully in the subject of this book.

The main problem is how to distinguish two partial orders as to their "niceness" or simplicity. What is a nice or simple partial order? A total order is maximally nice, this seems clear, everything grows monotonically. It seems questionable if (partial) equivalences make orders nice. Is $A < B < C$ better or worse than $A = B < C$? The AGS approach says yes, in above example we have just two points, $A = B$ and C, the counting approach (see below, Sect. 12.5) leaves the question open, and does not commit.

We now give the formal definition of the problem, and then turn to the AGS approach.

Definition 12.2.1
A quality matrix is a two-dimensional matrix $a_{i,j}$, $1 \leq i \leq m$, $1 \leq j \leq n$, $=$ with $a_{i,j} = 0$ or 1, where columns indicate different qualities, and rows different instances. For instance, for computer screens, columns might be screen size, reaction time, dot size, etc. where 1 is always better than 0 in overall quality. A bigger screen is better, a screen with faster reaction is better, and with smaller dot size is better too.

Column j is considered at least as good as column j' iff $\forall 1 \leq i \leq m$ $a_{i,j} \geq a_{i,j'}$.

Computer screens

	P	C	I	R	D	S
	price over 450	self collection	screen bigger than 24 in.	reaction time below 4 ms	dot size less than 0.275	stereo-phonic
Screen 1	0	1	0	1	0	1
Screen 2	0	0	1	1	0	1
Screen 3	0	0	0	0	1	?
Screen 4	1	1	0	0	1	1

1 = yes; 0 = no;? = no data

In an actual matrix, the order need not be total, as different manufacturers might emphasise different features (in our example).

We are now ready to formulate the reasoning problem.

Suppose one (not more) $a_{i,j}$ is unknown, it is 0 or 1, but we do not know which, we write the value as "?". Can we guess what it probably is? To obtain an answer, we try the matrix with 0, say $M(0)$, in place of "?", and the matrix with 1, say $M(1)$, in place of "?", and see if one of them has a nicer order of the columns than the other. The one with the nicer order gives then the better guess.

In the Kal Vachomer context, two quite different ways to measure "nice" of a partial order have been suggested.

The first is the AGS approach, Sect. 12.3, the second is the extremely simple arrow counting approach, see Sect. 12.5. As they are quite different, it is not surprising that they do not always give the same results. It is more surprising that they give the same result in all cases discussed in Abraham et al. (2009), and which do not involve equalities. This might be an argument for the simplistic counting approach.

12.2.2 Historical Origin

D. Gabbay and co-authors looked at the Talmudic rule of Kal Vachomer, see Abraham et al. (2009), and also pointed out that the same or similar rules were discussed in Arab and Indian philosophy, Qiyas and Kaimutika Nyaya, respectively.

12.3 The AGS Approach

The material in this Sect. 12.3 is taken from Abraham et al. (2009).

12.3.1 *Description*

The AGS approach considers four qualities of a partial order (or, equivalently, a directed graph, where an arrow points to the bigger element)

(1) the number of nodes,
(2) the number of connected parts,
(3) the maximal number of necessary direction changes in paths, see below,
(4) the dimension in a multiset representation, see Sect. 12.3.1.2. This is the most complicated part of their approach.

The number of nodes is trivial, so is the number of connected parts, e.g. Diagram 12.5.3 and e.g. Diagram 12.5.4 have four connected parts, Diagram 12.5.5 and Diagram 12.5.6 have two connected parts. (Note that in all diagrams, the lines are supposed to be upward pointing arrows.)

12.3.1.1 Direction Changes

Consider Diagrams 12.5.3 and 12.5.4. When we want to go from G to I, we have one direction change. We first follow the arrow from G to H, then go against the arrow from H to I. When we want to go from G to K_0, we have three direction changes. If we add now a new point, say P, with $G < P$ and $K_0 < P$, we have a new path $G - P - K_0$ with only one direction change. For any two points X, Y, let $\rho(X, Y)$ be the minimal number of direction changes on all paths from X to Y, and let ρ be the $max\{\rho(X, Y) : X, Y$ points in the graph $\}$, this is the maximal number of direction changes in the graph (or partial order). (Thus, without P, $\rho(G, K_0) = 3$, with P added $\rho(G, K_0) = 1$).

The definition of the dimension is more complicated and demands some preparation.

12.3.1.2 Definition of the Dimension

Definition 12.3.1
A multiset is a set, where one or more elements may occur several times, e.g. $\{a, a, b, c, c, c\}$. We will write such sets $\{2a, b, 3c\}$, or $\{a^2, b, c^3\}$ etc. if this increases readability. A 1-multiset is a multiset where at most one element occurs more than once, e.g. $\{3a, b, c\}$. To simplify notation, we write sets like $\{a, b, c\}$ also abc, mul-

tisets like $\{3a, b, c\}$ also 3abc, etc. If $\{m * a, b_1, \ldots, b_n\}$ is a 1-multiset, then m will be called its index, and $n + 1$ its dimension.

$\{m * a, b_1, \ldots, b_n\} \subseteq \{m' * a, c_1, \ldots, c_{n'}\}$ (multiset inclusion) iff $m \leq m'$ and $\{b_1, \ldots b_n\} \subseteq c_1, \ldots, c_{n'}\}$

Definition 12.3.2
Let $(S, <)$ be a partially ordered set, and L a set (of "labels"). A function f from S into the set of 1-multisets of L is called an (L, f) representation of $(S, <)$ iff $x \leq y$ iff $f(x) \subseteq f(y)$, where $\subseteq$ is multiset inclusion.

Fact 12.3.1
For every $(S, <)$ there is a multiset representation.

Proof
Let $L := S$, and $f(x) := \{y : y \leq x\}$. □

The idea behind the use of 1-multisets is to represent the longest chain in the order by increasing the number of a's and code the rest only with additional elements. Thus, a total order (considered best) would need no additional elements, so the dimension would be 1. The lower the dimension, the "better" the partial order.

Example 12.3.1
Consider the following partial order: $\{A < B, A < C\}$. As B, C are incomparable, we need at least two elements, e.g. $A = a,\ B = 2a,\ C = ab$.

(To simplify notation, we write $A = a$, instead of $\lambda(A) = a$, as in Abraham et al. (2009), we also write $A = a$ instead of $A = \{a\}$ etc., for the same reason.)

Consider now $\{A < B, C\}$, i.e. C is isolated. Again, we need at least two elements, e.g. $A = a,\ B = 2a,\ C = b$.

So we cannot distinguish the two orders by the dimension alone.

We give a few more examples to make the reader familiar with the concept.

Example 12.3.2
The relations are to be closed under transitivity.

(1) For the multiset case, all the following are equivalent, and have the form $\{na, b\}$:

(1.1) A, B
$A = a,\ B = b.$

(1.2) $A < B,\ A < C$
$A = \emptyset,\ B = a,\ C = b.$

(1.3) $A < B < C < D,\ A < E < F$
$A = \emptyset,\ E = b,\ F = ab,\ B = 2a,\ C = 3a,\ D = 4a.$

(1.4) $A < B < C,\ A < D < C$

(1.5) $A < B < C,\ D < E < F$
$A = 3a,\ D = b,\ E = ab,\ F = 2ab,\ B = 4a,\ C = 5a.$

(1.6) $A < B < C < D < E,\ B < F < D$
$A = \emptyset,\ B = a,\ F = ab,\ C = 2a,\ D = 2ab,\ E = 3ab.$

The following is more complicated, and has the form $\{na, b, c\} : A < B < C < D, A < E < C < F$
$A = \emptyset,\ B = a,\ E = b,\ C = ab,\ D = 2ab,\ F = abc$

(2) Multisets and number of maximal chains do not coincide:
A, B, C, D has four maximal chains, and needs $\{na, b, c\}$:
$A = 2a,\ B = ab,\ C = ac,\ D = bc.$
$A < B < C < D, A < E < F < G, B < F, C < G$ has 4 maximal chains, but needs only $\{na, b\}$:
$A = a,\ B = 2a,\ C = 3a,\ D = 4a,\ E = ab,\ F = 2ab,\ G = 3ab$

12.3.2 A Problem with the Original AGS Algorithm

The AGS algorithm is quite complicated, and we refer the reader to the original article Abraham et al. (2009).

Example 12.3.3
This example shows that the algorithm presented in Abraham et al. (2009) does not always give the best result.

We have six points, $A < A', B < B', C < C'$.

The algorithm works as follows: Choose, e.g. $A < A'$ as T. Then $A = a$, $B = b$, $C = g$. $A' = 2a$, $B' = bd$, $C' = ge$, using five labels.

But $A = a,\ B = b,\ C = g,\ A' = 2a,\ B' = bd,\ C' = gd$ uses four labels only, and $A = 2a,\ A' = 3a,\ B = b,\ B' = ab,\ C = c,\ C' = ac$ uses even three.

12.4 There Is No Straightforward Inductive Algorithm for the AGS Approach

In this section, we show that the problem is not so simple, in particular, it cannot be solved by a straightforward induction. This is shown by giving examples which demonstrate that one has to “look forward”, or, alternatively, one has to backtrack. What seemed to be a good solution can turn out a bad one, but we see this only by looking at points “higher up”.

12.4.1 Even the Case with Simple (Not Multi) Sets Is Quite Complicated

We consider here graphs generated by a subset of some powerset (with the natural ordering by inclusion), and show that we need a certain number of atomic labels to represent them.

The examples show that it is probably quite difficult to come up with a minimal number of elements—let alone working with multisets. Example 12.4.3 shows that an inductive construction is impossible.

Example 12.4.1
Consider a flat graph, $X_1, \ldots, X_6$, six points, mutually incomparable, we have one maximal antichain of six elements.
Four labels suffice:
ab, ac, ad, bc, bd, cd.

Example 12.4.2
But we can also force the antichain up, necessitating at least five elements:

Consider the graph generated by abe, ace, ade, bce, bde, cde, ae, be, ce, a, e. As a and e are incomparable, they cannot be empty, so they contain at least one element each. Thus, ae etc. contain at least two elements, and the top elements at least three. But there are not six mutually incomparable subsets of size at least three in a four-element set.

12.4.1.1 There Is No Inductive Algorithm by the Natural Ordering for the Simple Set Case

Example 12.4.3
This example shows that, in general, an inductive procedure is impossible.

Recall that, for a given set of n elements, the number of subsets of size $m < n$ is $\frac{n!}{(n-m)!*m!}$.

Take now a structure consisting of one antichain with 20 elements, and nothing else. This can be represented with six elements and subsets of size 3, as $\frac{6!}{3!*3!} = 20$.

Take a structure with four antichains, each of size 20, and one above the other. Thus, the size of the representing sets will increase at least by 1 from the lowest antichain to the next, etc.

As we are not allowed to look ahead, we begin again with subsets of size 3 of a set of six atomic labels for the lowest antichain. So the next antichain must consist of sets of at least four elements, the next of five, the final of six. Thus, we need at least eight elements for representation, as $\frac{8!}{6!*2!} = 28$, 7 elements will not do.

If, however, we had begun with subsets of size 2 of a seven-element set, by $\frac{7!}{5!*2!} = 21$, we could have in the top layer only five element subsets, and this is again possible, so seven elements will do. But we have to look at the whole structure to see this.

12.4.2 The Multiset Case

We work with a set of atomic labels $L := \{a\} \cup B$, where a may occur several times, this will sometimes be written a^n for n times a, instead of na, etc.

We have the following trivial fact:

Fact 12.4.1
(1) Let $B' \subseteq B$, then $a^n B'$ and $a^m B'$ are comparable, as $n \leq m$ or $m \leq n$.
(2) Let $B' \subset B'' \subseteq B$, then $a^n B'$ is not comparable to $a^m B''$ iff $n > m$.

Corollary 12.4.2
(1) To code an antichain of size 2^n, *we need* B *of size at least* n.
(2) We can code an antichain of size 2^n *with* B *of size* n.

Proof
(1) Suppose B is smaller, then $card(\mathcal{P}(B)) < 2^n$, so two elements of the antichain are coded by the same $B' \subseteq B$, contradicting Fact 12.4.1, (1).
(2) Let (B') be the exponent of the (unique by Fact 12.4.1, (1)) $a^{(B')}B'$. Code the elements of the antichain by $\{a^{(B')}B' : B' \subseteq B\}$, where $B' \subset B''$ implies $(B'') < (B')$. Then the codes are pairwise incomparable by Fact 12.4.1, (2). Note that $(\emptyset)$ is the biggest exponent, and (B) the smallest. (The idea is that, if the B-part of two codes is comparable, we make the a-part comparable in the other direction, so the whole codes are incomparable.) □

Example 12.4.4
Consider the structure $X < Y$ and an isolated Z.
Obviously, we need at least one b. We may code this by $a < a^2$, b, or, by $b < ab$, a^2, and we have two, non-isomorphic, codings.

In the following, we will code a bottom antichain of size 2^n by $\{a^{(B')}B' : B' \subseteq B\}$, where B has n elements.

12.4.2.1 There Is No Inductive Algorithm by the Natural Ordering for the Multiset Case

We now show that an upward inductive algorithm, using the natural ordering, is impossible. For this, we discuss progressively more complicated examples. The last one, Example 12.4.7, is perhaps the most interesting, as it shows that we have to consider an arbitrarily deep and wide substructure (with non-trivial interior nodes), to see that a decision taken lower down cannot be upheld.

Example 12.4.5
Consider an antichain of four elements at the bottom, say A, B, C, D.
We might code this with the labels abc, a^2b, a^2c, a^3.
In the next level, we have an antichain of two elements, say X, Y, and they have both A, B as predecessors.
Suppose A was coded by a^3, B by a^2b. Then we can code X by a^5b, Y by a^4bc, and need no new label.
Suppose now that A was coded by abc, B by a^2b. Then X and Y must include a^2bc, we may for instance make X a^4bc, but now we have to introduce a new variable, say d, and make Y a^3bcd, by Fact 12.4.1, (1).

So, we have to look ahead. But it can be much more complicated. Take again above example. Suppose we have now two antichains, X, Y, and X', Y', one is above A, B, the other above C, D. Which one will have the abc? If X, Y is higher than X', Y', then we might have needed d already elsewhere, so we can use it without additional cost. But X might also be higher than X', Y' higher than Y. What shall we do?

Example 12.4.6
This example shows that even the initial step of coding four elements with three labels, as done above, might not always work: again, we have to look ahead.

Consider again an antichain of four elements at the bottom, say A, B, C, D. Again, we might code this with the labels abc, a^2b, a^2c, a^3.

Suppose we have in the second layer one new point above exactly each pair from A, B, C, D. One of the bottom nodes will be coded by $a^n bc$, another by $a^{n'} b$, another by $a^{n''} c$. Suppose $n' \geq n''$. Let X be above the bottom elements coded by $a^{n'} b$ and $a^n bc$. Then it will also be above the bottom element coded by $a^{n''} c$. But this is not wanted.

Thus, in this situation, we need a new label, say d, to code the element coded by $a^n bc$.

(We could also put the second layer nodes X on "stilts", so they will have arbitrary height, like $a^n b < a^{n+1} b < a^{n+2} b < \cdots < X$, etc., so we have to climb up arbitrarily high to see the problem.)

Fact 12.4.3
Consider a bottom antichain with elements $a^{(B')} B'$, where $B' \subseteq B$.

Fix now $D \subseteq B$, let $D' := B - D$, and consider $X := \{a^{(D'E)} D'E : E \subseteq D\}$. Then, of course, $(D') = (D'\emptyset) > (D'E)$ for all $E \neq \emptyset$, by Fact 12.4.1, (2).

Let $X = X' \cup X''$, where X', X'' are disjoint and have the same cardinality, and introduce two new nodes, C' and C'', such that $C' > A'$ for all $A' \in X'$, $C'' > A''$ for all $A'' \in X''$, but for no $A'' \in X''$ $C' > A''$, and for no $A' \in X'$ $C'' > A'$.

Suppose without loss of generality $a^{(D')} D' \in X'$. Then there is $b \in D$ such that for no $a^{(D'E)} D'E \in X'$, $b \in E$. (Otherwise, by maximality of (D'), all $x \in X$ would be below C', as the code of C' would contain all $b \in D$, and the highest exponent of a.)

On the other hand, for cardinality reasons, and by $card(X') = card(X'')$, there cannot be two such $b \in D$, as we would not have enough coding possibilities left. □

Example 12.4.7
Using $n + 1$ atomic labels, $L := \{a, b_0, \ldots, b_{n-1}\}$, we can code a bottom antichain $X_{0,0}, \ldots, X_{0,2^n-1}$ as follows: Work in the binary system. Set $B := \{b_0, \ldots, b_{n-1}\}$, and code $B' \subseteq B$ by $c(B') := \Sigma\{2^i : b_i \in B'\}$. This gives a natural total order on $\mathcal{P}(B)$, and we use the inverse of this order for the exponent of a. Thus, as it should be, $(\emptyset) = (0, \ldots, 0)$ is the biggest exponent, and $(B) = (1, \ldots, 1)$ the smallest exponent.

In more detail, code $X_{0,i}$ by $a^{(i)}i$, where i is written in binary, i coding as above a subset of B. Thus, $X_{0,0}$ is coded by $a^{(\emptyset)}\emptyset$, $X_{0,1}$ by $a^{(0\ldots1)}0\ldots1 = a^{(b_0)}b_0$, $X_{0,2}$ by $a^{(0\ldots10)}0\ldots10 = a^{(b_1)}b_1$, etc., up to $X_{0,2^n-1} = a^{(b_{n-1}\ldots b_0)}b_{n-1}\ldots b_0$.

Then create new nodes above the bottom level, etc. always grouping successive lower nodes together, as follows:
$X_{1,0},\ldots,X_{1,2^{n-1}-1}$
$X_{1,i} > X_{0,i*2}, X_{0,i*2+1}$
$X_{k,0},\ldots,X_{k,2^{n-k}-1}$
$X_{k,i} > X_{k-1,i*2}, X_{k-1,i*2+1}$
up to $k = n-1$ (included).

The labelling of the new nodes is made by taking the union of lower labels. Our ordering of the exponents shows that this is possible, with exactly the relations as defined. See Diagram 12.4.1 for an example with $n = 3$. In the last line, the left bit codes d, the middle one c, the right one b. The exponent of a is the bitwise complement of the binary figure in the last line.

Diagram 12.4.1

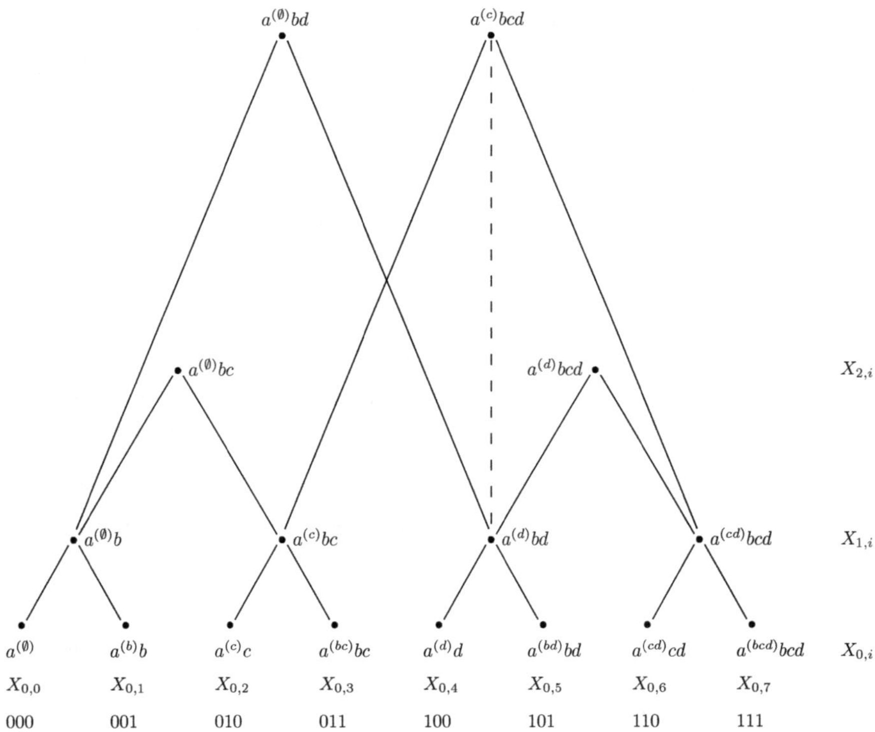

For instance, (d) is the highest exponent in the right half, but all exponents on the left half are bigger than (d). Thus, all nodes on the right half are below $a^{(d)}bcd$, and none on the left is below $a^{(d)}bcd$. For $a^{(\emptyset)}bc$, all nodes on the right half contain d, so they are not below $a^{(\emptyset)}bc$, etc. We add now two additional nodes, $a^{(\emptyset)}bd$, and $a^{(c)}bcd$.

The latter will have more nodes below it than intended—see the broken line in the diagram. Consider first the node labelled $a^{(\emptyset)}bd$. The nodes below $a^{(cd)}bcd$ contain c, so they are not concerned, the same holds for those below $a^{(c)}bc$. But it is impossible to add the node $a^{(c)}bcd$: By $(c) > (d) > (bd)$, we see that $a^{(d)}d < a^{(c)}bcd$ and $a^{(bd)}bd < a^{(c)}bcd$, a contradiction.

This is no accident, it does not depend on the specific choice and distribution of the base labels, as we show now. "(labelled …)" refers to the example for $n = 3$, described in Diagram 12.4.1.

Consider, for an arbitrary labelling, $X_{n-1,0}$ (labelled by $a^{(\emptyset)}bc$) and $X_{n-1,1}$ (labelled by $a^{(d)}bcd$) (these are all which are on level n). One of them has to be above $a^{(\emptyset)}$, without loss of generality, let this be $X_{n-1,0}$. Note that $(\emptyset)$ has to be the strictly biggest exponent, otherwise we have no antichain. One of the atomic labels, say b_j (d in the diagram) does not occur in the labelling of $X_{n-1,0}$, otherwise, all bottom nodes would be below $X_{n-1,0}$. For cardinality reasons, all others have to occur in the labelling of $X_{n-1,0}$, see Fact 12.4.3. Moreover, b_j occurs in all labels of the bottom nodes below $X_{n-1,1}$, and all combinations of the other b_k occur below $X_{n-1,1}$. In particular, we have $a^{(b_j)}b_j$ and $a^{(b_{n-1}\ldots b_0)}b_{n-1}\ldots b_0$ below $X_{n-1,1}$, and, by the same reasoning, (b_j) is the strictly biggest exponent below $X_{n-1,1}$.

We split now $X_{n-1,0}$ into $X_{n-2,0}$ (labelled $a^{(\emptyset)}b$) and $X_{n-2,1}$ (labelled $a^{(c)}bc$) and repeat the argument, using again Fact 12.4.3.

Suppose, without loss of generality, $a^{(\emptyset)}\emptyset$ is below $X_{n-2,0}$, so there must be some $a^{(b_{j'})}b_{j'}$ (labelled $a^{(c)}c$) below $X_{n-2,1}$. As $a^{(b_{j'})}b_{j'}$ is not below $X_{n-1,1}$, $(b_{j'}) > (b_j)$. Split now $X_{n-1,1}$ into $X_{n-2,2}$ (labelled $a^{(d)}bd$) and $X_{n-2,3}$ (labelled $a^{(cd)}bcd$), and suppose without loss of generality $a^{(b_j)}b_j$ is below $X_{n-2,2}$. Create a new node X (labelled $a^{(c)}bcd$) above $X_{n-2,1}$ and $X_{n-2,3}$. Then it is bigger than $a^{(b_{j'})}b_{j'}$, so its label has the exponent $(b_{j'})$, but it is also above $a^{(b_{n-1}\ldots b_0)}b_{n-1}\ldots b_0$ (labelled $a^{(bcd)}bcd$), so it is also above $a^{(b_j)}b_j$, a contradiction by $(b_{j'}) > (b_j)$. But we detect this only at level $n-2$, and we have to look at arbitrarily big subsets of the construction (in width and depth!) to find a contradiction. Thus, in a strong sense, a recursion is impossible.

Note that we may modify above example, e.g. introduce a smallest node with label $\emptyset$, and then lift the whole construction by adding everywhere a new set of labels, so we can embed it into an arbitrary diagram. Thus, the problem is not only with the base level.

12.5 The Arrow Counting Approach

12.5.1 *Definition and Discussion*

The arrow counting approach is based on the intuition that a total order is the best—it describes a simple law, there is a direction in which things get ever better. Conversely, the empty order is the worst. We have to generalise this to intermediate cases.

Fix now the number of points, say $1, \ldots, n$. The total relation (closed under transitivity, but without loops!) has $n - 1$ pairs involving point n, $n - 2$ new pairs involving point $n - 1$, etc. The empty relation has 0 pairs. So, if the total order relation has k pairs, and the partial order at hand m pairs, m/k gives the quality of the relation, the bigger, the better.

This is, of course, extremely simple, and one can find arguments against it, as it does not look at finer graph properties. It is all the more astonishing that it often—but not always—gives the same results as the much more complicated AGS approach.

We assume that there are no loops in the partial order, i.e. all points are different. We do not think there is a universal answer whether loops simplify a partial order, by making the number of points smaller, or the best picture is just one straight chain without complications. We do not commit, but underline that our approach is intended for the loop-free case only.

We emphasise that this approach favours long chains over several small chains, by transitivity, see e.g. Diagrams 12.5.1, and 12.5.2, and "punishes" unconnected parts, as there are no arrows going in or out. Moreover, and again by transitivity, direction changes in paths are bad.

Of course, one can add other properties to improve on this approach, but we chose explicitly to keep it simple, and code everything into the number of pairs.

A very basic algorithm to calculate the number of arrows We make no claims for efficiency or elegance, this is just a rough sketch for completeness' sake. For n points, make an nxn matrix, M. In M, row A, column B, note 1 iff B is a direct predecessor of A. Close under transitivity by starting with those B which have no direct predecessors. This exist, as we have no cycles. Add the predecessors of B to the predecessors of A (mark the elements, no counting). Mark the A which are done—where no changes occur. Such A can be treated next. Details are left to the interested reader.

12.5.2 Comparison of the AGS and the Arrow Counting Approach

The approaches (AGS and counting) differ considerably, so it is rather surprising that they quite often agree.

12.5.2.1 Agreements

Both approaches agree on all diagrams without equality in Abraham et al. (2009) (Fig. (29/30), 39, 40, (44/45), 47, 55, 59, 63).

Thus both approaches chose the same variant ("?" = 0 or "?" = 1) as better.

12.5.2.2 Disagreements

We consider three examples which show the difference between the two approaches.

Each compares two diagrams, which originate from changing one bit from 0–1 in one vector in a set of vectors. This is the situation of KAL. We will call them (Dia) and (Dib), where $i = 1, 2, 3$.

We show how to represent the diagrams by such vectors, show that each pair is equivalent in the matrix dimension (and the number of nodes, of course), but they may differ in direction changing and number of arrows (considering the transitive closure).

The number of connected components is also the same in each case (2, 4 and 3).

For simplification of the diagrams: all lines are meant to be upward pointing arrows.

The codings by bit sequences and by multisets were checked with a small computer program.

Example 12.5.1
(See Diagrams 12.5.1 and 12.5.2.)
(D1a) $A_0 < B < C < D, E < F$
versus
(D1b) $B < C < D, E < F < A_1$
Vector representation:
$A_0 = (01\ 000\ 11)$, $B = (01\ 001\ 11)$, $C = (01\ 011\ 11)$, $D = (01\ 111\ 11)$, $E = (10\ 000\ 00)$, $F = (10\ 000\ 01)$, $A_1 = (11\ 000\ 11)$,
So A_0 changes to A_1 by $0 \rightarrow 1$ at place 1.
Arrows: (D1a) 7, (D1b) 6, so (D1a) is better by the arrow criterion.
Multiset representation (for better readability, the number of a's is written as exponent):
(D1a):
$A_0 = a^2,\ B = a^3,\ C = a^4,\ D = a^5,\ E = b,\ F = ab$
(D1b):
$B = a^3,\ C = a^4,\ D = a^5,\ E = b,\ F = ab,\ A_1 = a^2b$
So no difference by the multiset criterion.
Direction changes: no difference
So by the AGS criterion, they are equivalent, but not by counting.

Example 12.5.2
(See Diagrams 12.5.3 and 12.5.4.)
(D2a) $G < H,\ I < H,\ I < J,\ K_0 < J,\ L,\ M,\ N$
versus
(D2b) $G < H,\ I < H,\ I < J,\ L < K_1,\ M,\ N$

(Note: M, N are only used to create a sufficiently big antichain, and thus force $1+3$ elements in the multiset representation. D2a has a maximal antichain of size 6, D2b of size 5.)
Vector representation
$G = (0\ 100\ 000)$, $H = (0\ 110\ 000)$, $I = (0\ 010\ 000)$, $J = (0\ 011\ 000)$, $K_0 = (0\ 001\ 000)$, $L = (1\ 000\ 000)$, $K_1 = (1\ 001\ 000)$, $M = (0\ 000\ 010)$, $N = (0\ 000\ 001)$,
Arrows: both 4, so no difference by the arrow criterion.

Direction changes: (D2a) has 1 change more.
Multiset:
(D2a):
$G = ab,\ H = abc,\ I = ac,\ J = a^2c,\ K_0 = a^2,\ L = ad,\ M = bd,\ N = cd$
(D2b):
$G = ab,\ H = abc,\ I = ac,\ J = a^2c,\ L = a^3,\ K_1 = a^4,\ M = ad,\ N = bd$
So they are equivalent by arrow counting, and (D2a) is worse than (D2b) for AGS.

Example 12.5.3
(See Diagrams 12.5.5 and 12.5.6.)
(D3a) $G < H,\ I < H,\ I < J,\ A_0 < J,\ A_0 < B < C < D,\ E < F,\ L$
versus
(D3b) $G < H,\ I < H,\ I < J,\ B < C < D,\ E < F < A_1,\ L < A_1$
Vector representation:
$G = (010\ 00\ 000\ 00)$, $H = (011\ 00\ 000\ 00)$, $I = (001\ 00\ 000\ 00)$,
$J = (001\ 11\ 000\ 11)$, $A_0 = (000\ 11\ 000\ 11)$, $B = (000\ 11\ 001\ 11)$,
$C = (000\ 11\ 011\ 11)$, $D = (000\ 11\ 111\ 11)$, $E = (100\ 10\ 000\ 00)$,
$F = (100\ 10\ 000\ 01)$, $A_1 = (100\ 11\ 000\ 11)$, $L = (100\ 01\ 000\ 10)$,
Multiset representation:

In both cases, the maximal antichain has size 5, so we need at least three elements apart from a.
(D3a):
$G = ab, H = abc, I = ac, J = a^3c, A_0 = a^3, B = a^4, C = a^5, D = a^6, E = ad, F = a^2d,\ L = bd$
(D3b):
$G = a^3b,\ H = a^3bc,\ I = ac,\ J = a^4c,\ B = a^6,\ C = a^7,\ D = a^8,\ E = ad,\ F = a^2d,\ L = bd,\ A_1 = a^2bd$
(D3a) has four direction changes, (D3b) two direction changes, (D3a) has 11 arrows, (D3b) has 10 arrows, so for AGS (D3b) is better, for arrow counting (D3a) is better

Diagram 12.5.1

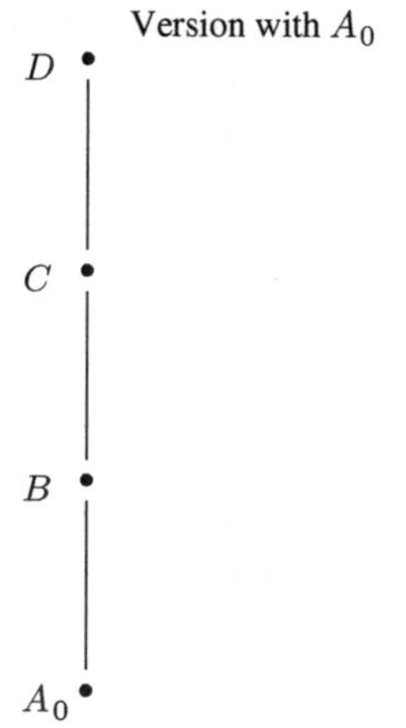

Diagram 12.5.2

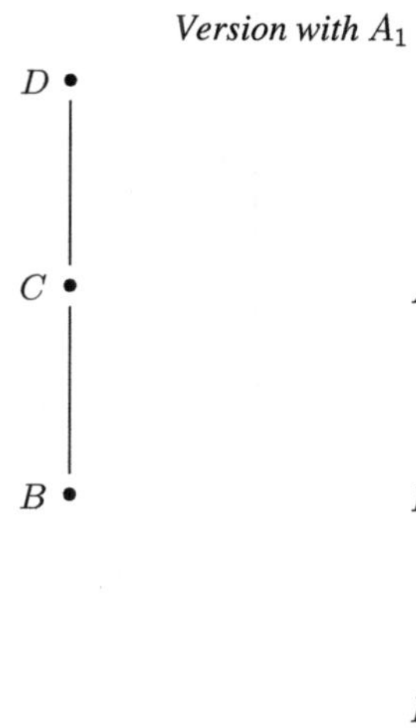

Diagram 12.5.3

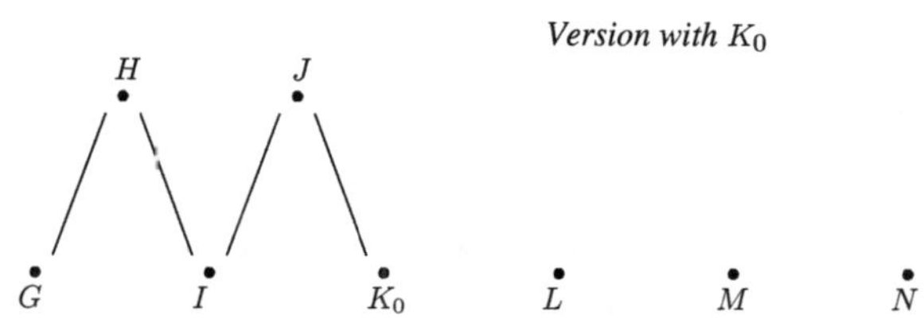

Diagram 12.5.4

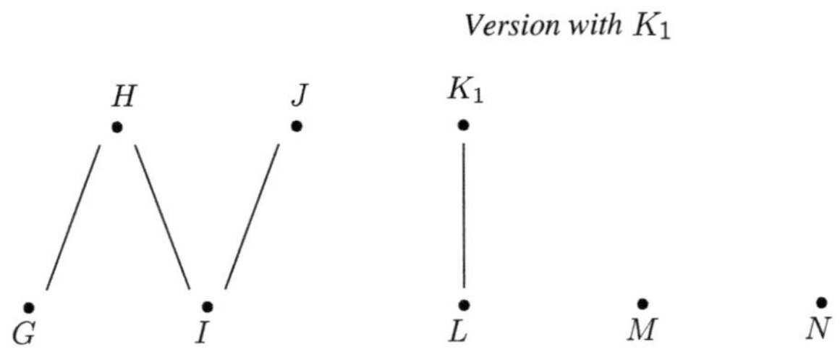

Diagram 12.5.5

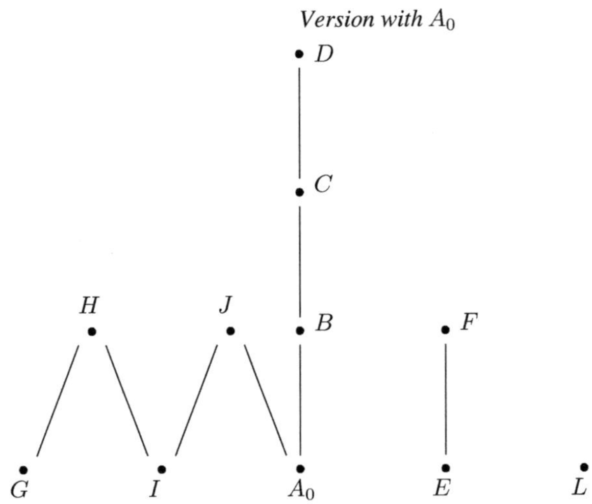

Diagram 12.5.6

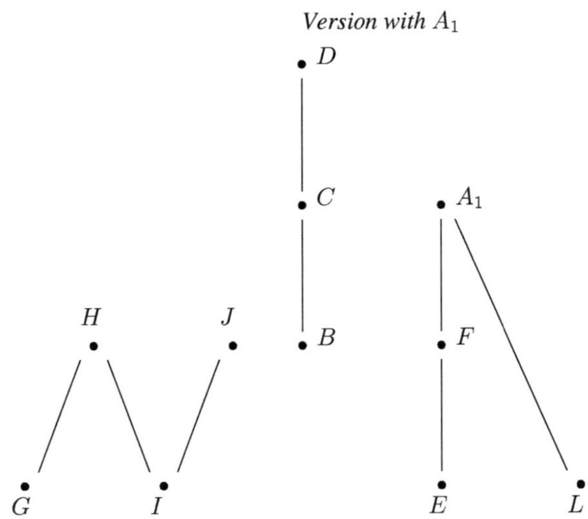

Chapter 13
Equational CTD

13.1 Summary

We apply the equational approach to logic to define numerical equational semantics and consequence relations for contrary-to-duty obligations, thus avoiding some of the traditional known paradoxes in this area. We also discuss the connection with abstract argumentation theory. Makinson and Torre's input output logic and Governatori and Rotolo's logic of violation.

13.2 Methodological Orientation

This chapter gives equational semantics to contrary to duty obligations (CTDs) and thus avoids some of the known CTD paradoxes. The chapter's innovation is on three fronts.

(1) Extend the equational approach from classical logic and from argumentation (Gabby 2012, 2013) to deontic modal logic and contrary-to-duty obligations (Carmo and Jones 2002).
(2) Solve some of the known CTD paradoxes by providing numerical equational semantics and consequence relation to CTD obligation sets.
(3) Have a better understanding of argumentation semantics.

Our starting point in this section is classical propositional logic, a quite familiar logic to all readers. We give it equational semantics and define equational consequence relation. This will explain the methodology and concepts behind our approach and prepare us to address CTD obligations. We then, in Sect. 13.2, present some theory and problems of CTD obligations and intuitively explain how we use equations to represent CTD sets.

Section 13.3 deals with technical definitions and discussions of the equational approach to CTD obligations, Sect. 13.6 compares with input output logic, Sect. 13.7

D.M. Gabbay and K. Schlechta, *A New Perspective on Nonmonotonic Logics*,
DOI 10.1007/978-3-319-46817-4_13

compares with the logic of violation and we conclude in Sect. 13.8 with general discussion and future research.

Let us begin.

13.2.1 Discussion and Examples

Definition 13.2.1 Classical propositional logic has the language of a set of atomic propositions Q (which we assume to be finite for our purposes) and the connectives $\neg$ and $\wedge$. A classical model is an assignment $h : Q \mapsto \{0, 1\}$. h can be extended to all wffs by the following clauses:

- $h(A \wedge B) = 1$ iff $h(A) = h(B) = 1$
- $h(\neg A) = 1 - h(A)$

The set of tautologies are all wffs A such that for all assignments h, $h(A) = 1$.

The other connectives can be defined as usual

$$a \to b = \text{def. } \neg(a \wedge \neg b)$$
$$a \vee b = \neg a \to b = \neg(\neg a \wedge \neg b)$$

Definition 13.2.2 (1) A numerical conjunction is a binary function $\mu(x, y)$ from $[0, 1]^2 \mapsto [0, 1]$ satisfying the following conditions:

(1.1) μ is associative and commutative
$\mu(x, \mu(y, z)) = \mu(\mu(x, y), z)$
$\mu(x, y) = \mu(y, x)$
(1.2) $\mu(x, 1) = x$
(1.3) $x < 1 \Rightarrow \mu(x, y) < 1$
(1.4) $\mu(x, y) = 1 \Rightarrow x = y = 1$
(1.5) $\mu(x, 0) = 0$
(1.6) $\mu(x, y) = 0 \Rightarrow x = 0$ or $y = 0$

(2) We give two examples of a numerical conjunction

$$\mathbf{n}(x, y) = \min(x, y)$$
$$\mathbf{m}(x, y) = xy$$

For more such functions see the Wikipedia entry on t-norms (http://en.wikipedia.org/wiki/T-norm). However, not all t-norms satisfy condition (f) above.

Definition 13.2.3 (1) Given a numerical conjunction μ, we can define the following numerical (fuzzy) version of classical logic.

(1.1) An assignment is any function $\mathbf{h}$ from wff into $[0, 1]$.
(1.2) $\mathbf{h}$ can be extended to $\mathbf{h}_\mu$ defined for any formula by using μ by the following clauses:

- $\mathbf{h}_\mu(A \wedge B) = \mu(\mathbf{h}_\mu(A), \mathbf{h}_\mu(B))$
- $\mathbf{h}_\mu(\neg A) = 1 - \mathbf{h}_\mu(A)$

(2) We call μ-tautologies all wffs A such that for all $\mathbf{h}$, $\mathbf{h}_\mu(A) = 1$.

Remark 13.2.1 Note that on $\{0, 1\}$, $\mathbf{h}_\mu$ is the same as h. In other words, if we assign to the atoms value in $\{0, 1\}$, then $\mathbf{h}_\mu(A) \in \{0, 1\}$ for any A. This is why we also refer to μ as "semantics".

The difference in such cases is in solving equations, and the values they give to the variables $0 < x < 1$.

Consider the equation arising from $(x \to x) \leftrightarrow \neg(x \to x)$. We want

$$\mathbf{h}_\mathbf{m}(x \to x) = \mathbf{h}_\mathbf{m}(\neg(x \to x))$$

We get

$$(1 - \mathbf{m}(x))\mathbf{m}(x) = [1 - \mathbf{m}(x) \cdot (1 - \mathbf{m}(x))]$$

or equivalently

$$\mathbf{m}(x)^2 - \mathbf{m}(x) + \frac{1}{2} = 0.$$

Which is the same as

$$\left(\mathbf{m}(x) - \frac{1}{2}\right)^2 + \frac{1}{4} = 0.$$

There is no real numbers solution to this equation.

However, if we use the $\mathbf{n}$ semantics we get

$$\mathbf{h}_\mathbf{n}(x \to x) = \mathbf{h}_\mathbf{n}(\neg(x \to x))$$

or

$$\min(\mathbf{n}(x), (1 - \mathbf{n}(x)) = 1 - \min(\mathbf{n}(x), 1 - \mathbf{n}(x))$$

$\mathbf{n}(x) = \frac{1}{2}$ is a solution.

Note that if we allow $\mathbf{n}$ to give values to the atoms in $\{0, \frac{1}{2}, 1\}$, then all formulae A will continue to get values in $\{0, \frac{1}{2}, 1\}$. I.e. $\{0, \frac{1}{2}, 1\}$ is closed under the function $\mathbf{n}$, and the function $\nu(x) = 1 - x$.

Also all equations with $\mathbf{n}$ can be solved in $\{0, \frac{1}{2}, 1\}$.

This is not the case for $\mathbf{m}$. Consider for the example the equation corresponding to $x \equiv x \wedge \cdots \wedge x$, $(n + 1$ times).

The equation is $x = x^{n+1}$. We have the solutions $x = 0, x = 1$ and all roots of unity of $x^n = 1$.

Definition 13.2.4 Let I be a set of real numbers $\{0, 1\} \subseteq I \subseteq [0, 1]$. Let μ be a semantics. We say that I supports μ iff the following holds:

(1) For any $x, y \in I$, $\mu(x, y)$ and $\nu(x) = 1 - x$ are also in I.

(2) By a μ expression we mean the following

(2.1) x is a μ expression, for x atomic

(2.2) If X and Y are μ expressions then so are $\nu(X) = (1 - X)$ and $\mu(X, Y)$

(3) We require that any equation of the form $E_1 = E_2$, where E_1 and E_2 are μ expressions has a solution in I, if it is at all solvable in the real numbers.

Remark 13.2.2 Note that it may look like we are doing fuzzy logic, with numerical conjunctions instead of t-norms. It looks like we are taking the set of values $\{0, 1\} \subseteq I \subseteq [0, 1]$ and allowing for assignments **h** from the atoms into I and assuming that I is closed under the application of μ and $\nu(x)$. For $\mu = \mathbf{n}$, we do indeed get a three-valued fuzzy logic with the following truth table, Fig. 13.1.

Note that we get the same system only because our requirement for solving equations is also supported by $\{0, \frac{1}{2}, 1\}$ for **n**.

The case for **m** is different. The values we need are all solutions of all possible equations. It is not the case that we choose a set I of truth values and close under **m**, and ν.

It is the case of identifying the set of zeros of certain polynomials (the polynomials arising from equations). This is an algebraic geometry exercise.

Remark 13.2.3 The equational approach allows us to model what is considered traditionally inconsistent theories, if we are prepared to go beyond $\{0, 1\}$ values. Consider the liar paradox $a \leftrightarrow \neg a$. The equation for this is (both for **m** for **n**) $a = 1 - a$ (we are writing 'a' for '$\mathbf{m}(a)$' or '$\mathbf{n}(a)$' f). This solves to $a = \frac{1}{2}$.

Fig. 13.1 3 valued fuzzy truth table

A	B	$\neg A$	$A \wedge B$	$A \vee B$	$A \to B$
0	0	1	0	0	1
0	$\frac{1}{2}$	1	0	$\frac{1}{2}$	1
0	1	1	0	1	1
$\frac{1}{2}$	0	$\frac{1}{2}$	0	$\frac{1}{2}$	$\frac{1}{2}$
$\frac{1}{2}$	$\frac{1}{2}$	$\frac{1}{2}$	$\frac{1}{2}$	$\frac{1}{2}$	$\frac{1}{2}$
$\frac{1}{2}$	1	$\frac{1}{2}$	$\frac{1}{2}$	1	1
1	0	0	0	1	0
1	$\frac{1}{2}$	0	$\frac{1}{2}$	1	$\frac{1}{2}$
1	1	0	1	1	1

13.2.2 Theories and Equations

The next series of definitions will introduce the methodology involved in the equational point of view.

Definition 13.2.5 (1) (1.1) A classical equational theory has the form

$$\Delta = \{A_i \leftrightarrow B_i \mid i = 1, 2, \ldots\},$$

where A_i, B_i are wffs.

(1.2) A theory is called a B-theory[1] if it has the form

$$x_i \leftrightarrow A_i,$$

where x_i are atomic, and for each atom y there exists at most one i such that $y = x_i$.

(2) (2.1) A function **f**: wff $\rightarrow [0, 1]$ is an μ model of the theory if we have that **f** is a solution of the system of equations $\mathbf{Eq}(\Delta)$.

$$\mathbf{h}_\mu(A_i) = \mathbf{h}_\mu(B_i), i = 1, 2, \ldots$$

(2.2) Δ is μ consistent if it has an μ model

(3) We say that a theory Δ μ semantically (equationally) implies a theory Γ if every solution of $\mathbf{Eq}(\Delta)$ is also a solution of $\mathbf{Eq}(\Gamma)$.
We write

$$\Delta \vDash_\mu \Gamma.$$

Let $\mathbb{K}$ be a family of functions from the set of wff to $[0, 1]$. We say that $\Delta \vDash_{(\mu,\mathbb{K})} \Gamma$ if every μ solution **f** of $\mathbf{Eq}(\Delta)$ such that $\mathbf{f} \in \mathbb{K}$ is also an μ solution of $\mathbf{Eq}(\Gamma)$.

(4) We write

$$A \vDash_\mu B$$

iff the theory $\top \leftrightarrow A$ semantically (equationally) implies $\top \leftrightarrow B$.

Similarly we write $A \vDash_{(\mu,\mathbb{K})} B$. In other words, if for all suitable solutions $\mathbf{f}, \mathbf{f}(A) = 1$ implies $\mathbf{f}(B) = 1$.

[1] B for Brouwer, because we are going to use Brouwer's fixed point theorem to show that theories always have models.

Example 13.2.1 (1) Consider $A \wedge (A \to B)$ does it $\mathbf{m}$ imply B? The answer is yes. Assume $\mathbf{m}(A \wedge (A \to B)) = 1$ then $\mathbf{m}(A)(1 - \mathbf{m}(A)(1 - \mathbf{m}(B))) = 1$. Hence $\mathbf{m}(A) = 1$ and $\mathbf{m}(A)(1 - \mathbf{m}(B)) = 0$. So $\mathbf{m}(B) = 1$.
We now check whether we always have that $\mathbf{m}(A \wedge (A \to B) \to B) = 1$.
We calculate $\mathbf{m}(A \wedge (A \to B) \to B) = [1 - \mathbf{m}(A \wedge (A \to B))(1 - \mathbf{m}(B))]$.

$$= [1 - \mathbf{m}(A)(1 - \mathbf{m}(A)(1 - \mathbf{m}(B))x(1 - \mathbf{m}(B))]$$

Let $\mathbf{m}(A) = \mathbf{m}(B) = \frac{1}{2}$. we get

$$= \left[1 - \frac{1}{2}\left(1 - \frac{1}{2} \times \frac{1}{2}\right) \cdot \frac{1}{2}\right] = 1 - \frac{3}{16} = \frac{13}{16}.$$

Thus the deduction theorem does not hold. We have

$$A \wedge (A \to B) \vDash B$$

but

$$\nvDash A \wedge (A \to B) \to B.$$

(2) (2.1) Note that the theory $\neg a \leftrightarrow a$ is not $(\{0, 1\}, \mathbf{m})$ consistent while it is $(\{0, \frac{1}{2}, 1\}, \mathbf{m})$ consistent.

(2.2) The theory $(x \to x) \leftrightarrow \neg(x \to x)$ is not $([0, 1], \mathbf{m})$ consistent but it is $(\{0, \frac{1}{2}, 1\}, \mathbf{n})$ consistent, but not $(\{0, 1\}, \mathbf{n})$ consistent.

Remark 13.2.4 We saw that the equation theory $x \wedge \neg x \leftrightarrow \neg(x \wedge \neg x)$ has no solutions (no $\mathbf{m}$-models) in $[0, 1]$. Is there a way to restrict $\mathbf{m}$ theories so that we are assured of solutions? The answer is yes. We look at B-theories of the form $x_i \leftrightarrow E_i$ where x_i is atomic and for each x there exists at most one clause in the theory of the form $x \leftrightarrow E$. These we called B-theories. Note that if $x = \top$, we can have several clauses for it. The reason is that we can combine

$$\top \leftrightarrow E_1$$
$$\top \leftrightarrow E_2$$

into

$$\top \leftrightarrow E_1 \wedge E_2.$$

The reason is that the first two equations require

$$\mathbf{m}(E_i) = \mathbf{m}(\top) = 1$$

which is the same as

$$\mathbf{m}(E_1 \wedge E_2) = \mathbf{m}(E_1) \cdot \mathbf{m}(E_2) = 1.$$

If x is atomic different from $\top$, this will not work because

$$x \leftrightarrow E_i$$

requires $\mathbf{m}(x) = \mathbf{m}(E_i)$ while $x \leftrightarrow E_1 \wedge E_2$ requires $\mathbf{m}(x) = \mathbf{m}(E_1)\mathbf{m}(E_2)$.

The above observation is important because logical axioms have the form $\top \leftrightarrow A$ and so we can take the conjunction of the axioms and that will be a theory in our new sense.

In fact, as long as our μ satisfies

$$\mu(A \wedge B) = 1 \Rightarrow \mu(A) = \mu(B) = 1$$

we are OK.

Theorem 13.2.5 *Let Δ be a B-theory of the form*

$$x_i \leftrightarrow E_i.$$

Then for any continuous μ, Δ has a $([0, 1], \mu)$ *model.*

Proof Follows from Brouwer's fixed point theorem, because our equations have the form

$$\mathbf{f}(\vec{x}) = \mathbf{f}(\vec{E}(\vec{x}))$$

in $[0, 1]^n$, where $\vec{x} = (x_1, \ldots, x_n)$ and $\vec{E} = (E_1, \ldots, E_n)$. ■

Remark 13.2.6 If we look at B-theories, then no matter what μ we choose, such theories have μ-models in $[0, 1]$. We get that all theories are μ-consistent. A logic where everything is consistent is not that interesting.

It is interesting, therefore, to define classes of μ models according to some meaningful properties. For example, the class of all $\{0, 1\}$ models. There are other classes of interest. The terminology we use is intended to parallel semantical concepts used and from argumentation theory.

Definition 13.2.6 Let Δ be a B-theory. Let $\mathbf{f}$ be a μ-model of Δ. Let A be a wff.

(1) We say $\mathbf{f}(A)$ is crisp (or decided) if $\mathbf{f}(A)$ is either 0 or 1. Otherwise we say $\mathbf{f}(A)$ is fuzzy or undecided.
(2) (2.1) $\mathbf{f}$ is said to be crisp if $\mathbf{f}(A)$ is crisp for all A.

(2.2) We say that $\mathbf{f} \leq \mathbf{g}$, if for all A, if $\mathbf{f}(A) = 1$ then $\mathbf{g}(A) = 1$, and if $\mathbf{f}(A) = 0$ then $\mathbf{g}(A) = 0$.
We say $\mathbf{f} < \mathbf{g}$ if $\mathbf{f} \leq \mathbf{g}$ and for some A, $\mathbf{f}(A) \notin \{0, 1\}$ but $\mathbf{g}(A) \in \{0, 1\}$.
Note that the order relates to crisp values only.

(3) Define the μ-crisp (or μ-stable) semantics for Δ to be the set of all crisp μ-model of Δ.
(4) Define the μ-grounded semantics for Δ to be the set of all μ-models $\mathbf{f}$ of Δ such that there is no μ-model $\mathbf{g}$ of Δ such that $\mathbf{g} < \mathbf{f}$.
(5) Define the μ-preferred semantics of Δ to be the set of all μ-models $\mathbf{f}$ of Δ such that there is no μ-model $\mathbf{g}$ of Δ with $\mathbf{f} < \mathbf{g}$.
(6) If $\mathbb{K}$ is a set of μ models, we therefore have the notion of $\Delta \vDash_{\mathbb{K}} \Gamma$ for two theories Δ and Γ.

13.2.3 Generating B-Theories

Definition 13.2.7 Let S be a finite set of atoms and let R_a and R_s be two binary relations on S. We use $\mathcal{A} = (S, R_a, R_s)$ to generate a B-theory which we call the argumentation network theory generated on S from the attack relation R_a and the support relation R_s.

For any $x \in S$, let $y_1, \ldots, y_m$ be all the elements y of S such that yR_ax and let $z_1, \ldots, z_n$ be all the elements z of S such that xR_sz (of course m, n depend on x). Write the theory $\Delta_{\mathcal{A}}$.

$$\{x \leftrightarrow \bigwedge z_j \wedge \bigwedge \neg y_i \mid x \in S\}$$

We understand the empty conjunction as $\top$.

These generate equations

$$x = \min(z_j, 1 - y_i)$$

using the $\mathbf{n}$ function or

$$x = (\Pi_j z_j)(\Pi_i(1 - y_i))$$

using the $\mathbf{m}$ function.

Remark 13.2.7 (1) If we look at a system with attacks only of the form $\mathcal{A} = (S, R_a)$ and consider the $\mathbf{n}$(min) equational approach for $[0, 1]$ then $\mathbf{n}$ models of the corresponding B-theory $\Delta_{\mathcal{A}}$ correspond exactly to the complete extensions of (S, R_a). This was extensively investigated in Gabby (2012, 2013). The semantics defined in Definition 13.2.6, the stable, grounded a preferred $\mathbf{n}$-semantics corresponding to the same named semantics in argumentation, when restricted to B-theories arising from argumentation.
If we look at μ other than $\mathbf{n}$, for example, we look at $\mu = \mathbf{m}$, we get different

Fig. 13.2 Example of loops

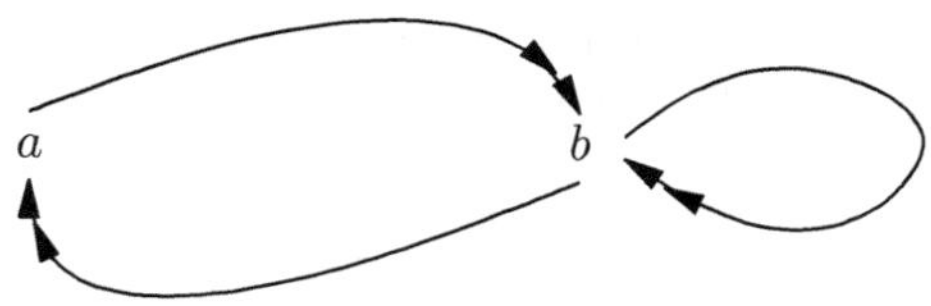

semantics and extensions for argumentation networks. For example, the network of Fig. 13.2 has the **n** extensions $\{a = 1, b = 0\}$ and $\{a = b = \frac{1}{2}\}$ while it has the unique **m** extension $\{a = 1, b = 0\}$.

(2) This correspondence suggests new concepts in the theory of abstract argumentation itself. Let $\Delta_{\mathcal{A}}$, $\Delta_{\mathcal{B}}$ be two B-theories arising from two abstract argumentation system $\mathcal{A} = (S, R_{\mathcal{A}})$ and $\mathcal{B} = (S, R_{\mathcal{B}})$ based on the same set S. Then the notion of $\Delta_{\mathcal{A}} \vDash_{\mathbb{K}} \Delta_{\mathcal{B}}$ as defined in Definition 13.2.5 suggest the following consequence relation for abstract argumentation theory.

- $\mathcal{A} \vDash_{\mathbb{K}} \mathcal{B}$ iff any $\mathbb{K}$-extension ($\mathbb{K}$=complete, grounded, stable, preferred) of $\mathcal{A}$ is also a $\mathbb{K}$-extension of $\mathcal{B}$.

So, for example, the network of Fig. 13.3(a) semantically entails the network of Fig. 13.3(b).

Remark 13.2.8 We can use the connection of equational B-theories with argumentation networks to export belief revision and belief merging from classical logic into argumentation. There has been considerable research into merging of argumentation networks. Classical belief merging offers a simple solution. We only hint here, the full study is elsewhere (Gabbay and Rodrigues 2012).

Let $\mathcal{A}_i = (S, R_i), i = 1, \ldots, n$, be the argumentation networks to be merged based on the same S. Let Δ_i be the corresponding equational theories with the corresponding semantics, based on **n**. Let $\mathbf{f}_i$ be respective models of Δ_i and let μ be a merging function, say $\mu = \mathbf{m}$.

Let $\mathbf{f} = \mu(\mathbf{f}_1, \ldots, \mathbf{f}_n)$. Then the set of all such $\mathbf{f}$s is the semantics for the merge result. Each such an $\mathbf{f}$ yields an extension.

Remark 13.2.9 The equational approach also allows us to generate more general abstract argumentation networks. The set S in (S, R_a) need not be a set of atoms. It can be a set of wffs.

Fig. 13.3 Example for network semantic entailment

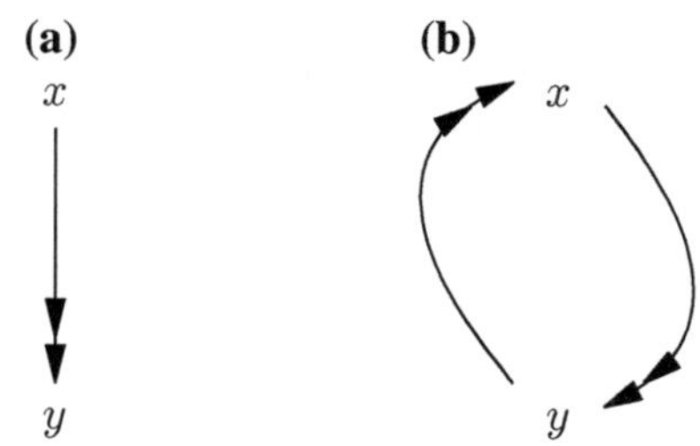

Thus following Definition 13.2.7 and Remark 13.2.7, we get the equations (for each A, B_j and where B_j are all the attackers of A:

$$\mathbf{f}(A) = \mu(\mathbf{f}(\neg B_1), \ldots,).$$

There may not be a solution.

13.3 Equational Modelling of Contrary-to-Duty Obligations

This section will use our μ-equational logic to model contrary-to-duty (CTD) sets of obligations. So far such modelling was done in deontic logic and there are difficulties involved. Major among them is the modelling of the Chisholm set (Chisholm 1963).

We are going to use our equational semantics and consequence of Sect. 13.1 and view the set of contrary-to-duty obligations as a generator for an equational theory. This will give an acceptable paradox free semantics for contrary-to-duty sets.

We shall introduce our semantics in stages. We start with the special case of the generalised Chisholm set and motivate and offer a working semantical solution. Then we show that this solution does not work intuitively well for more general sets where there are loops. Then we indicate a slight mathematical improvement which does work. Then we also discuss a conceptual improvement.

The reader might ask why not introduce the mathematical solution which works right from the start? The answer is that we do not do this for reasons of conceptual motivation, so we do not appear to be pulling a rabbit out of a hat!

We need to first introduce the contrary-to-duty language and its modelling problems.

13.3.1 Contrary-to-Duty Obligations

Consider a semi-formal language with atomic variables $Q = \{p, q, r, \ldots\}$ the connective $\to$ and the unary operator $\bigcirc$. We can write statements like

(1) $\bigcirc\neg$ fence
You should not have a fence
(2) fence $\to \bigcirc$ whitefence
If you do have a fence it should be white.
(3) Fact: fence

We consider a generalised Chisholm set of contrary-to-duty obligations (CTD) of the form

$$Oq_0$$

and for $i = 0, \dots, n$ we have the CTD as

$$q_i \to Oq_{i+1}$$
$$\neg q_i \to O\neg q_{i+1}$$

and the facts $\pm q_j$ for some $j \in J \subseteq \{0, 1, \dots, n+1\}$. Note that for the case of $n = 1$ and fact $\neg q_0$ we have the Chisholm paradox.

13.3.2 Standard Deontic Logic and Its Problems

A logic with modality $\Box$ is **KD** modality if we have the axioms

K0 All substitution instances of classical tautologies
K1 $\Box(p \wedge q) \equiv (\Box p \wedge \Box q)$
K2 $\vdash A \Rightarrow \vdash \Box A$
D $-\Box\bot$

It is complete for frames of the form (S, R, a), where $S \neq \varnothing$ is a set of possible worlds, $a \in S$, $R \subseteq S \times S$ and $\forall x \exists y(xRy)$.

Standard Deontic Logic **SDL** is a **KD** modality O. We read $u \vDash Op$ as saying p holds in all ideal worlds relative to u, i.e. $\forall t(uRt \Rightarrow t \vDash p)$. So the set of ideal worlds relative to u is the set $I(u) = \{t \mid uRt\}$.

The **D** condition says $I(x) \neq \varnothing$ for $x \in S$.

Following (Jones and Poern 1985), let us quickly review some of the difficulties facing **SDL** in formalizing the Chisholm paradox.

The Chisholm Paradox

A. Consider the following statements:

1. It ought to be that a certain man go to the assistance of his neighbour.
2. It ought to be that if he does go he tell them he is coming.
3. If he does not go then he ought not to tell them he is coming.
4. He does not go.

It is agreed that intuitively (1)–(4) of Chisholm set A are consistent and totally independent of each other. Therefore it is expected that their formal translation into logic **SDL** should retain these properties.

B. Let us semantically write the Chisholm set in semi-formal English, where p and q as follows, p means HELP and q means TELL.

1. Obligatory p.
2. $p \to$ Obligatory q.
3. $\neg p \to$ Obligatory $\neg q$.
4. $\neg p$.

Consider also the following:

5. p.
6. Obligatory q.
7. Obligatory $\neg q$. q.

We intuitively accept that (1)–(4) of B are consistent and logically independent of each other. Also we accept that (3) and (4) imply (7), and that (2) and (5) imply (6). Note that some authors would also intuitively expect to conclude (6) from (1) and (2).

Now suppose we offer a logical system **L** and a translation τ of (1), (2), (3), (4) of Chisholm into **L**.

For example, **L** could be Standard Deontic Logic or **L** could be a modal logic with a dyadic modality $O(X/Y)$ (X is obligatory in the context of Y). We expect some coherence conditions to hold for the translation, as listed in Definition 13.3.1.

Definition 13.3.1 (*Coherence conditions for representing contrary-to-duty obligations set in any logic*) We now list coherence conditions for the translation τ and for **L**.

We expect the following to hold:

(a) "Obligatory X" is translated the same way in (1), (2) and (3).
Say $\tau(\text{Obligatory } X) = \varphi(X)$.

(b) (2) and (3) are translated the same way, i.e. we translate the form:
(23): $X \to$ Obligatory Y
to be $\psi(X, Y)$ and the translation does not depend on the fact that we have (4) $\neg p$ as opposed to (5) p.
Furthermore, we might, but not necessarily, expect $\psi(X/\top) = \varphi(X)$.

(c) if X is translated as $\tau(X)$ then (4) is translated as $\neg\tau(X)$, the form (23) is translated as $\psi(\tau(\text{X}), \tau(Y))$ and (1) is translated as $\varphi(\tau(X))$.

(d) the translations of (1)–(4) remain independent in **L** and retain the connections that the translations of (2) and (5) imply the translation of (6), and the translations of (3) and (4) imply the translation of (7).

(e) the translated system maintains its properties under reasonable substitution in **L**.
The notion of reasonable substitution is a tricky one. Let us say for the time being that if we offer a solution for one paradox, say $\Pi_1(p, q, r, \ldots)$ and by substitution for $p, q, r, \ldots$ we can get another well-known paradox Π_2, then we would like to have a solution for Π_2. This is a reasonable expectation from mathematical reasoning. We give a general solution to a general problem which yields specific solutions to specific problems which can be obtained from the general problem.

(f) the translation is essentially linguistically uniform and can be done item by item in a uniform way depending on parameters derived from the entire database. To explain what we mean consider in classical logic the set

(1) p
(2) $p \to q$.

To translate it into disjunctive normal form we need to know the number of atoms to be used. Item (1) is already in normal form in the language of $\{p\}$ but in the language of $\{p, q\}$ its normal form is $(p \wedge q) \vee (p \wedge \neg q)$. If we had another item

(3) r

then the normal form of p in the language of $\{p, q, r\}$ would be
$(p \wedge q \wedge r) \vee (p \wedge q \wedge \neg r) \vee (p \wedge \neg q \wedge r) \vee (p \wedge \neg q \wedge \neg r)$.

The moral of the story is that although the translation of (1) is uniform algorithmically, we need to know what other items are in the database to set some parameters for the algorithm.

Jones and Pörn, for example, examine in Jones and Poern (1985) possible translations of the Chisholm (1)–(4) into **SDL**. They make the following points:

(1) If we translate according to, what they call, option a:

    (1a) Op
    (2a) $O(p \to q)$
    (3a) $\neg p \to O\neg q$
    (4a) $\neg p$

    then we do not have consistency, although we do have independence

(2) If we translate the Chisholm item (2) according to what they call option b:

    (2b) $p \to Oq$

    then we have consistency but not independence, since (4a) implies logically (2b).

(3) If (3a) is replaced by

    (3b) $O(\neg p \to \neg q)$

    then we get back consistency but lose independence, since (1a) implies (3b).

(4) Further, if we want (2) and (5) to imply (6), and (3) and (4) to imply (7) then we cannot use (3b) and (2a).

The translation of the Chisholm set is a "paradox" because known translations into Standard Deontic Logic (the logic with O only) are either inconsistent or dependent.

All the above statements together are logically independent and are consistent. Each statement is independent of all the others. If we want to embed the (model them) in some logic, we must preserve these properties and correctly get all intuitive inferences from them.

Remark 13.3.1 We remark here that the Chisholm paradox has a temporal dimension to it. The $\pm$tell comes before the $\pm$go. In symbols, the $\pm q$ is temporally before the $\pm p$. This is not addressed in the above discussion.

Consider a slight variation:

(1) It ought to be that a certain man go to the assistance of his neighbour.
(2) It ought to be that if he does not go he should write a letter of explanation and apology.
(3) If he does go, then he ought not write a letter of explanation and apology.
(4) He does not go.

Here $p =$ he does go and $q =$ he does not write a letter. Here q comes after p.

It therefore makes sense to supplement the Chisholm paradox set with a temporal clause as follows:

(1) p comes temporally before q.

In the original Chisholm paradox the supplement would be:

(1) Tell comes temporally before go.

13.3.3 The Equational Approach to CTD

We are now ready to offer equational semantics for CTD. Let us summarise the tools we have so far.

(1) We have μ semantics for the language of classical logic.
(2) Theories are sets of equivalences of the form $E_1 \leftrightarrow E_2$.
(3) We associate equations with such equivalences.
(4) Models are solutions to the equations.
(5) Using models, we define consequence between theories.
(6) Axioms have the for $\top \leftrightarrow E$
(7) B-theories have the form $x \leftrightarrow E$, where x is atomic and E is unique to x.
(8) We always have solutions for equations corresponding to B-theories.

Our strategy is therefore to associate a B-theory $\Delta(\mathbb{C})$ with any contrary-to-duty set $\mathbb{C}$ and examine the associated μ-equations for a suitable μ. This will provide semantics and consequence for the CTD sets and we will discuss how good this representation is.

The perceptive reader might ask, if Obligatory q is a modality, how come we hope to successfully model it in μ classical logic? Don't we need modal logic of it? This is a good question and we shall address it later. Of course modal logic can be translated into classical logic, so maybe the difficulties and paradoxes are "lost in translation". See Remark 13.3.8.

Definition 13.3.2 (1) Consider a language with atoms, the semi-formal $\rightarrow$ and $\neg$ and a semi-formal connective O.
A contrary-to-duty expression has the form $x \rightarrow Oy$ where x and y are literals, i.e. either atoms q or negations of atoms $\neg q$, and where we also allow for x not to appear. We might write $\top \rightarrow Oy$ in this case, if it is convenient.

(2) Given a literal x and a set $\mathbb{C}$ of CTD expressions, then the immediate neighbourhood of x in $\mathbb{C}$. is the set $\mathbb{N}_x$ of all expressions from $\mathbb{C}$ of the form

$$z \to Ox$$

or the form

$$x \to Oy.$$

(3) A set $\mathbb{F}$ of facts is just a set of literals.
(4) A general CTD system is a pair $(\mathbb{C}, \mathbb{F})$
(5) A Chisholm CTD set $\mathbb{CH}$ has the form

$$\begin{aligned} &x_i \to Ox_{i+1} \\ &\neg x_i \to O\neg x_{i+1} \\ &Ox_1 \end{aligned}$$

where $1 \leq i \leq m$ and x_i are literals (we understand that $\neg\neg x$ is x).

Example 13.3.1 Figure 13.4 shows a general CTD set

$$\mathbb{C} = \{a \to Ob, b \to O\neg a\}$$

Figure 13.5 shows a general Chisholm set. We added an auxiliary node x_0 as a starting point.

Figure 13.6 shows a general neighbourhood of a node x.

We employed in the figures the device of showing, whenever $x \to Oy$ is given, two arrows, $x \to y$ and $x \twoheadrightarrow \neg y$. The single arrow $x \to y$ means "from x go to y" and the double arrow $x \twoheadrightarrow \neg y$ means "from x do not go to $\neg y$".

Remark 13.3.2 In Figs. 13.4, 13.5 and 13.6 we understand that an agent is at the starting point x_0 and he has to go along the arrows $\to$ to follow his obligations. He should not go along any double arrow, but if he does, new obligations (contrary to duty) appear.

This is a mathematical view of the CTD. The obligations have no temporal aspect to them but mathematically there is an obligation progression $(\pm x_0, \pm x_1, \pm x_2, \ldots)$.

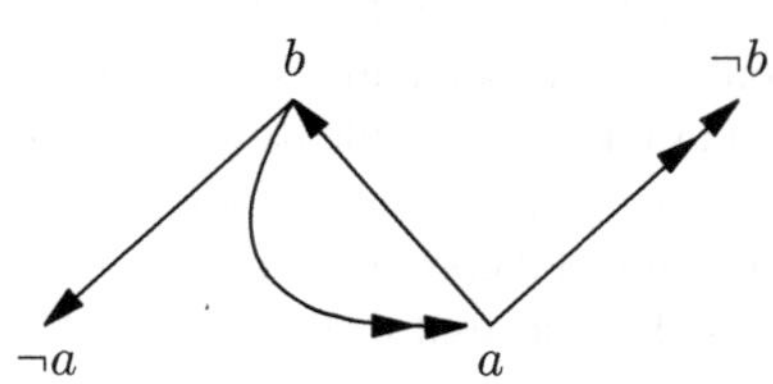

Fig. 13.4 Example for illustrating the Equational Approach

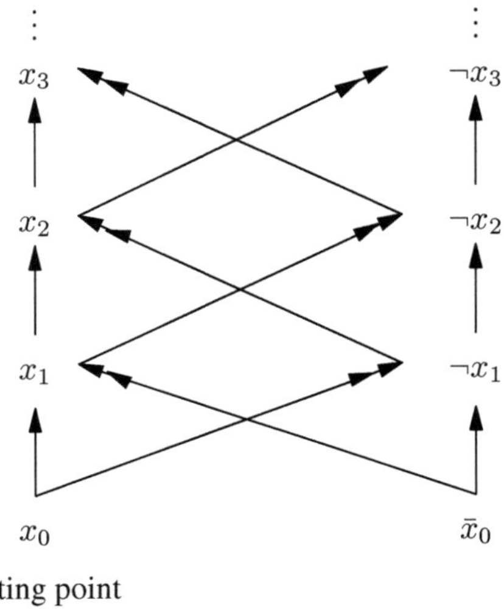

Fig. 13.5 Illustrating a CTD system

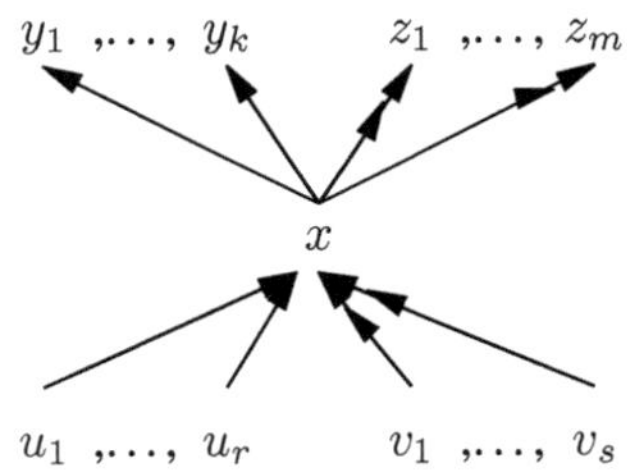

Fig. 13.6 General condition of a point x in a network

In the Chisholm example, the obligation progression is (± go, ±tell), while the practical temporal real life progression is (±tell, ±go). We are modelling the obligation progression.

To be absolutely clear about this we give another example where there is similar progression. Take any Hilbert axiom system for classical logic. The consequence relation $A \vdash B$ is timeless. It is a mathematical relation. But in practice to show $A \vdash B$ from the axioms, there is a progression of substitutions and uses of modus ponens. This is a mathematical progression of how we generate the consequence relations.

Remark 13.3.3 We want to associate equations with a given CTD set. This is essentially giving semantics to the set. To explain the methodology of what we are doing, let us take an example from the modal logic **S4**. This modal logic has wffs of the form $\Box q$. To give semantics for $\Box q$ we need to agree on a story for "$\Box$" which respects the logical theorems which "$\Box$" satisfies (completeness theorem). The following are possible successful stories about "$\Box$" for which there is completeness.

(1) Interpret $\Box$ to mean provability in Peano arithmetic.
(2) $\Box q$ means that q holds in all possible accessible situations (Kripke models).

(3) $\Box$ means topological interior in a topological space.
(4) $\Box$ means the English progressive tense:
$\Box$ eat = "is eating"
(5) $\Box$ means constructive provability.

For the case of CTD we need to adopt a story respecting the requirement we have on CTD.

Standard deontic logic **SDL** corresponds to the story that the meaning of OA in a world is that A holds in all accessible relative ideal worlds. It is a good story corresponding to the intuition that our obligations should take us to a better worlds. Unfortunately, there are difficulties with this story, as we have seen.

Our story is different. We imagine we are in states and our obligations tell us where we can and where we cannot go from our state. This is also intuitive. It is not descriptive as the ideal world story is, but it is operational, as real life is.

Thus in Fig. 13.6 an agent at node x wants to say that he is a "good boy". So at x he says that he intends to go to one of $y_1, \ldots, y_k$ and that he did not come to x from $v_1, \ldots, v_k$, where the obligation was not to go to x.

Therefore the theory we suggest for node x is

$$x \leftrightarrow \left(\bigwedge_i y_i \wedge \bigwedge_j \neg v_j \right)$$

We thus motivated the following intuitive, but not final, definition.

Let $\mathbb{C}$ be a CTD set and for each x let $\mathbb{N}_x$ be its neighbourhood as in Fig. 13.6.

We define the theory $\Delta(\mathbb{C})$ to be

$$\left\{ x \leftrightarrow \left(\bigwedge_i y_i \wedge \bigwedge_j \neg v_j \right) \mid \text{for all } \mathbb{N}_x \right\}. \qquad (*1)$$

This definition is not final for technical reasons. We have literals "$\neg q$" and we do not want equivalences of the form $\neg q \leftrightarrow E$. So we introduce a new atom $\bar{q}$ to represent $\neg q$ with the theory $\bar{q} \leftrightarrow \neg q$.

So we take the next more convenient definition.

Definition 13.3.3 (1) Let $\mathbb{C}$ be a CTD set using the atoms Q. Let $Q^* = Q \cup \{\bar{q} \mid q \in Q\}$, where $\bar{q}$ are new atoms.

Consider $\mathbb{C}^*$ gained from $\mathbb{C}$ by replacing any occurrence of $\neg q$ by $\bar{q}$, for $q \in Q$. Using this new convention Fig. 13.5 becomes Fig. 13.7.

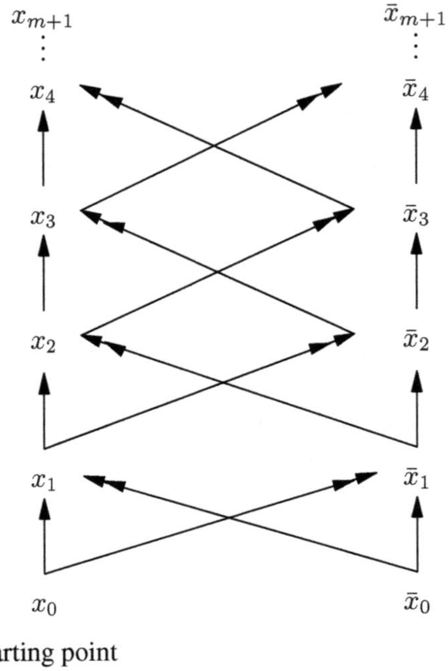

Fig. 13.7 Figure 13.5 after intervention

(2) The theory for the CTD set represented by Fig. 13.7 is therefore

$$
\begin{array}{l}
x_0 \leftrightarrow \top, \bar{x}_0 \leftrightarrow \bot \\
x_0 \leftrightarrow x_1, \bar{x}_0 \leftrightarrow \bar{x}_1 \\
x_i \leftrightarrow x_{i+1} \wedge \bar{x}_{i-1} \\
\bar{x}_i \leftrightarrow \bar{x}_{i+1} \wedge x_{i-1} \\
\bar{x}_i \leftrightarrow \neg x_i \\
x_{m+1} \leftrightarrow \bar{x}_m \\
\bar{x}_{m+1} \leftrightarrow x_m \\
\text{for } 1 \le i \le m
\end{array}
$$

The above is not a B-theory. The variable $\bar{x}_i$ has two clauses associated with it. (x_0 is OK because the second equation is $\top$). So is $\bar{x}_0$.
It is convenient for us to view clause $\bar{x}_i = \neg x_i$ as an integrity constraint. So we have a B-theory with some additional integrity constraints.
Note also that we regard all x_i and $\bar{x}_i$ as different atomic letters. If some of them are the same letter, i.e. $x_i = x_j$ then we regard that as having further integrity constraints of the form $x_i \leftrightarrow x_j$.

(3) The equations corresponding to this theory are

$$\begin{aligned} & x_0 = 1, \bar{x}_0 = 0 \\ & x_0 = x_1, \bar{x}_0 = \bar{x}_1 \\ & x_i = \min(x_{i+1}, 1 - \bar{x}_{i-1}) \\ & \bar{x}_i = \min(\bar{x}_{i+1}, 1 - x_{i-1}) \\ & \bar{x}_i = 1 - x_i \\ & x_{m+1} = 1 - \bar{x}_m \\ & \bar{x}_{m+1} = 1 - x_m \\ & \text{for } 1 \leq i \leq m \end{aligned}$$

Remember we regard the additional equation

$$\bar{x}_i = 1 - x_i$$

as an integrity constraint.
Note also that we regard all x_i and $\bar{x}_i$ as different atomic letters. If some of them are the same letter, i.e. $x_i = x_j$ then we regard that as having further integrity constraints of the form $x_i \leftrightarrow x_j$. The rest of the equations have a solution by Brouwer's theorem. We look at these solutions and take only those which satisfy the integrity constraints. There may be none which satisfy the constraints, in which case the system overall has no solution!

(4) The dependency of variables in the equations of Fig. 13.7 is described by the relation $x \Rightarrow y$ reading (x depends on y), where

$$x \Rightarrow y = \text{def. } (x \rightarrow y) \vee (y \twoheadrightarrow x).$$

Figure 13.8 shows the variable dependency of the equations generated by Fig. 13.7 up to level 3.

Lemma 13.3.4 *(1) The equations associated with the Chisholm set of Fig. 13.7 have the following unique solution, and this solution satisfies the integrity constraints:*

$$x_0 = 1, x_i = 1, \bar{x}_i = 0, \textit{ for } 0 \leq i \leq m+1$$

(2) All the equations are independent.

Proof

(1) By substitution we see the proposed solution is actually a solution. It is unique because $x_0 = 1$ and the variable dependency of the equations, as shown in Fig. 13.8, is acyclic.
(2) Follows from the fact that the variable dependency of the equations is acyclic. The variable x_i can depend only on the equations governing the variables below

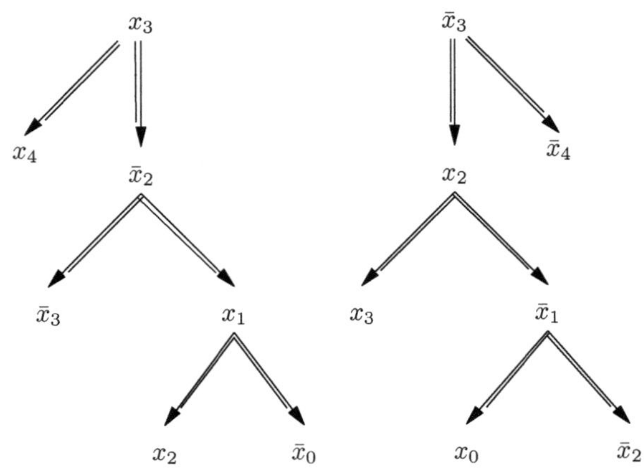

Fig. 13.8 Variable dependency

it in the dependency graph. Since it has the last equation in the tree, it cannot be derived from the equations below it. ∎

Remark 13.3.5 We mentioned before that the theory (*1) and its equations above do not work for loops. Let us take the set $a \to \bigcirc\neg a$.

The graph for it, according to our current modelling would be Fig. 13.9.

The equations for this figure would be

$$a = \min(1 - a, \bar{a})$$
$$a = 1 - \bar{a}$$

which reduces to

$$a = 1 - a$$
$$a = \frac{1}{2}$$

It does not have a consistent {0, 1} solution.

We can fix the situation by generally including the integrity constraints $\bar{x} = 1 - x$ in the graph itself.

So Fig. 13.9 becomes Fig. 13.10, and the equations become

$$a = \min(\bar{a}, 1 - a, 1 - \bar{a})$$
$$\bar{a} = 1 - a$$

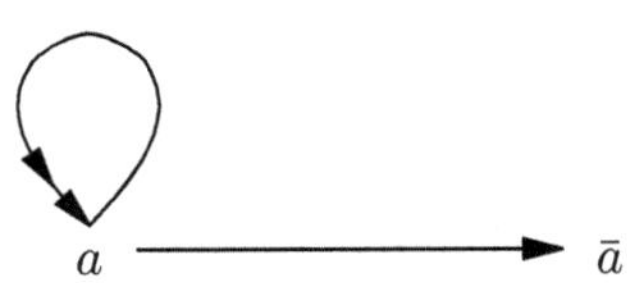

Fig. 13.9 Graph for the CTD $a \to \bigcirc\neg a$

Fig. 13.10 Graph fixing Fig. 13.9

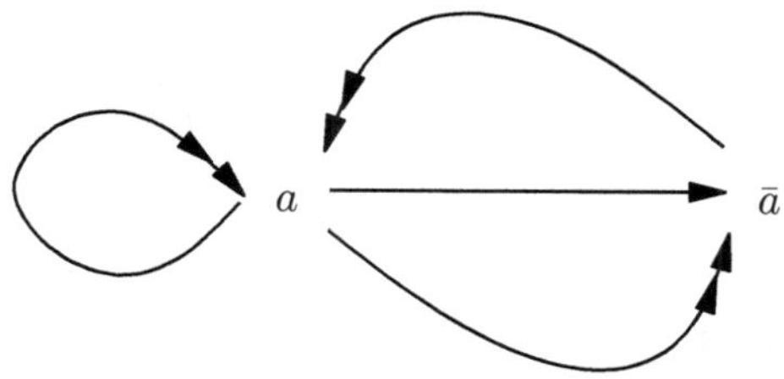

The two equations reduce to

$$a = \min(a, 1 - a)$$

which has the solution

$$a = 0, \bar{a} = 1$$

which fits our intuition.

Let us call this approach, (namely the approach where we do not view the equations $\bar{x} = 1 - x$ as integrity constraints but actually insert instead double arrow in the graph itself) the mathematical approach. What we have done here is to incorporate the integrity constraints $\bar{x} = 1 - x$ into the graph. Thus Fig. 13.7 would become Fig. 13.11, and the equations for the figure would become

Fig. 13.11 Graph modifying Fig. 13.7

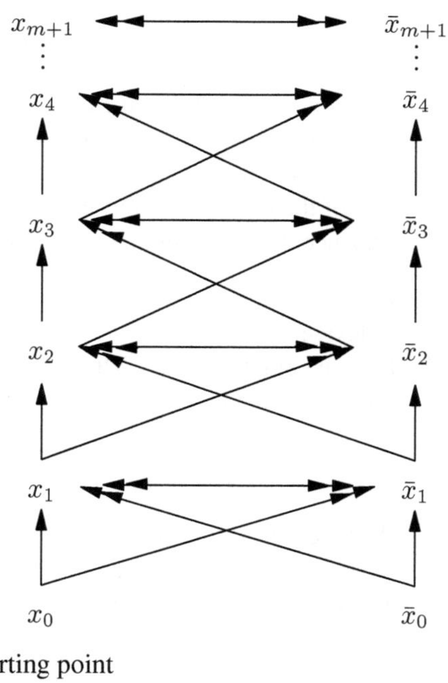

$$\begin{aligned} x_i &= \min(x_{i+1}, 1 - \bar{x}_i, 1 - \bar{x}_{i-1}) \\ \bar{x}_i &= \min(\bar{x}_{i+1}, 1 - x_i, 1 - x_{i-1}) \\ x_0 &= 1, \bar{x}_0 = 0 \\ x_{m+1} &= \min(1 - \bar{x}_{m+1}, 1 - \bar{x}_m) \\ \bar{x}_{m+1} &= min(1 - x_{m+1}, 1 - x_m) \end{aligned}$$

for $1 \leq i \leq m$.

For the Chisholm set, we still get the same solution for these new equations, namely

$$\begin{aligned} x_0 &= x_1 = \cdots x_{m+1} = 1 \\ \bar{x}_0 &= \bar{x}_1 = \cdots = \bar{x}_{m+1} = 0 \end{aligned}$$

The discussion that follows in Definition 13.3.4 onwards applies equally to both graphs.

We shall discuss this option in detail in Sect. 2.4.

The reader should note that we used here a mathematical trick. In Fig. 13.11, there are two conceptually different double arrows. The double arrow $x_i \twoheadrightarrow x_{i+1}$ comes from an obligation $x_i \rightarrow \bigcirc x_{i+1}$, while the double arrows $x \twoheadrightarrow \bar{x}$ and $\bar{x} \twoheadrightarrow x$ come from logic (because $\bar{x} = \neg x$). We are just arbitrarily mixing them in the graph!

Definition 13.3.4 Consider Fig. 13.7. Call this graph by $\mathbb{G}(m + 1)$. We give some definitions which analyse this figure.

First note that this figure can be defined analytically as a sequence of pairs

$$((x_0, \bar{x}_0), (x_1, \bar{x}_1), \ldots, (x_{m+1}, \bar{x}_{m+1})).$$

The relation $\rightarrow$ can be defined between nodes as the set of pairs $\{(x_i, x_{i+1})$ and $(\bar{x}_i, \bar{x}_{i+1})$ for $i = 0, 1, \ldots, m\}$. The relation $\twoheadrightarrow$ can be defined between nodes as the set of pairs $\{(x_i, \bar{x}_{i+1})$ and $(\bar{x}_i, x_{i+1})$ for $i = 0, 1, \ldots, m\}$. The starting point is a member of the first pair, in this case it is x_0, the left hand element of the first pair in the sequence, but we could have chosen $\bar{x}_0$ as the starting point.

(1) Let xRy be defined as $(x \rightarrow y) \vee (x \twoheadrightarrow y)$ and let R^* be the transitive and reflexive closure of R.
(2) Let z be either x_i or $\bar{x}_i$. The truncation of $\mathbb{G}(m + 1)$ at z is the subgraph of all points above z including z and $\bar{z}$ and all the arrow connections between them.

$$\mathbb{G}_z = \{y | zR^*y\} \cup \{\bar{z}\}$$

We take z as the starting point of $\mathbf{G}(m + 1)_z$. Note that $\mathbb{G}(m + 1)_z$ is isomorphic to $\mathbb{G}(m + 1 - i)$. It is the same type of graph as $\mathbb{G}(m + 1)$, only it starts at z. The corresponding equations for $\mathbb{G}_z$ will require $z = 1$.

(3) A path in the graph is a full sequence of points $(x_0, z_1, \ldots, z_{m+1})$ where z_i is $\bar{x}_i$ or x_i.[2]
(4) A set of "facts" $\mathbb{F}$ in the graph is a set of nodes choosing at most exactly one of each pair $\{x_i, \bar{x}_i\}$.
(5) A set of facts $\mathbb{F}$ restricts the possible paths by stipulating that the paths contain the nodes in the facts.

Example 13.3.2 Consider Fig. 13.7. The following is a path Π in the graph

$$\Pi = (x_0, x_1, x_2, x_3, \ldots, x_{m+1})$$

If we think in terms of an agent going along this path, then this agent committed two violations. Having gone to $\bar{x}_1$ instead of to x_1, he committed the first violation. From $\bar{x}_1$, the CTD says he should have gone to $\bar{x}_2$, but he went to x_2 instead. This is his second violation. After that he was OK.

Now look at the set of facts $= \{\bar{x}_1, x_2\}$. This allows for all paths starting with $(x_0, \bar{x}_1, x_2, \ldots)$. So our agent can still commit violations after x_2. We need more facts about his path.

Suppose we add the fact $\bar{x}_4$. So our set is now $\mathbb{F} = \{\bar{x}_1, x_2, \bar{x}_4\}$.

We know now that the agent went from x_2 onto $\bar{x}_4$. The question is, did he pass through $\bar{x}_3$? If he goes to x_3, there is no violation and from there he goes to $\bar{x}_4$, and now there is violation.

If he goes to x_3, then the violation is immediate but when he goes from $\bar{x}_3$ to $\bar{x}_4$, there is no violation.

The above discussion is a story. We have to present it in terms of equations, if we want to give semantics to the facts.

Example 13.3.3 Let us examine what is the semantic meaning of facts. We have given semantic meaning to a Chisholm set $\mathbb{C}$ of contrary to duties; we constructed the graph, as in Fig. 13.7 and from the graph we constructed the equations and we thus have equational semantics for $\mathbb{C}$.

We now ask what does a fact do semantically?

We know what it does in terms of our story about the agent. We described it in Example 13.3.2. What does a fact do to the graph? Let us take as an example the fact $\bar{x}_3$ added to the CTD set of Fig. 13.7. What does it do? The answer is that it splits the figure into two figures, as shown in Figs. 13.12 and 13.13.

Note that Fig. 13.13 is the truncation of Fig. 13.7 at $\bar{x}_3$, and Fig. 13.12 is the complement of this truncation.

[2]Note that the facts are sets of actual nodes. We can take the conjunction of the actual nodes as a formula faithfully representing the set of facts. Later on in this chapter we will look at an arbitrary formula ϕ as generating the set of facts $\{y|y$ is either x_i or $\neg x_i$, unique for each i, such that $\phi| - y\}$.

According to this definition, $\phi = x_1 \vee x_2$, generates no facts. We will, however, find it convenient later in the chapter, (in connection with solving the Miner's Paradox, Remark 13.4.2 below) to regard a disjunction as generating several possible sets of facts, one for each disjunct. See also Remark 13.4.1 below.

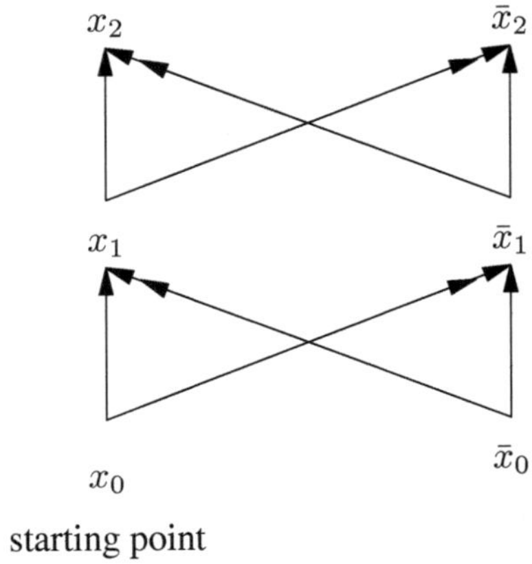

Fig. 13.12 Complement of the truncation of Fig. 13.7

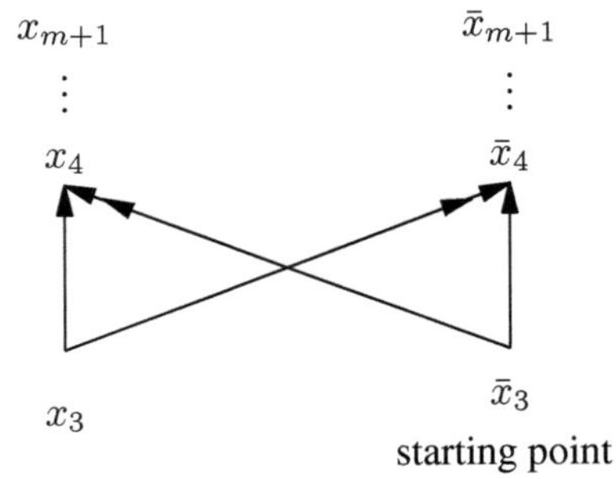

Fig. 13.13 The truncation of Fig. 13.7

Thus the semantical graphs and equations associated with $(\mathbb{C}, \{\bar{x}_3\})$ are the two figures, Figs. 13.12 and 13.13 and the equations they generate.

The "facts" operation is associative. Given another fact, say z it will be in one of the figures and so that figure will further split into two.

Definition 13.3.5 Given a Chisholm system $(\mathbb{C}, \mathbb{F})$ as in Definition 13.3.2 we define its semantics in terms of graphs and equations. We associate with it with following system of graphs (of the form of Fig. 13.7) and these graphs will determine the equations, as in Definition 13.3.3.

The set $\mathbb{C}$ has a graph $\mathbb{G}(\mathbb{C})$. The set $\mathbb{F}$ can be ordered according to the relation R in the graph $\mathbb{G}(\mathbb{C})$ as defined in Definition 13.3.4. Let $(z_1, \ldots, z_k)$ be the ordering of $\mathbb{F}$.

We define by induction the following graphs:

(1) (1.1) Let $\mathbb{G}_k^+$ be $\mathbb{G}(\mathbb{C})_{z_k}$, (the truncation of $\mathbb{G}(\mathbb{C})$ at z_k). item Let $\mathbb{G}_k^-$ be $\mathbb{G}(\mathbb{C})-\mathbb{G}_k^+$ (the remainder graph after deleting from it the top part $\mathbb{G}_k^+$).

(1.2) The point z_{k-1} is in the graph $\mathbb{G}_k^-$.

(2) Assume that for z_i, $1 < i \leq k$ we have defined $\mathbb{G}_i^+$ and $\mathbb{G}_i^-$ and that $\mathbb{G}_i^+$ is the truncation of $\mathbb{G}_{i+1}^-$ at point z_i, and that $\mathbb{G}_i^- = \mathbb{G}_{i+1}^- - \mathbb{G}_i^+$. We also assume that

z_{i-1} is in $\mathbb{G}_i^-$.

Let $\mathbb{G}_{i-1}^+ = (\mathbb{G}_i^-)_{z_{i-1}}$, (i.e. the truncation of $\mathbb{G}_i^-$ at point z_{i-1}).
Let $\mathbb{G}_{i-1}^- = \mathbb{G}_i^- - \mathbb{G}_{i-1}^+$.

(3) The sequence of graphs $\mathbb{G}, \mathbb{G}_1^-, \mathbb{G}_1^+, \mathbb{G}_2^+, \ldots, \mathbb{G}_k^+$ is the semantical object for $(\mathbb{C}, \mathbb{F})$. They generate equations which are the equational semantics for $(\mathbb{C}, \mathbb{F})$.

Example 13.3.4 Consider a system $(\mathbb{C}, \mathbb{F})$ where $\mathbb{F}$ is a maximal path, i.e. $\mathbb{F}$ is the sequence $(z_1, \ldots, x_{m+1})$. The graph system for it will be as in Fig. 13.14.

Remark 13.3.6 The nature of the set of facts $\mathbb{F}$ is best understood when the set $\mathbb{C}$ of Chisholm CTDs is represented as a sequence. Compare with Definition 13.3.5.

$\mathbb{C}$ has the graph $\mathbb{G}(\mathbb{C})$. The graph can be represented as a sequence

$$\mathbf{E} = ((x_0, \bar{x}_0), (x_1, \bar{x}_1), \ldots, (x_{m+1}, \bar{x}_{m+1}))$$

together with the starting point (x_0).

When we get a set of facts $\mathbb{F}$ and arrange it as a sequence $(z_1, \ldots, z_k)$ in accordance with the obligation progression, we can add x_0 to the sequence and look at $\mathbb{F}$ as

$$\mathbb{F} = (x_0, z_1, \ldots, z_k).$$

We also consider $(\mathbb{E}, \mathbb{F})$ as a pair, one the sequence $\mathbb{E}$ and the other as a multiple sequence of starting points. The graph $\mathbb{G}_i$ is no more and no less than the subsequence $\mathbb{E}_i$, beginning from the pair $(z_i, \bar{z}_i)$ up to the pair $(z_{i+1}, \bar{z}_{i+1})$ but *not* including $(z_{i+1}, \bar{z}_{i+1})$.

This way it is easy to see how $\mathbb{G}$ is the sum of all the $\mathbb{G}_i$, strung together in the current progression order. Furthermore, we can define the concept of "the fact z_j is in violation of the CTD of z_i", for $i < j$. To find out if there was such a violation, we solve the equations for

$$\mathbb{E}_i = ((z_i, \bar{z}_i), \ldots, (x_{m+1}, \bar{x}_{m+1}))$$

and if the equation solves with $z_j = 0$ then putting $z_j = 1$ is a violation.

Remark 13.3.7 Let us check whether our equational modelling of the Chisholm CTD set satisfies the conditions set out in Definition 13.3.1.

Fig. 13.14 General graph for *example* 13.3.4

starting point	z_{m+1}	$\bar{z}_{m+1}$	graph $\mathbb{G}_{m+1}^+$
	$\vdots$	$\vdots$	$\vdots$
starting point	z_1	$\bar{z}_1$	graph $\mathbb{G}_1^+$
starting point	x_0	$\bar{x}_0$	graph $\mathbb{G}_1^-$

Consider Fig. 13.15a, b:

(a) Obligatory x must be translated the same way throughout.
This holds because we use a variable x in a neighbourhood generated equation.

(b) The form $X \to OY$ must be translated uniformly no matter whether $X = q$ or $X = \neg q$.
This is true of our model.

(c) This holds because "X" is translated as itself.

(d) The translation of the clauses must be all independent.
Indeed this holds by Lemma 13.3.4.
It is also true that (see Fig. 13.15a)

(2) $p \to Oq$

and

(5) p

imply

(6) Oq

This holds because (5) p is a fact. So this means that Fig. 13.15b truncated at the point p.
The truncated figure is indeed what we would construct for Oq.
A symmetrical argument shows that (4) and (3) imply (7).

Fig. 13.15 Graph for *Remark* 13.3.7

(a)

1.	Op	5.	p
2.	$p \to Oq$	6.	Oq
3.	$\neg p \to Oq$	7.	$O\neg q$
4.	$\neg p$		

(b)

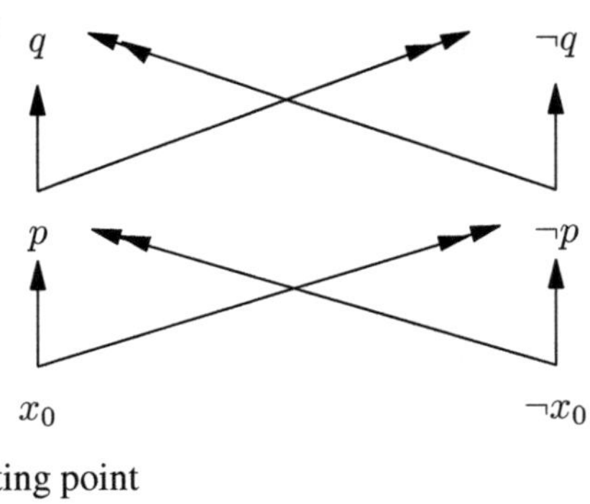

(e) The system is required to be robust with respect to substitution.
This condition arose from criticism put forward in de Boer et al. (2012) against the solution to the Chisholm paradox offered in Jones and Poern (1985). Jones and Poern (1985) relies on the fact that p, q are independent atoms.
The solution does not work when $q \vdash p$, e.g. substituting for "q" the wff "$r \wedge p$" (like $p =$ fence and $q =$ white fence).
In our case we use equations and if we substitute "$r \wedge p$" for "q" we get the equations

$$r \wedge p = 1 - p$$
$$p = r \wedge p$$

Although this type of equation is not guaranteed a solution, there is a solution in this case; $p = r = 1$.

If we add the fact $\neg p$, i.e. $1 - p = 1$, $p = 0$, (there is no fence) the equation solves to $\neg q = \neg p \vee \neg r$, which is also $= 1$ because of $\neg p$. So we have no problem with such substitution. In fact we have no problem with any substitution because the min function which we use always allows for solutions.

(f) The translation must be uniform and it to be done item by item.
Yes. Indeed, this is what we do!

Remark 13.3.8 We can now explain how classical logic can handle CTD, even though the CTD $x \rightarrow Oy$ involves a modality. The basic graph representation such as Fig. 13.7 can be viewed as a set of possible worlds where the variables x and y act as nominals (i.e. atoms naming worlds by being true exactly at the world they name). x is a world, y is a world and $x \rightarrow y$ means y is ideal for x. $x \twoheadrightarrow \bar{y}$ means that $\bar{y}$ is sub-ideal for x. Let $\Box_1$ be the modality for $\rightarrow$ and $\Box_2$ the modality for $\twoheadrightarrow$. Then we have a system with two disjoint modalities and we can define

$$OA \equiv \Box_1 A \wedge \Box_2 \neg A.$$

Now this looks familiar and comparable to Jones and Poern (1985), and especially to Gabbay et al. (2014). The perceptive reader might ask, if we are so close to modal logic, and in the modal logic formulation there are the paradoxes, why is it that we do not suffer from the paradoxes in the equational formulation?

The difference is because of how we interpret the facts! The equational approach spreads and inserts the facts into different worlds according to the obligation progression. Modal logic cannot do that because it evaluates formulae in single worlds. With equations, each variable is a nominal for a different world but is also is natural to substitute values to several variables at the same time!

Evaluating in several possible worlds at the same time in modal logic would solve the paradox but alas, this is not the way it is done.

Another difference is that in modal logic we can iterate modalities and write for example,

$$O(x \to Oy).$$

We do not need that in Chisholm sets. This simplifies the semantics.

13.3.4 Looping CTDs

So far we modelled the Chisholm set only. Now we want to expand the applicability of the equational approach and deal with looping CTDs, as in the set in Fig. 13.4. Let us proceed with a series of examples.

Example 13.3.5 Consider the CTD set of Fig. 13.4. If we write the equations for this example we get

(1) $a = \min(b, 1 - b)$
(2) $b = \neg a$
(3) $\neg b = 1 - a$

and the constants

(4) $\neg b = 1 - b$
(5) $\neg a = 1 - a$.

The only solution here is $a = b = \frac{1}{2}$. In argumentation and in classical logic terms this means the theory of Fig. 13.4 is $\{0, 1\}$ inconsistent.

This is mathematically OK, but is this the correct intuition? Consider the set $\{b, \neg a\}$. The only reason this is not a solution is because we have $a \twoheadrightarrow \neg b$ and if $a = 0$, we get $\neg b = 1$ and so we cannot have $b = 1$.

However, we wrote $a \twoheadrightarrow \neg b$ because of the CTD $a \to Ob$, which required us to go from a to b (i.e $a \to b$) and in this case we put in the graph $a \twoheadrightarrow \neg b$ to stress "do not go to $\neg b$".

However, if $a = 0$, why say anything? We do not care in this case whether the agent goes from a to b!

Let us look again at Fig. 13.6. We wrote the following equation for the node x

$$x = \min(u_i, 1 - v_j).$$

The rationale behind it was that we follow the rules, so we are going to u_i as our obligations say, and we came to x correctly, not from v_j, because $v_j \to O\neg x$ is required. Now if $v_j = 1$ (in the final solution) then the equation is correct. But if $v_j = 0$, then we do not care if we come to x from v_j, because $v_j \to O\neg x$ is not activated. So somehow we need to put into the equation that we care about v_j only when $v_j = 1$.

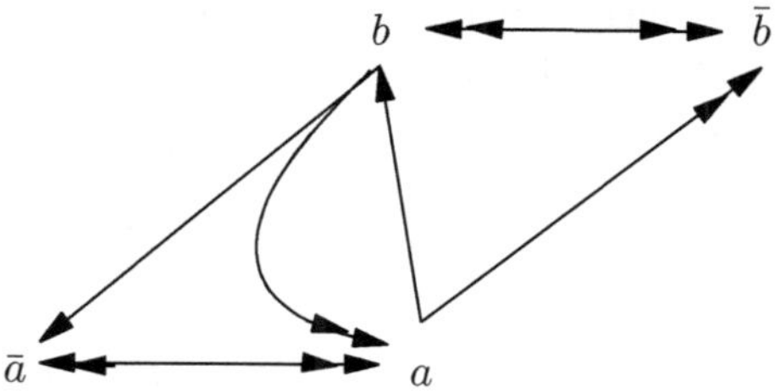

Fig. 13.16 Converting Fig. 13.4

Remark 13.3.9 Let us develop the new approach mentioned in example 13.3.5 and call it the soft approach. We shall compare it with the mathematical approach of Remark 13.3.5.

First we need a δ function as follows:

$$\delta(w) = \bot \text{ if } w = \bot$$

and

$$\delta(w) = \top \text{ if } w \neq \bot.$$

$\delta(w) = w$, if we are working in two valued $\{0, 1\}$ logic. Otherwise it is a projective function

$$\delta(0) = 0 \text{ and } \delta(w) = 1 \text{ for } w > 0.$$

We can now modify the equivalences (∗1) (based on Fig. 13.6) as follows:

Let $v_1, \ldots, v_s$ be as displayed in the configuration of in Fig. 13.6. Let $J, K \subseteq \{1, \ldots, s\}$ be such that $J \cap K = \varnothing$ and $J \cup K = \{1, \ldots, s\}$. Consider the expression

$$\varphi_{J,K} = \bigwedge_{j \in J} \delta(v_j) \wedge \bigwedge_{k \in K} \neg\delta(v_k).$$

This expression is different from 0 (or $\bot$), exactly when K is the set of all indices k for which $v_j = \bot$.

Replace (∗1) by the following group of axioms for each pair J, K and for each x

$$x \wedge \varphi_{J,K} \leftrightarrow \varphi_{J,K} \wedge \bigwedge_r u_r \wedge \bigwedge_{j \in J} \neg v_j. \qquad (*2)$$

Basically what (∗2) says is that the value of x should be equal to

$$\min\{u_r, 1 - v_j \text{ for those } j \text{ whose value is } \neq 0\}.$$

Note that this is an implicit definition for the solution of the equations. It is clear when said in words but looks more complicated when written mathematically. Solutions may not exist.

Example 13.3.6 Let us now look again at Fig. 13.4.

The soft equations discussed in Remark 13.3.9 are

$$\begin{aligned}\delta(a) \wedge \bar{b} &= \delta(a)(1-a)\\ \delta(b) \wedge a &= \delta(b)\min(b, 1-b)\\ b &= \bar{a}\\ \bar{b} &= 1-b\\ \bar{a} &= 1-a.\end{aligned}$$

For these equations $\bar{a} = 1, \bar{b} = a = 0, b = 1$ is a solution.

Note that $\bar{a} = \bar{b} = 1$ and $a = b = 0$ is *not* a solution!

Let us now examine and discuss the mathematical approach alternative, the one mentioned in Remark 13.3.5. The first step we take is to convert Fig. 13.4 into the right form for this alternative approach by adding double arrows between all x and $\bar{x}$. We get Fig. 13.16.

The equations are the following:

$$\begin{aligned}a &= \min(b, 1-\bar{a}, 1-b)\\ \bar{a} &= 1-a\\ b &= \min(\bar{a}, 1-\bar{b})\\ \bar{b} &= \min(1-a, 1-b).\end{aligned}$$

Let us check whether $a = \bar{b} = 0$ and $b = \bar{a} = 1$ is a solution. We get respectively by substitution

$$\begin{aligned}0 &= \min(1, 0, 0)\\ 1 &= 1-0\\ 1 &= \min(1, 1-0)\\ 0 &= min(1-0, 1-1).\end{aligned}$$

Indeed, we have a solution. Let us try the solution $\bar{b} = \bar{a} = 1$ and $a = b = 0$. Substitute in the equations and get

$$\begin{aligned}0 &= \min(0, 0, 1)\\ 1 &= 1-0\\ 0 &= \min(1, 1-1)\\ 1 &= \min(1-0, 1-0).\end{aligned}$$

Again we have a solution.

This solution also makes sense. Note that this is not a solution of the previous soft approach!

We need to look at more examples to decide what approach to take, and which final formal definition to give.

Example 13.3.7 Consider the following two CTD sets, put forward by two separate security advisors D and F.

D1: you should have a dog
Od

D2: If you do not have a dog, you should have a fence
$\neg d \to Of$

D3: If you have a dog you should not have a fence
$d \to O\neg f$

F1: You should have a fence
Of

F2: If you do not have a fence you should have a dog
$\neg f \to Od$

F3: If you do have a fence you should not have a dog.
$f \to O\neg d$

If we put both sets together we have a problem. They do not agree, i.e. {D1, D2, D3, F1, F2, F3}. However, we can put together both D1, D2 and F1, F2. They do agree, and we can have both a dog and a fence.

The mathematical equational modelling of D1 and D2 also models D3, i.e. D1, D2 $\vDash$ D3 and similarly F1, F2 $\vDash$ F3. So according to this modelling {D1, D2, F1, F2} cannot be consistently together. Let us check this point. Consider Fig. 13.17.

The equations for Fig. 13.17 are:

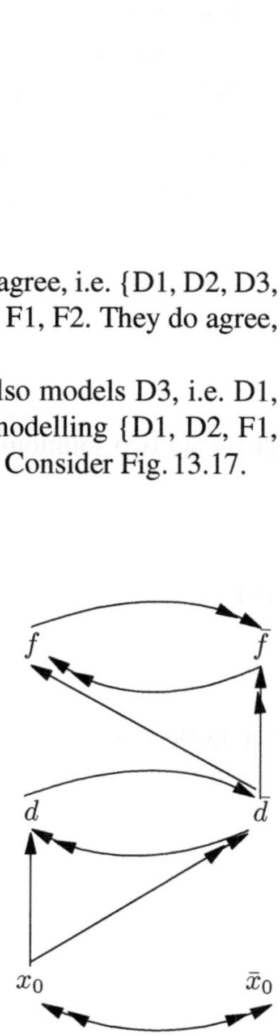

Fig. 13.17 Figure for illustrating consistency problems

$$\begin{aligned}
x_0 &= 1\\
x_0 &= d\\
\bar{x}_0 &= 1 - x_0\\
d &= 1 - \bar{d}\\
\bar{d} &= \min(1 - d, 1 - x_0)\\
\bar{d} &= f\\
f &= 1 - \bar{f}\\
\bar{f} &= \min(1 - f, 1 - \bar{d})
\end{aligned}$$

The only solution is

$$\begin{aligned}
x_0 = d = \bar{f} = 1\\
\bar{x}_0 = \bar{d} = f = 0.
\end{aligned}$$

The important point is that $\bar{f} = 1$, i.e. no fence.

Thus D1,D2 $\vdash \bar{f}$.

By complete symmetry beget that F1, F2 $\vdash \bar{d}$. Thus we cannot have according to the mathematical approach that having both a dog and a fence is consistent with {D1, D2, F1, F2}.

Let us look now at the soft approach. Consider Fig. 13.18.

The soft equations for Fig. 13.18 are:

$$\begin{aligned}
&x_0 = 1\\
&x_0 = d\\
&\min(x_0, \bar{d}) = \min(x_0, 1 - x_0)\\
&\min(\bar{d}, \bar{f}) = \min(\bar{d}, 1 - \bar{d})
\end{aligned}$$

There are two solutions

$$x_0 = 1, d = 1, \bar{d} = 0, \bar{f} = 1, f = 0$$

and

$$x_0 = 1, d = 1, \bar{d} = 0, \bar{f} = 0, f = 1.$$

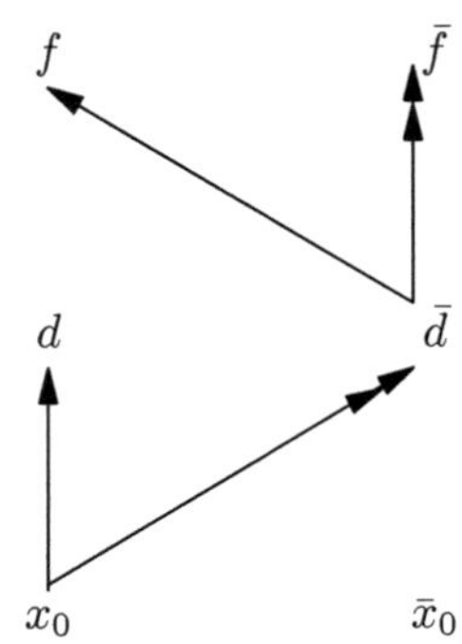

Fig. 13.18 Example for soft equations

The conceptual point is that since $\bar{d} = 0$, we say nothing about $\bar{f}$.

Now similar symmetrical solution is available for {F1, F2}. Since D1, D2 allow for $f = 1$ and F1, F2 allow for $d = 1$, they are consistent together. In view of this example, we should adopt the soft approach.

Remark 13.3.10 Continuing with the previous example 13.3.7, let us see what happens if we put together in the same CTD set the clauses {D1, D2, E1, E2} and draw the graph for them all together, in contrast to what we did before, where we were looking at two separate theories and seeking a joint solution. If we do put them together, we get the graph in Fig. 13.19.

If we use the mathematical equations, there will be no solution. If we use the soft approach equations, we get a unique solution

$$d = f = 1, \bar{d} = \bar{f} = 0$$

The reason for the difference, I will stress again, is in the way we write the equations for $\bar{d}$ and $\bar{f}$. In the mathematical approach we write

$$\begin{aligned} \bar{d} &= \min(f, 1 - \bar{f}) \\ \bar{f} &= \min(d, 1 - \bar{d}) \\ \bar{d} &= 1 - d \\ \bar{f} &= 1 - f \end{aligned}$$

In the soft approach we write

$$\begin{aligned} \min(\bar{d}, \bar{f}) &= \min(f, \bar{d}, 1 - \bar{f}) \\ \min(\bar{f}, \bar{d}) &= \min(d, \bar{f}, 1 - d) \end{aligned}$$

This example also shows how to address a general CTD set, where several single arrows can come out of a node (in our case x_0). The equations for x_0 in our example are:

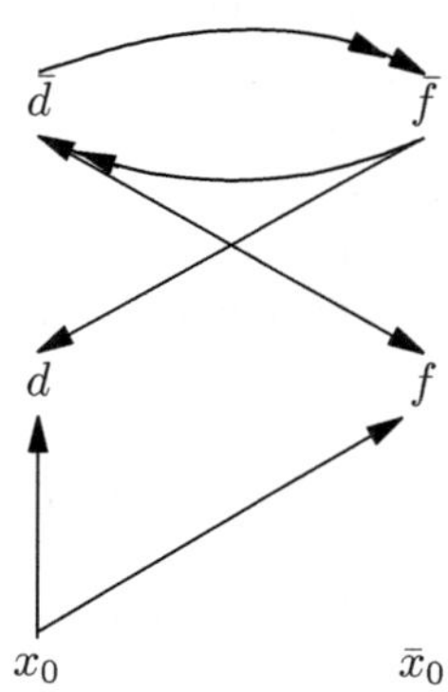

Fig. 13.19 A combination of the two graphs of *example* 13.3.7

$$x_0 = 1$$
$$x_0 = min(f, d)$$

which forces $d = f = 1$. We will check how to generalise these ideas in the next section.

13.3.5 Methodological Discussion

Following the discussions in the previous sections, we are now ready to give general definitions for the equational approach to general CTD sets. However, before we do that we would like to have a methodological discussion. We already have semantics for CTD. It is the soft equations option discussed in the previous subsection. So all we need to do now is to define the notion of a general CTD set (probably just a set of clauses of the form $\pm x \to O \pm y$) and apply the soft equational semantics to it. This will give us a consequence relation and a consistency notion for CTD sets and the next step is to find proof theory for this consequence and prove a completeness theorem.

We need to ask, however, to what extent is the soft semantics going to be intuitive and compatible with our perception of how to deal with conflicting CTD sets? So let us have some discussion about what is intuitive first, before we start with the technical definitions in the next section. Several examples will help.

Example 13.3.8 Consider the following CTD set:

(1) You should not have a dog
$O\neg d$
(2) If you have a dog you must keep it
$d \to Od$
(3) d: you have a dog

Here we have a problem. Is (1), (2), (3) a consistent set? In **SDL** we can derive from (2) and (3) $O\neg d$ and get a contradiction $Od \wedge O\neg d$.

However, in our semantics we produce a graph and write equations and if we have solutions, then the set is consistent. Let us do this.

The original graph for clauses (1)–(2) is Fig. 13.20. This graph generates equations.

The fact d splits the graph and we get the two graphs in Figs. 13.21 and 13.22.

The solution of the soft equations for the original graph (without the fact d) is $x_0 = \bar{d} = 1, d = 0$.

The solution for the two split graphs, after the fact d gives $d = 1$ for Fig. 13.21 and $\bar{d} = 0$ for Fig. 13.22.

There is no mathematical contradiction here. We can identify a violation from the graphs. However we may say there is something unintuitive, as the CTD proposal for a remedy for the violation $O\neg d$, namely $d \to Od$ violates the original obligation

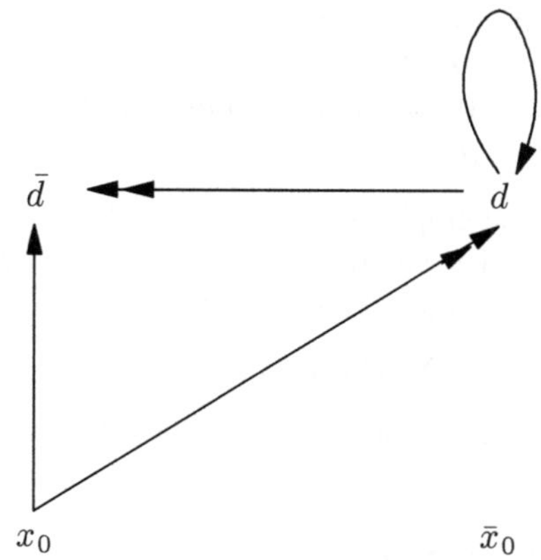

Fig. 13.20 Graph for *example* 13.3.8

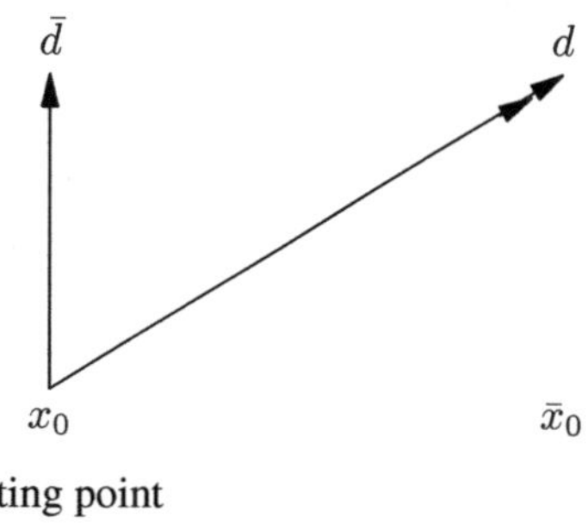

Fig. 13.21 First split from Fig. 13.20

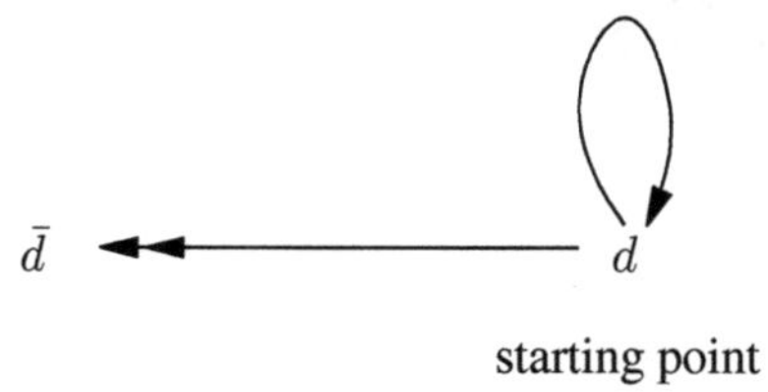

Fig. 13.22 Second split from Fig. 13.20

$O\neg d$, and actually perpetuates this violation. This we see on the syntactical level. No problem in the semantics.

We can explain and say that since the fact d violates $O\neg d$ then a new situation has arisen and $O\neg d$ is not "inherited" across a CTD. In fact, in the case of a dog it even makes sense. We should not have a dog but if we violate the obligation and get it, then we must be responsible for it and keep it.

The next example is more awkward to explain.

Example 13.3.9 This example is slightly more problematic. Consider the following:

(1) You should not have a dog
$O\neg d$

(2) you should not have a fence
$O\neg f$

(3) If you do have a dog you should have a fence
$d \to Of$

The graph for (1)–(3) is Fig. 13.23.

The solution is $\bar{d} = \bar{f} = 1, d = f = 1$.

Let us add the new fact

(4) d: You have a dog

The graph of Fig. 13.23 splits into two graphs, Figs. 13.24 and 13.25.

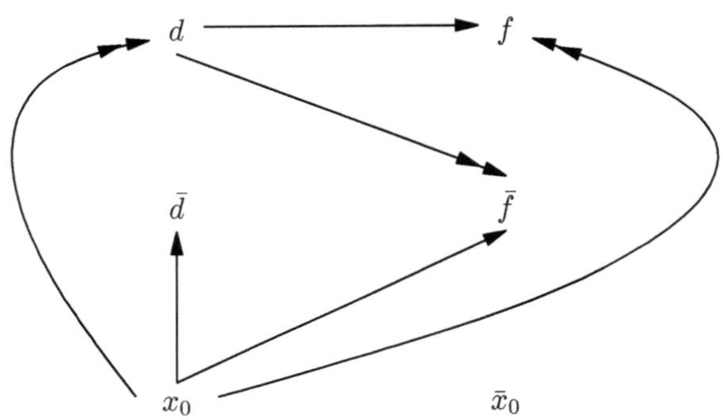

Fig. 13.23 Graph for *example* 13.3.9

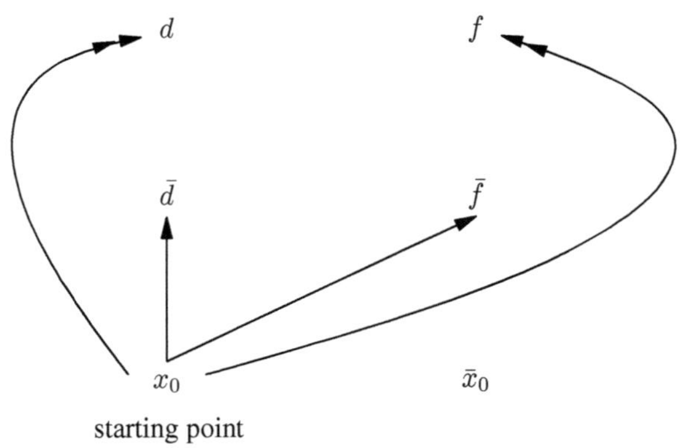

Fig. 13.24 First split for Fig. 13.23

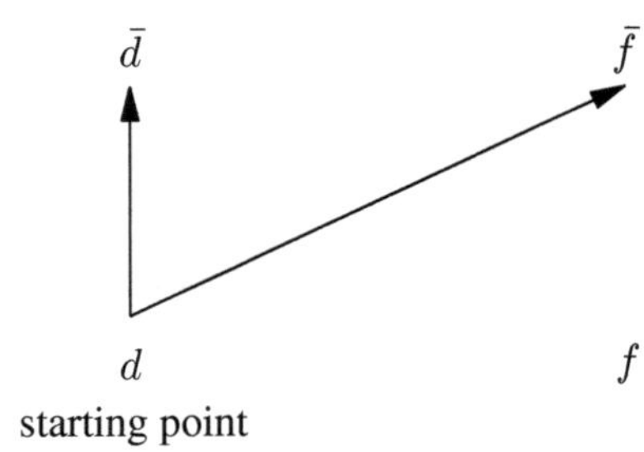

Fig. 13.25 Second split for Fig. 13.23

The equations for Fig. 13.24 solve to $\bar{d} = \bar{f} = 1, d = f = 0$. The equations for Fig. 13.25 solve to $d = f = 1, \bar{d} = \bar{f} = 0$.

There is no mathematical contradiction here because we have three separate graphs and their solutions. We can, and do, talk about violations, not contradictions.

Note that in **SDL** we can derive Of and $O\neg f$ from (1), (2), (3) and we do have a problem, a contradiction, because we are working in a single same system.

Still, even for the equational approach, there is an intuitive difficulty here. The original $O\neg f$ is contradicted by $d \to Of$. The "contradiction" is that we offer a remedy for the violation d namely Of by violating $O\neg f$.

You might ask, why offer the remedy Of? Why not say keep the dog chained? Oc? The Oc remedy does not violate $O\neg f$.

The explanation that by having a dog (violating $O\neg d$) we created a new situation is rather weak, because having a fence is totally independent from having a dog, so we would expect that the remedy for having a dog will not affect $O\neg f$!

The important point is that the equational approach can identify such "inconsistencies" and can add constraints to avoid them if we so wish.

Remark 13.3.11 Let us adopt the view that once a violation is done by a fact then any type of new rules can be given. This settles the problems raised in example 13.3.9. However, we have other problems. We still have to figure out a technical problem, namely how to deal with several facts together. In the case of the Chisholm set there were no loops and so there was the natural obligation progression. We turned the set of facts into a sequence and separated the original graph (for the set of CTD without the facts) into a sequence of graphs, and this was our way of modelling the facts. When we have loops there is a problem of definition, how do we decompose the original graph when we have more than one fact? The next example will illustrate.

Example 13.3.10 This example has a loop and two facts. It will help us understand our modelling options in dealing with facts. Consider the following clauses. This is actually the Reykjavik paradox, see for example, Governatori and Rotolo (2005):

(1) There should be no dog
    $O\neg d$
(2) There should be no fence
    $O\neg f$
(3) If there is a dog then there should be a fence
    $d \to Of$
(4) If there is a fence then there should be a dog
    $f \to Od$.

The figure for these clauses is Fig. 13.26.

The soft equations solve this figure into $x_0 = \bar{d} = \bar{f} = 1$. $f = d = 0$. We now add the input that there is a dog and a fence.

(5) d: dog, f: fence

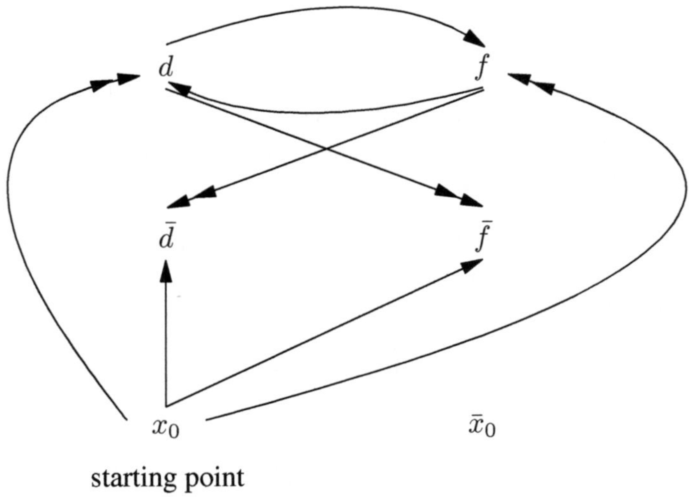

Fig. 13.26 Graph for *example* 13.3.10

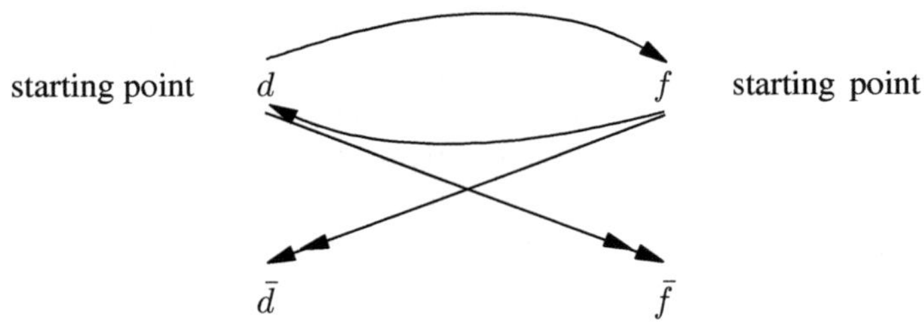

Fig. 13.27 First split for Fig. 13.26

The question is how to split Fig. 13.26 in view of this input.

If we substitute $d = 1$ and $f = 1$ together and split, we get Fig. 13.27, with two starting points.

Comparison with the original figure shows two violations of $O\neg d$ and $O\neg f$.

Let us now first add the fact d and then add the fact f.

When we add the fact d, Fig. 13.26 splits (actually is modified) into Fig. 13.28. This Figure happens to look just like Fig. 13.27 with only d as a starting point. (Remember that any starting point x gets the equation $x = 1$.)

Adding now the additional fact f changes Fig. 13.28 into Fig. 13.29. In fact we would have got Fig. 13.29 first, had we introduced the fact f first, and then added the fact d, we would have got Fig. 13.29.

The difference between the sequencing is in how we perceive the violations.

The following is a summary.

Option 1. Introduce facts $\{d, f\}$ simultaneously. Get Fig. 13.27, with two starting points. There are two violations, one of $O\neg f$ and one of $O\neg d$. This is recognised

Fig. 13.28 Second split for Fig. 13.26

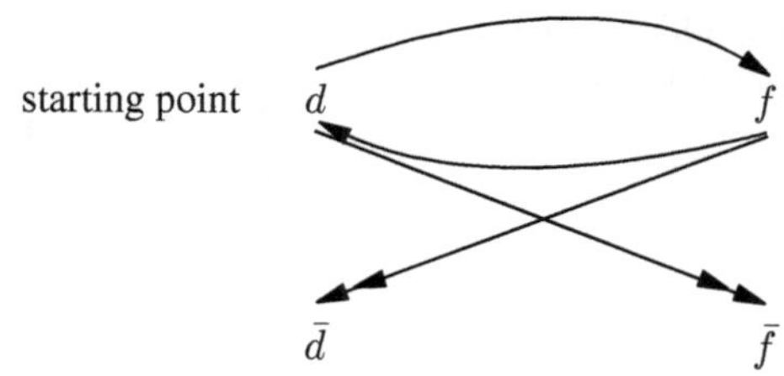

Fig. 13.29 Modifying Fig. 13.28

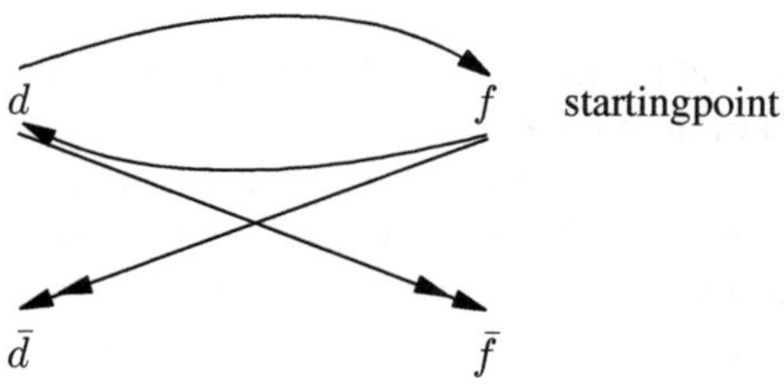

by comparing the solutions for the equations of Fig. 13.26 with those of Fig. 13.27.
Option 2. Introduce the fact d first. Figure 13.26 changes into Fig. 13.28. Solving the equations for these two figures shows a violation of $O\neg d$ and a vioation of $O\neg f$, because f also gets $f = 1$ in the equations of Fig. 13.28.
Option 2df. We now add to option 2d the further fact f. We get that Fig. 13.28 becomes Fig. 13.29. The solutions of the two figures are the same, $f = d = 1$. So adding f gives no additional violation.

We thus see that adding $\{d, f\}$ together or first d and then f or (by symmetry) first f and then d all essentially agree and there is no problem. So where is the problem with simultaneous facts? See the next example 13.3.11.

Example 13.3.11 (*Example* 13.3.10 *continued*) We continue the previous example 13.3.10:

Let us try to add the facts $\{d, \neg f\}$ to the CTD set of Fig. 13.26. Here we have a problem because we get Fig. 13.30. In this figure both d and $\bar{f}$ are starting points. These two must solve to $d = \bar{f} = 1$. This is impossible in the way we set up the system. This means that it is inconsistent from the point of view of our semantics

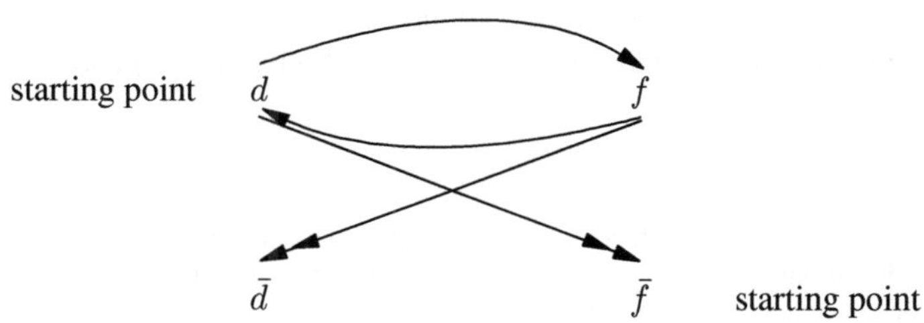

Fig. 13.30 Adding to Fig. 13.26

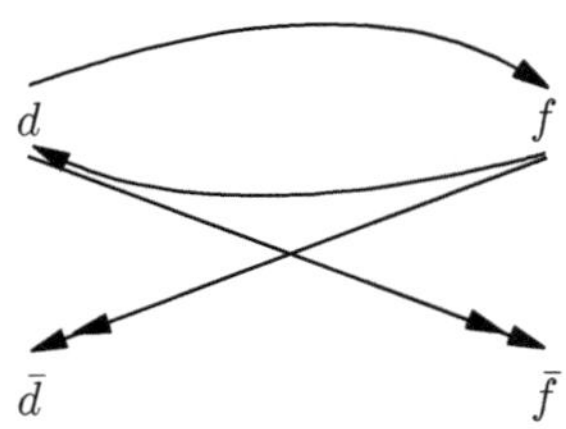

Fig. 13.31 Adding more and violating Fig. 13.26

to add the facts $\{d, \neg f\}$ simultaneously in the semantics, or technically to have two starting points!

But we know that it is consistent and possible in reality to have a dog and no fence. So where did we go wrong in our semantic modelling? Mathematically the problem arises with making two nodes starting points. This means that we are making two variables equal to 1 at the same time. The equations cannot adjust and have a solution.[3]

The obvious remedy is to add the facts one at a time. Option 3d first adds d and then takes option 3d $\neg f$ and add $\neg f$ and in parallel, option $4\neg f$ first adds the fact $\neg f$ and then take option $4\neg f d$ and add the fact d. Let us see what we get doing these options and whether we can make sense of it.

Option 3d¬f. Adding the fact d would give us Fig. 13.28 from Fig. 13.26. We now add fact $\neg f$. This gives us Fig. 13.31 from Fig. 13.28.

Figure 13.31 violates Fig. 13.28.

Option 3¬fd. If we add the fact $\neg d$ first, we get Fig. 13.31 from the original Fig. 13.26.

If we now add the fact d, we get Fig. 13.28.

The solution to the equations of this figure is $d = 1, f = 1, \bar{f} = 0$, but we already have the fact $\neg d$, so the $f = 1$ part cannot be accepted.

Summing up:

- facts $\{d, \neg f\}$ cannot be modelled simultaneously.
- First d then $\neg f$, we get that $\neg f$ violates $d \to Of$.
- first $\neg f$ then d, we get that $d \to Of$ cannot be implemented.

So the differences in sequencing the facts manifests itself as differences in taking a point of view of the sequencing of the violations.

The two views, when we have as additional data both d and $\neg f$, are therefore the following:

[3] Remember when we substitute a fact we split the graph into two and so the equations change. We are not just substituting values into equations (in which case the order simultaneous or not does not matter), we are also changing the equations.

Recall what you do in Physics: If we have, for example, the equation $y = \sin x$ and we substitute for x a very small positive value, then we change the equation to $y = x$.

we view $d \to Of$ as taking precedent and $\neg f$ is violating it

or

we view $O\neg f$ as taking precedence over $d \to Of$ and hence $d \to Of$ cannot be implemented.

13.4 Equational Semantics for General CTD Sets

We now give general definitions for general equational semantics for general CTD sets.

Definition 13.4.1 (1) Let Q be a set of distinct atoms. Let $\bar{Q}$ be $\{\bar{a} | a \in Q\}$. Let $Q^* = Q \cup \bar{Q} \cup \{\top, \bot\}$. For $x \in \bar{Q}$, let $\bar{x}$ be x (i.e. $\bar{\bar{x}} = x$). Let $\bar{\top} = \bot$ and $\bar{\bot} = \top$.

(2) A general CTD clause has the form $x \to Oy$, where $x, y \in Q^*$ and $x \neq \bot$, $y \neq \bot, \top$.

(3) Given a set $\mathbb{C}$ of general CTD clauses let $Q^*(\mathbb{C})$ be the set $\{x, \bar{x} | x$ appears in a clause of $\mathbb{C}\}$.

(4) Define two relations on $Q^*(\mathbb{C})$, $\to$ and $\twoheadrightarrow$ as follows:

- $x \to y$ if the clause $x \to Oy$ is in $\mathbb{C}$
- $x \twoheadrightarrow y$ if the clause $x \to O\bar{y}$ is in $\mathbb{C}$

(5) Call the system $\mathbb{G}(\mathbb{C}) = (Q^*(\mathbb{C}), \to, \twoheadrightarrow)$ the graph of $\mathbb{C}$.

(6) Let $x \in Q^*(\mathbb{C})$. Let

$$E(x \to) = \{y | x \to y\}$$
$$E(\twoheadrightarrow x) = \{y | y \twoheadrightarrow x\}$$

Definition 13.4.2 (1) Let $\mathbb{C}$ be a CTD set and let $\mathbb{G}(\mathbb{C})$ be its graph. Let x be a node in the graph.
Let $\mathbf{f}$ be a function from $Q^*(\mathbb{C})$ into [0, 1].
Define

$$E^+(\twoheadrightarrow x, \mathbf{f}) = \{y | y \twoheadrightarrow x \text{ and } \mathbf{f}(x) > 0\}.$$

(2) Let $\mathbf{f}$, x be as in (1). We say $\mathbf{f}$ is a model of $\mathbb{C}$ if the following holds:

(2.1) $\mathbf{f}(\top) = 1, \mathbf{f}(\bot) = 0$
(2.2) $\mathbf{f}(\bar{x}) = 1 - \mathbf{f}(x)$
(2.3) $\mathbf{f}(x) = \min(\{\mathbf{f}(y) | x \to y\} \cup \{1 - \mathbf{f}(z) | z \in E^+(\twoheadrightarrow x, \mathbf{f})\})$

(3) We say $\mathbf{f}$ is a $\{0, 1\}$ model of $\mathbb{C}$ if $\mathbf{f}$ is a model of $\mathbb{C}$ and $\mathbf{f}$ gives values in $\{0,1\}$.

Example 13.4.1 Consider the set

(1) $\top \to a$
(2) $a \to \bar{a}$

This set has no models. However, (2) alone has a model $\mathbf{f}(a) = 0, \mathbf{f}(\bar{a}) = 1$. The equations for (2) are: $a = 1 - \bar{a}, a = \min(1 - a, \bar{a})$.

The graph for (1) and (2) is Fig. 13.32

The graph for (2) alone is Fig. 13.32 without the node $\top$.

Definition 13.4.3 Let $\mathbb{C}$ be a CTD set. Let $\mathbb{G}(\mathbb{C}) = (Q^*, \to, \twoheadrightarrow)$ be its graph. Let $x \in Q^*$. We define the truncation graph $\mathbb{G}(\mathbb{C})_x$, as follows:

(1) Let R^* be the reflexive and transitive closure of R, where

$$xRy = \text{def}(\text{x} \to \text{y}) \vee (\text{x} \twoheadrightarrow \text{y}).$$

(2) Let Q^*_x be the set

$$\{z | xR^*z \vee \bar{x}R^*z\} \cup \{\top, \bot\}$$

Let $\to_x = (\to \restriction Q_x) \cup \{\top \to x\}$.

Then

$$\mathbb{G}(\mathbb{C})_x = (Q^*_x, \to_x, \twoheadrightarrow \restriction Q^*_x).$$

(3) In words: the truncation of the graph at x is obtained by taking the part of the graph of all points reachable from x or $\bar{x}$ together with $\top$ and $\bot$ and adding $\top \to x$ to the graph.

Example 13.4.2 Consider a m level Chisholm set as in Fig. 13.11. The truncation of this figure at point $\bar{x}_3$ is essentially identical with Fig. 13.13. It is Fig. 13.33. The difference is that we write "$\top \to \bar{x}_3$" instead of "$\bar{x}_3$ starting point". These two have the same effect on the equations namely that $\bar{x}_3 = 1$.

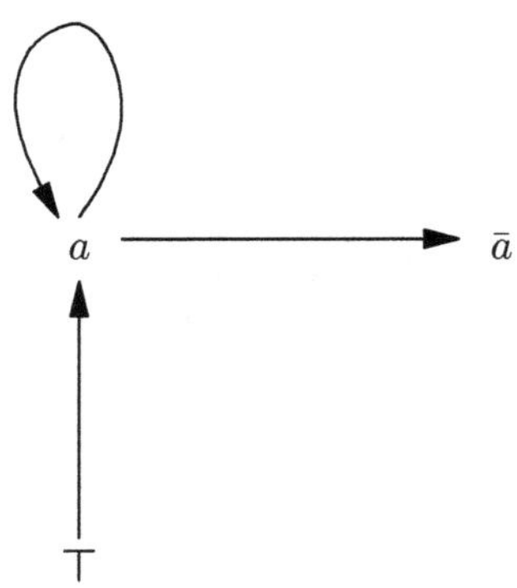

Fig. 13.32 Graph for *example* 13.4.1

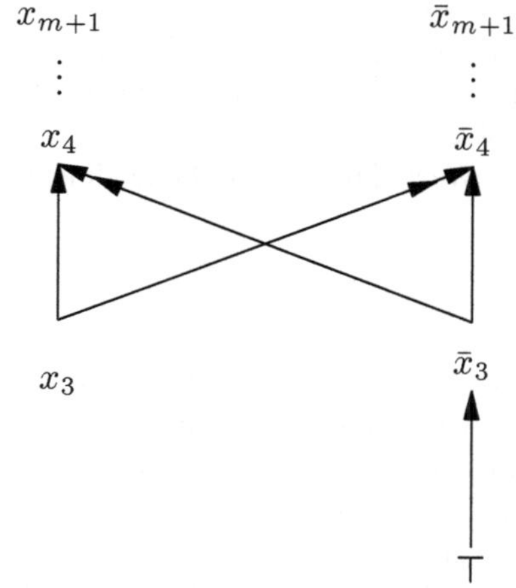

Fig. 13.33 Truncation from Fig. 13.11

Definition 13.4.4 Let $\mathbb{C}$ be a CTD set. Let $\mathbb{F}$ be a set of facts. We offer equational semantics for $(\mathbb{C}, \mathbb{F})$.

(1) Let Ω be any ordering of $\mathbb{F}$.

$$\Omega = (f_1, f_2, \ldots, f_k).$$

(2) Let $\mathbb{G}(\mathbb{C})$ be the graph of $\mathbb{C}$ and consider the following sequence of graphs and their respective equations.

$$(\mathbb{G}(\mathbb{C}), \mathbb{G}(\mathbb{C})_{f_1}, \mathbb{G}(\mathbb{C})_{f_1 f_2}, \ldots, \mathbb{G}(\mathbb{C})_{f_1,\ldots,f_k})$$

We call the above sequence

$$\mathbb{G}(\mathbb{C})_\Omega.$$

We consider Ω as a "point of view" of how to view the violation sequence arising from the facts.

(3) The full semantics for $(\mathbb{C}, \mathbb{F})$ is the family of all sequences $\{\mathbb{G}(\mathbb{C})_\Omega\}$ for all Ω orderings of $\mathbb{F}$.

Remark 13.4.1 The CTD sets considered so far had the form $\pm x \to O \pm y$, where x and y are atomic. This remark expands our language allowing x, y to be arbitrary propositional formulae. Our technical machinery of graphs and equations works just the same for this case. We can write in the graph $A, \bar{A}$ and then write the appropriate equations, and we use $\mathbf{h}_\mu$. For example, the equation $\bar{A} = 1 - A$ becomes $\mathbf{h}_\mu(\bar{A}) = 1 - \mathbf{h}_\mu(A)$. Starting points A must satisfy $\mathbf{h}_\mu(A) = 1$, all the same as before. The only difference is that since the equations become implicit on the atoms, we may not have a solution.

In practice, the way we approach such a CTD set is as follows: Let $\mathbb{C}_1$ be a set of CTD obligations of the form $\{A_i \to O B_i\}$. We pretend that A_i, B_i are all atomic.

We do this by adding a new atomic constant $y(A)$, associated with every wff A. The set $\mathbb{C}_1 = \{A_i \to OB_i\}$ becomes the companion set $\mathbb{C}_2\{y(A_i) \to Oy(B_i)\}$. We now apply the graphs and equational approach to $\mathbb{C}_2$ and get a set of equations to be solved.

We add to this set of equations the further constraint equations

$$y(A_i) = \mathbf{h}_\mu(A_i)$$
$$y(B_i) = \mathbf{h}_\mu(B_i).$$

We now solve for the atomic propositions of the language.

We need to clarify one point in this setup. What do we mean by facts $\mathbb{F}$? We need to take $\mathbb{F}$ as a propositional theory, it being the conjunction of some of the A_i. If we are given a set $\mathbb{C}_1$ of contrary to duty clauses of the form $A \to OB$ and facts $\mathbb{F}_1$, we check whether $\mathbb{F}_1 \vDash A$ in classical logic, (or in any other logic we use as a base. Note that if A_i are all atomic then it does not matter which logic we use as a base the consequence between conjunctions of atoms is always the same). If yes, then to the companion set $\mathbb{C}_2$ we add the fact $y(A)$. We thus get the companion set of facts $\mathbb{F}_2$ and we can carry on. This approach is perfectly compatible with the previous system where A, B were already atomic. The theory $\mathbb{F}$ is the conjunction of all the $\pm$ atoms in $\mathbb{F}$.

There is a slight problem here. When the formulae involved were atomic, a set of facts was a set of atoms $\mathbb{F}$, obtained by choosing one of each pair $\{+x, -x\}$. So $\mathbb{F}$ was consistent. When the formulae involved are not atomic, even if we choose one of each pair $\{A, \neg A\}$, we may end up with a set $\mathbb{F}$ being inconsistent. We can require that we choose only consistent sets of facts and leave this requirement as an additional constraint.

This remark is going to be important when we compare our approach to that of Makinson and Torre's input output logic approach.

Example 13.4.3 To illustrate what we said in Remark 13.4.1, let us consider Fig. 13.17.

The equations are listed in example 13.3.7. Let us assume that in the figure we replace d by $y(D)$, where $D = d \vee c$. We get the equations involving $y(D)$ instead of D and get the solution as in example 13.3.7, to be

$$x_0 = y(D) = \bar{f} = 1$$
$$\bar{x}_0 = y(\bar{D}) = f = 0.$$

Now we have the additional equation

$$\begin{aligned} y(D) &= \mathbf{h}_\mu(D) \\ &= \mathbf{h}_\mu(c \vee d) \\ &= \max(c, d) \end{aligned}$$

So we get $\max(c, d) = 1$ and we do have the solution with $d = 1, c = 1, ord = 0, c = 1, ord = 1, c = 0$.

Remark 13.4.2 (*Miner Paradox (Kolodny and MacFarlane 2010; Willer 2012)*) We begin with a quote from Malte Willer in Willer (2012)

> Every adequate semantics for conditionals and deontic ought must offer a solution to the miners paradox about conditional obligations..... Here is the miners paradox. Ten miners are trapped either in shaft A or in shaft B, but we do not know which one. Water threatens to flood the shafts. We only have enough sand bags to block one shaft but not both. If one shaft is blocked, all of the water will go into the other shaft, killing every miner inside. If we block neither shaft, both will be partially flooded, killing one miner [See Fig. 13.34].
>
> Lacking any information about the miners exact whereabouts, it seems to say that
>
> (1) We ought to block neither shaft.
>
> However, we also accept that
>
> (2) If the miners are in shaft A, we ought to block shaft A,
> (3) If the miners are in shaft B, we ought to block shaft B.
>
> But we also know that
>
> (4) Either the miners are in shaft A or they are in shaft B.
>
> And (2)–(4) seem to entail
>
> (5) Either we ought to block shaft A or we ought to block shaft B, which contradicts (1).
>
> Thus we have a paradox.

We formulate the Miners paradox as follows:

(1) $\top \to O\neg\text{Block}A$
$\quad\top \to O\neg\text{Block}B$
(2) Miners in $A \to O\text{Block}A$
(3) Miners in $B \to O\text{Block}B$
(4) Facts: Miners in $A \vee$ Miners in B.

The graph for (1)–(3) is Fig. 13.35.

The Miners paradox arises because we want to detach using (2), (3) and (4) and get (5).

Fig. 13.34 The Miners paradox

Action	if miners in A	if miners in B
Block A	All saved	All drowned
Block B	All drowned	All saved
Block neither shaft	One drowned	One drowned

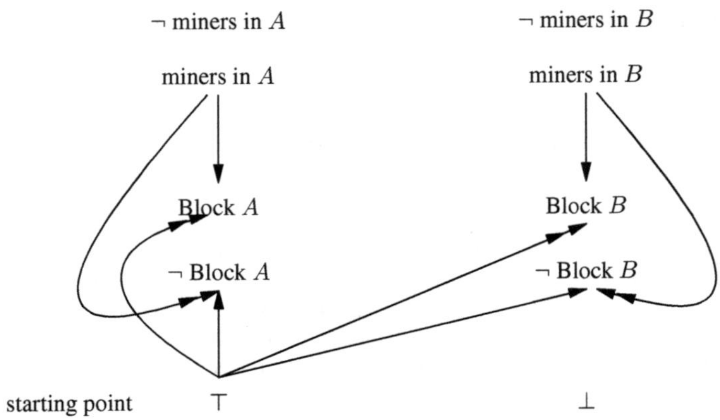

Fig. 13.35 First graph for the Miners paradox

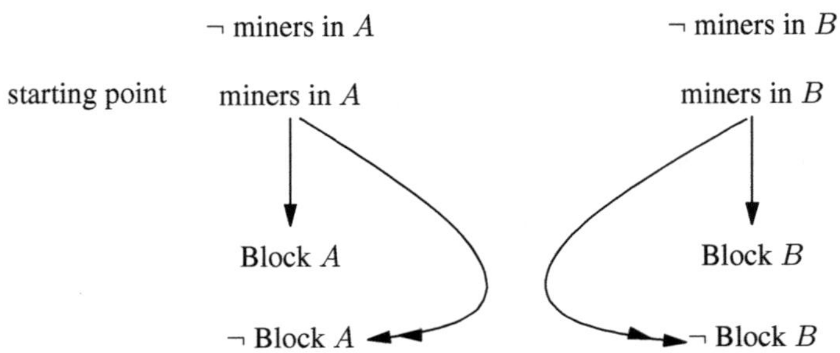

Fig. 13.36 Second graph for the Miners paradox

(5) $O\text{Block}A \vee O\text{Block}B$

which contradicts (1).

However, according to our discussion, facts simply choose new starting points in the figure. The fact (4) is read as two possible sets of facts. Either the set of the fact that miner in A or the other possibility, the set containing miner in B. We thus get two possible graphs, Figs. 13.36 and 13.37.

We can see that there is no paradox here.

We conclude with a remark that we can solve the paradox directly using H. Reichenbach (Reichenbach 1947) reference points, without going through the general theory of this chapter. See Baniasad et al. (2013).

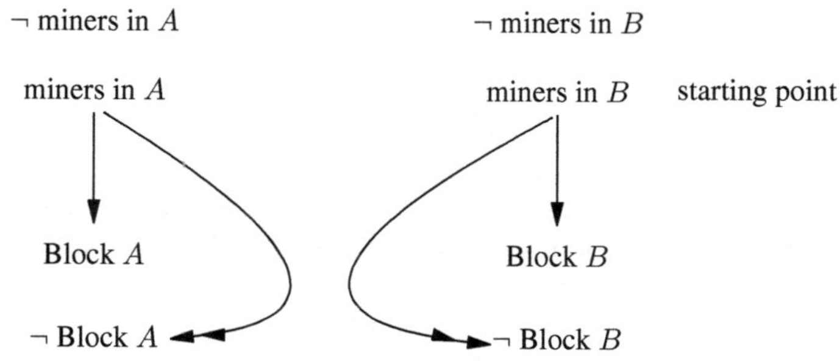

Fig. 13.37 Third graph for the Miners paradox

13.5 Proof Theory for CTDs

Our analysis in the previous sections suggest proof theory for sets of contrary-to-duty obligations. We use Gabbay's framework of labelled deductive systems (Gabbay 1996).

We first explain intuitively our approach before giving formal definitions. Our starting point is Definition 13.4.1. The contrary-to-duty obligations according to this definition have the form $x \to Oy$, where x, y are atoms q or their negation $\neg q$ and x may be $\top$ and y is neither $\top$ nor $\bot$.

For our purpose we use the notation $x \Rightarrow y$. We also use labels annotating the obligations, and we write

$$t : x \Rightarrow y.$$

The label we use is the formula itself

$$t = (x \Rightarrow y).$$

Thus our CTD data for the purpose of proof theory has the form

$$(x \Rightarrow y) : x \Rightarrow y.$$

Given two CTD data items of the form

$$t : x \Rightarrow y; s : y \Rightarrow z$$

we can derive a new item

$$t * s : x \Rightarrow z,$$

where $*$ is a concatenation of sequences. (Note that the end letter of t is the same as the beginning letter of s, so we can chain them.)

So we have the rule

$$\frac{(x \Rightarrow y) : x \Rightarrow y; (y \Rightarrow z) : y \Rightarrow z}{(x \Rightarrow y, y \Rightarrow z) : x \Rightarrow z}$$

It may be that we also have $(x \Rightarrow z) : x \Rightarrow z$ (i.e. the CTD set contains $x \to Oy, y \to Oz$ and $x \to Oz$), in which case $x \Rightarrow z$ will have two different labels, namely

$$t_1 = (x \Rightarrow y, y \Rightarrow z)$$
$$t_2 = (x \Rightarrow z).$$

We thus need to say that the proof theory allows for lables which are sets of chained labels (we shall give exact definitions later). So the label for $x \Rightarrow z$ would be $\{t_1, t_2\}$. There may be more labels $t_3, t_4, \ldots$ for $x \Rightarrow z$ depending on the original CTD set.

Suppose that in the above considerations $x = \top$. This means that our CTD set described above has the form $\{Oy(\text{being } \top \to Oy), y \to Oz \text{ and } Oz\}$. By using the chaining rule we just described (and not mentioning any labels) we also get

$$\top \Rightarrow z, \top \Rightarrow y.$$

We can thus intuitively detach with $\top$ and get that our CTD set proves $\{y, z\}$. Notations $\vdash \{y, z\}$.

Alternatively, even if x were arbitrary, not necessarily $\top$, we can detach with x and write $x \vdash \{y, z\}$.

Of course when we use labels we will write

$$t : x \vdash \{s_1 : y, s_2 : z\}$$

the labels s_1, s_2 will contain in them the information of how y, z were derived from x.

To be precise, if for example,

$$\mathbb{C} = \{(x \Rightarrow y) : x \Rightarrow y, (y \Rightarrow z) : y \Rightarrow z, (x \Rightarrow z) : x \Rightarrow z\}.$$

We get

$$(x) : x \vdash_{\mathbb{C}} \{(x, x \Rightarrow y, y \Rightarrow z) : z, (x, x \Rightarrow z) : z, (x, x \Rightarrow y) : y\}.$$

Definition 13.5.1 (1) Let Q be a set of atoms. Let $\neg$ be a negation and let $\Rightarrow$ be a CTD implication symbol.
A clause has the form $x \Rightarrow y$, where x is either $\top$ or atom q or $\neg q$ and y is either atom a or $\neg a$.
(2) A basic label is either $(\top)$ or (q) or $(\neg q)$. (q atomic) or a clause $(x \Rightarrow y)$.
(3) A chain label is a sequence of the following form

$$(x_0 \Rightarrow x_1, x_1 \Rightarrow x_2, \ldots, x_n \Rightarrow x_{n+1})$$

where $x_i \Rightarrow x_{i+1}$ are clauses. x_0 is called the initial element of the sequence and x_{n+1} is the end element.

(4) A set label is a set of chain labels.

(5) A labelled CTD dataset $\mathbb{C}$ is a set of elements of the form $(x \Rightarrow y) : (x \Rightarrow y)$, where $x \Rightarrow y$ is a clause and $(x \Rightarrow y)$ is a basic label.

(6) A fact has the form $(x) : x$, where x is either atom q or $\neg q$ or $\top$.

A fact set $\mathbb{F}$ is a set of facts.

Definition 13.5.2 Let $\mathbb{C}$ be a CTD dataset. We define the notions of

$$\mathbb{C} \vdash_n t : x \Rightarrow y,$$

where $n \geq 0$, t a basic or chain label. This we do by induction on n.

We note that we may have $\mathbb{C} \vdash_n t : x \Rightarrow y$ hold for several different ns and different ts all depending on $\mathbb{C}$.

Case $n = 0$
$\mathbb{C} \vdash_0 t : x \Rightarrow y$ if $t = (x \Rightarrow y)$ and $(x \Rightarrow y) : x \Rightarrow y \in \mathbb{C}$.
Case $n = 1$
$\mathbb{C} \vdash_1 t : x \Rightarrow y$ if for some $x \Rightarrow w$ we have $(x \Rightarrow w) : x \Rightarrow w$ in $\mathbb{C}$ and $(w \Rightarrow y) : (w \Rightarrow y)$ in $\mathbb{C}$ and $t = (x \Rightarrow w, w \Rightarrow y)$.

Note that the initial element of t is x and the end element is y.

Case $n = m + 1$
Assume that $\mathbb{C} \vdash_m t : x \Rightarrow y$ has been defined and that in such cases the end element of t is y and the initial element of t is x.

Let $\mathbb{C} \vdash_{m+1} t : x \Rightarrow y$ hold if for some $t' : x \Rightarrow w$ we have $\mathbb{C} \vdash_m t' : x \Rightarrow w$ (and therefore the end element of t' is w and the initial element of t is x) and $(w \Rightarrow y) : w \Rightarrow y \in \mathbb{C}$ and $t = t' * (w \Rightarrow y)$, where $*$ is concatenation of sequences.

Definition 13.5.3 Let $\mathbb{C}$ be a dataset and let $(x) : x$ be a fact. We write $\mathbb{C} \vdash^x_{n+1} t : y$ if for some $t : x \Rightarrow y$ we have $\mathbb{C} \vdash_n t : x \Rightarrow y$.

We may also use the clearer notation

$$\mathbb{C} \vdash_{n+1} (x, t) : y.$$

Example 13.5.1 Let $\mathbb{C}$ be the set

$$(x \Rightarrow y) : x \Rightarrow y$$
$$(y \Rightarrow z) : y \Rightarrow z$$
$$(z \Rightarrow y) : z \Rightarrow y$$

Then

$$\begin{aligned}&\mathbb{C} \vdash_0 (x \Rightarrow y) : x \Rightarrow y\\&\mathbb{C} \vdash_2 (x \Rightarrow y, y \Rightarrow z, z \Rightarrow y) : x \Rightarrow y\\&\mathbb{C} \vdash_1^x (x \Rightarrow y) : y\\&\mathbb{C} \vdash_3^x (x \Rightarrow y, y \Rightarrow z, z \Rightarrow y) : y\end{aligned}$$

or using the clearer notation

$$\begin{aligned}&\mathbb{C} \vdash_1 (x, (x \Rightarrow y)) : y\\&\mathbb{C} \vdash_3 (x, (x \Rightarrow y, y \Rightarrow z, z \Rightarrow y)) : y.\end{aligned}$$

Note also that y can be proved with different labels in different ways.

Definition 13.5.4 Let $\mathbb{C}$ be a dataset and let $\mathbb{F}$ be a set of facts. We define the notion of $\mathbb{C}, \mathbb{F} \vdash_n (z, t) : x$ where x, z are atomic or negation of atomic and z also possibly $z = \top$, as follows:

Case $n = 0$
$\mathbb{C}, \mathbb{F} \vdash_0 (z, t) : x$ if $(x) : x \in \mathbb{F}$ and $(z, t) = (x)$.

Case $n = m + 1$
$\mathbb{C}, \mathbb{F} \vdash_{m+1} (z, t) : x$ if $\mathbb{C} \vdash_n t : z \Rightarrow x$ and $z = \top$ or $(z) : z \in \mathbb{F}$.

Example 13.5.2 We continue example 13.5.1. We have

$$\begin{aligned}&\mathbb{C}, \{(z) : z\} \vdash_0 (z) : z\\&\mathbb{C} \vdash_2 (z, (z \Rightarrow y)) : y\\&\mathbb{C} \vdash_4 (x, (x \Rightarrow y, y \Rightarrow z, z \Rightarrow y)) : y\end{aligned}$$

Example 13.5.3 To illustrate the meaning of the notion of $\mathbb{C}, \mathbb{F} \vdash t : x$ let us look at the CTD set of Fig. 13.26 (this is the Reykajavic set) with $d =$ dog and $f =$ fence:

(1) $O\neg d$
written as $(\top \Rightarrow \neg d) : \top \Rightarrow \neg d$.
(2) $O\neg f$
written as $(\top \Rightarrow \neg f) : \top \Rightarrow \neg f$.
(3) $d \to Of$
written as $(d \Rightarrow f) : d \Rightarrow f$.
(4) $f \to Od$ written as $(f \Rightarrow d) : f \Rightarrow d$.

The above defines $\mathbb{C}$. Let the facts $\mathbb{F}$ be $(d) : d$ and $(\neg f) : \neg f$. We can equally write the facts as

$$\begin{aligned}&(\top \Rightarrow d) : \top \Rightarrow d)\\&(\top \Rightarrow \neg f) : \top \Rightarrow \neg f.\end{aligned}$$

(a) CTD point of view
Let us first look at the contrary-to-duty set and the facts intuitively from the deontic point of view. The set says that we are not allowed to have neither a dog

d nor a fence f. So good behaviour must "prove" from $\mathbb{C}$ the two conclusions $\{\neg d, \neg f\}$. This is indeed done by

$$\mathbb{C} \vdash_1 (\top, (\top \Rightarrow \neg d)) : \neg d$$
$$\mathbb{C} \vdash_1 (\top, (\top \Rightarrow \neg f) : \neg f$$

The facts are that we have a dog (in violation of $\mathbb{C}$) and not a fence

$$\mathbb{F} = \{(d) : d, (\neg f) : \neg f\}.$$

So we can prove

$$\mathbb{C}, \mathbb{F} \vdash_0 (d) : d$$
$$\mathbb{C}, \mathbb{F}) \vdash_0 (\neg f) : \neg f$$

but we also have

$$\mathbb{C}, \mathbb{F} \vdash_1 (d, (d \Rightarrow f)) : f.$$

We can see that we have violations, and the labels tell us what violates what. Let us take the facts as a sequence. First we have a dog and then not a fence. Let $\mathbb{F}_d = \{(d) : d\}$ and $\mathbb{F}_{\neg f} = \{(\neg f) : \neg f\}$.
then

$$\mathbb{C}, \mathbb{F}_d \vdash_0 (d) : d$$

$$\mathbb{C}, \mathbb{F}_d \vdash_1 (d, (d \Rightarrow f)) : f$$

which violates

$$\mathbb{C} \vdash_0 (\top, (\top \Rightarrow \neg f)) : \neg f$$

but $\mathbb{F} = \mathbb{F}_d \cup \mathbb{F}_{\neg f}$, and so $\mathbb{F}$ viewed in this sequence (first d then $\neg f$) gives us a choice of points of view. Is the addition $\neg f$ a violation of the CTD dog $\rightarrow O$ fence or is it in accordance with the original $O\neg f$?
The problem here is that the remedy for the violation of $O\neg d$ by the fact d is $d \rightarrow Of$, which is a violation of another CTD namely $O\neg f$. One can say the remedy wins or one can say this remedy is wrong, stick to $O\neg f$.
The important point about the proof system $\mathbb{C}, \mathbb{F} \vdash_n t : A$ is that we can get exactly all the information we need regarding facts and violations.

(b) Modal point of view
To emphasise the mechanical uninterpreted nature of the proof system let us give it a modal logic interpretation. We regard the labels as possible worlds and regard $*$ as indicating accessibility. We read $\mathbb{C}, \mathbb{F} \vdash_n t : A$ as $t \vDash A$, in the model $\mathbf{m}$ defined by $\mathbb{C}, \mathbb{F}$, i.e. $\mathbf{m} = \mathbf{m}(\mathbb{C}, \mathbb{F})$). The model of (a) above is shown in Fig. 13.38.
What holds at node t in Fig. 13.38 is the end element of the sequence t.

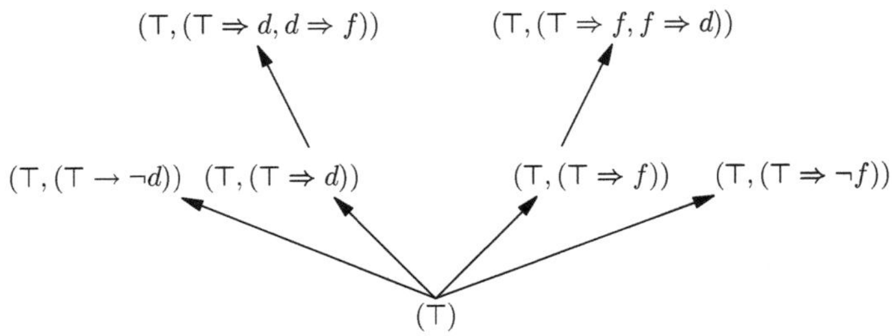

Fig. 13.38 A modal modelling of CTD

The facts give us no contradiction, because they are true at different worlds. At $(\top, (\top \Rightarrow d))$ we have dog and so at $(\top, (\top \Rightarrow d, d \Rightarrow f))$ we have a fence while at $(\top, (\top \Rightarrow \neg f))$, we have no fence.

Inconsistency can only arise if we have $(x) : z$ and $(x) : \neg z$ or $t : x \Rightarrow y$ and $t : x \Rightarrow \neg y$ but we cannot express that in our language.

(c) The deductive view This is a labelled deductive system view. We prove all we can from the system and to the extent that we get both A and $\neg A$ with different labels, we collect all labels and implement a *flattening policy*, to decide whether to adopt A or adopt $\neg A$.

Let us use, by way of example, the following flattening policy:

FP1 Longer labels win over shorter labels. (This means in CTD interpretation that once an obligation is violated, the CTD has precedence.)

FP2 In case of same length labels, membership in $\mathbb{F}$ wins. This means we must accept the facts!

So according to this policy we have

$$(\top, (\top \Rightarrow d, d \Rightarrow f)) : f \text{ wins over } (\top, (\top \Rightarrow \neg f)) : \neg f$$

and

$$(\top, (\top \Rightarrow d)) : d \text{ wins over } (\top, (\top \Rightarrow \neg d)) : \neg d.$$

So we get the result $\{d, f\}$.

We can adopt the input–output policy of Makinson–Torre. We regard $\mathbb{C}$ as a set of pure mathematical input output pairs.

We examine each rule in $\mathbb{C}$ against the input $\mathbb{F}$. If it yields a contradictory output, we drop the rule.

The final result is obtained by closing the input under the remaining rules. So let us check:

$$\text{Input} : \{d, \neg f\}.$$

Rules in $\mathbb{C}$
$\top \Rightarrow \neg f$, OK
$\top \Rightarrow \neg d$, drop rule
$d \Rightarrow f$, drop rule
$f \Rightarrow d$, not applicable.

Result of closure: $\{d, \neg f\}$.

The input–output approach is neither proof theory not CTD. To see this add another rule

$$\neg d \Rightarrow b,$$

where b is something completely different, consistent with $\pm d$, $\pm f$. This rule is not activated by the input $\{d, \neg f\}$. In the labelled approach we get b in our final set.

Problems of this kind have already been addressed by our approach of *compromise revision*, in 1999, see Gabbay (1996).

Example 13.5.4 Let us revisit the miners paradox of Remark 13.4.2 and use our proof theory.

We have the following data:

(1) $\top \Rightarrow \neg$ Block A
(2) $\top \Rightarrow \neg$ Block B
(3) Miners in $A \Rightarrow$ Block A
(4) Miners in $B \Rightarrow$ Block B
(5) Fact: Miners in $A \vee$ miners in B

using ordinary logic.

We get from (2), (3) and (4)

(5) Block $A \vee$ Block B

(5) contradicts (1).

Let us examine how we do this in our labelled system.

We have

1*. $(\top \Rightarrow$ Block $A) : \top \Rightarrow$ Bock A
$(\top \Rightarrow$ Block $B) : \top \Rightarrow$ Block B
2*. (miners in $A \Rightarrow$ Block A): miners in $A \Rightarrow$ Block A.
3*. (miners in $B \Rightarrow$ Block B): miners in $B \Rightarrow$ Block B
4*. (miners in $A \vee$ miners in B)': miners in $A \vee$ miners in B.

To do labelled proof theory we need to say how to chain the labels of disjunctions.

We do the obvious, we chain each disjunct. So, if t is a label with end element $x \vee y$ and we have two rules $x \Rightarrow z$ and $y \Rightarrow w$ then we can chain

$$(t, (x \Rightarrow z, y \Rightarrow w))$$

So we have the following results using such chaining:
(1*), (2*), (3*), (4*) $\vdash_1 (\top, (\top \Rightarrow \neg$ Block $A)) : \neg$ Block A

(1*), (2*), (3*), (4*) $\vdash_1$ $(\top, (\top \Rightarrow \neg \text{ Block } B)) : \neg \text{ Block } B$
(1*), (2*), (3*), (4*) $\vdash_3$ $((\text{miners} A \vee \text{ miners } B), ((\text{miners } A \Rightarrow \text{ Block } A), (\text{miners} B \Rightarrow \text{ Block } B))) : \text{ Block } A \vee \text{ Block } B$

Clearly we have proofs of $\neg$ Block A, $\neg$ Block B and Block $A \vee$ Block B but with different labels! The labels represent levels of knowledge. We can use a flattening process on the labels, or we can leave it as is.

There is no paradox, because the conclusions are on different levels of knowledge.

This can be seen also if we write a classical logic like proof.

To prove Block $A \vee$ Block B from (2), (3), (4), we need to use subproofs. Whenever we use a subproof we regard the subproof as a higher level of knowledge. To see this consider the attempt to prove

$$\frac{(l1.1) \quad A \Rightarrow B}{(l1.2) \quad \neg B \Rightarrow \neg A}$$

We get $\neg B \Rightarrow \neg A$ from the subproof in Fig. 13.39.

Let us now go back to the miners problem. The proof rules we have to use are

$$\textbf{MP} \quad \frac{A, A \Rightarrow B}{B}$$

Fig. 13.39 Box proof for *example* 13.5.4

Outer Box

(11.2.1) Assume $\neg B$, show $\neg A$

To show $\neg A$ use subproof in Inner Box

Inner Box

(1 1.2.1.1) Assume A, show $\bot$

(11.2.1.2) We want to reiterate (11.1) $A \Rightarrow B$ and bring it here to do modus ponens and get B

(1 1.2.1.3) We want to reiterate (1 1.2.1) $\neg B$ and bring it here to get a contradiction.

To do these actions we need proof theoretic permissions and procedures, because moving assumptions across levels of knowledge, from outer box to inner box
Such procedures are part of the definition of the logic.

$$\mathbf{DE}\quad \frac{\begin{array}{c} A \vee B \\ A \text{ proves } C \\ B \text{ proves } D \end{array}}{C \vee D}$$

RI We can reiterate positive (but not negative) wffs into subproofs.

The following is a proof using these rules:

Level 0

0.1a $\neg$ Block A. This is a negative assumption
0.1b $\neg$ Block B, negative assumption
0.2 miners in $A \Rightarrow$ Block A, assumption
0.3 miners in $B \Rightarrow$ Block B, assumption
0.4 miners in $A \vee$ miners in B
0.5 Block $A \vee$ Block B, would have followed from the proof in Fig. 13.40, if there were no restriction rule RI. As it is the proof is blocked.

Fig. 13.40 Box proof for the Miners paradox

Box 1

1.1	Miners in $A\vee$ miners in B, reiteration of 0.4 into Box 1
1.2	Miners in $A \Rightarrow$ Block A, reiteration of 0.2 into Box 1.
1.3	Miners in $B \Rightarrow$ Block B, reiteration of 0.3 into Box 1
1.4	Block $A\vee$ Block B, from 1.1, 1.2, and 1.3 using DE
1.5	To get a contradiction we need to bring 0.1a $\neg$ Block A 0.1.b $\neg$ Block B as reiterations into Box 1. However, we cannot do so because these are negative information assumptions and cannot be reiterated

13.6 Comparing with Makinson and Torre's Input Output Logic

This section compares our work with Input Output logic, I/O, of Makinson and Torre. Our starting point is Makinson and van der Torre (2001). Let us introduce I/O using Makinson and Torre own words from Makinson and van der Torre (2001).

> Input/output logic takes its origin in the study of conditional norms. These may express desired features of a situation, obligations under some legal, moral or practical code, goals, contingency plans, advice, etc. Typically they may be expressed in terms like: In such-and-such a situation, so-and-so should be the case, or ...should be brought about, or ...should be worked towards, or ...should be followed — these locutions corresponding roughly to the kinds of norm mentioned. To be more accurate, input/output logic has its source in a tension between the philosophy of norms and formal work of deontic logicians...
> Like every other approach to deontic logic, input/output logic must face the problem of accounting adequately for the behaviour of what are called 'contrary-to-duty' norms. The problem may be stated thus: given a set of norms to be applied, how should we determine which obligations are operative in a situation that already violates some among them. It appears that input/output logic provides a convenient platform for dealing with this problem by imposing consistency constraints on the generation of output.
> We do not treat conditional norms as bearing truth values. They are not embedded in compound formulae using truth-functional connectives. To avoid all confusion, they are not even treated as formulae, but simply as ordered pairs (a, x) of purely boolean (or eventually first-order) formulae.
>
> Technically, a normative code is seen as a set G of conditional norms, i.e. a set of such ordered pairs (a, x). For each such pair, the body a is thought of as an input, representing some condition or situation, and the head x is thought of as an output, representing what the norm tells us to be desirable, obligatory or whatever in that situation. The task of logic is seen as a modest one. It is not to create or determine a distinguished set of norms, but rather to prepare information before it goes in as input to such a set G, to unpack output as it emerges and, if needed, coordinate the two in certain ways. A set G of conditional norms is thus seen as a transformation device, and the task of logic is to act as its 'secretarial assistant'.

Makinson and Torre adapt an example from Prakken and Sergot (Prakken and Sergot 1996) to illustrate their use of input/output logic. We shall use the same example to compare their system with ours.

Example 13.6.1 We have the following two norms:

(1) The cottage should not have a fence or a dog;
$O\neg(f \vee d)$
or equivalently

(1.1) $O\neg f$
(1.2) $O\neg d$

(2) If it has a dog it must have both a fence and a warning sign.
$d \to O(f \wedge w)$
or equivalently

(2.3) $d \to Of$

(2.4) $d \to Ow$

In the notation of input/output logic the above data is written as

(2.5) $(\top, \neg(f \vee d))$
(2.6) $(d, f \wedge w)$.

Suppose further that we are in the situation that the cottage has a dog, in other words we have the fact:

(3) Fact: d
thus violating the first norm.

The question we ask is: what are our current obligations? or in other words, how are we going to model this set? We know from our analysis in the previous section that a key to the problem is modelling the facts and that deontic logic gets into trouble because it does not have the means to pay attention to what we called the obligation progression.

Figures 13.41, 13.42 and 13.43 describe our model, which is quite straight forward.

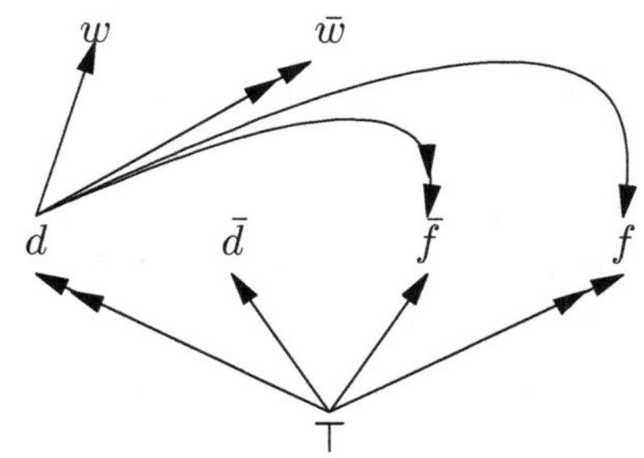

Fig. 13.41 First figure for describing a model for *example* 13.6.1

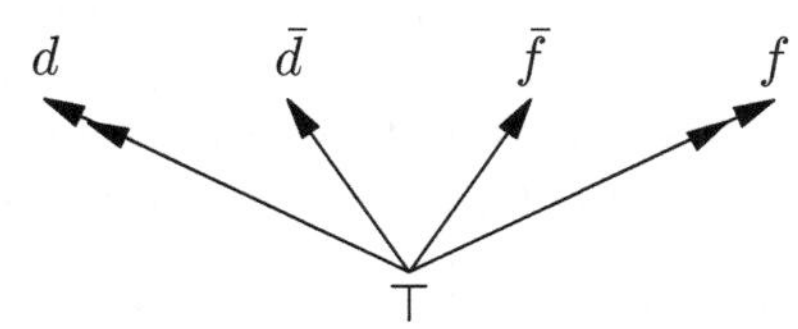

Fig. 13.42 Second figure for describing a model for *example* 13.6.1

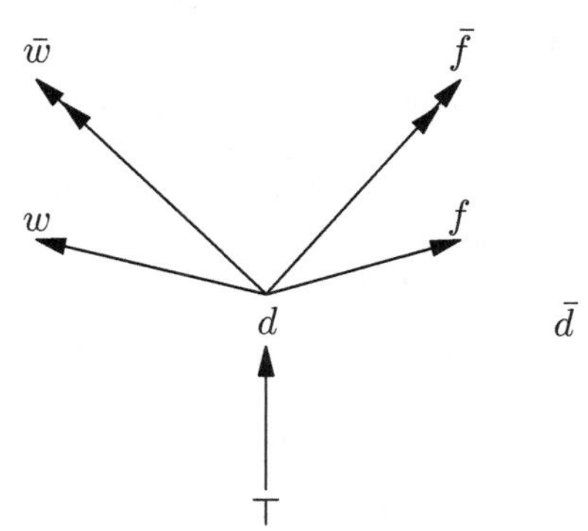

Fig. 13.43 Third figure for describing a model for *example* 13.6.1

Let us see how Makinson and Torre handle this example.

The input output model will apply the data as input to the input output rules (f) and (e). This is the basic idea of Makinson and Torre for handling CDT obligations with facts.

Makinson and Torre realse that, and I quote again

> Unrestricted input/output logic gives
>
> f: the cottage has a fence
>
> and
>
> w: the cottage has a warning sign.
>
> Less convincingly, because unhelpful in the supposed situation, it also gives
>
> $\neg d$: the cottage does not have a dog.
>
> Even less convincingly, it gives
>
> $\neg f$: the cottage does not have a fence,
>
> which is the opposite of what we want. These results hold even for simple-minded output, …

Makinson and Torre propose as a remedy to use constraints, namely to apply to the facts only those I/O rules which outputs are consistent with the facts. They say, and I quote again

> Our strategy is to adapt a technique that is well known in the logic of belief change cut back the set of norms to just below the threshold of making the current situation contrary to duty. In effect, we carry out a contraction on the set G of given norms. Specifically, we look at the maximal subsets G' of G such that out(G', A) is consistent with input A. To illustrate this consider the cottage example, where $G = \{(t, \neg(f \vee d), (d, f \wedge w)\}$, with the contrary-to-duty input d. Using just simple-minded output, G' has just one
>
> element $(d, f \wedge w)$ and so the output is just $f \wedge w$.

We note that this output corresponds to our Fig. 13.43.

Makinson and Torre continue to say, a key paragraph showing the difference between our methods and theirs:

> Although the …strategy is designed to deal with contrary-to-duty norms, its application turns out to be closely related to belief revision and nonmonotonic reasoning when the underlying input/output operation authorises throughput
>
> More surprisingly, there are close connections with the default logic of Reiter, falling a little short of identity…

Let us, for the sake of comparison, consider the CTD sets of Fig. 13.26 (this is actually the Reykajavik set of CTDs) and the facts as considered in example 13.3.11. We have the following CTD (or equivalently the input output rules):

(1) $(\top, \neg(d \vee f))$

(2) (d, f)
(3) (f, d)

The input is $A = d \wedge \neg f$.

In this example, the only rules (x, y) for which $A \vDash x$ are rules (1) and (2), but neither of their output is consistent with the input. So nothing can be done here. This corresponds to the lack of solution of our equations where we want to make both d and $\neg f$ the starting points. Our analysis in example 13.3.11 however, gives a different result, because we first input d then $\neg f$ and in parallel, put $\neg f$ and then d.

So apart from the difference that input output logic is based on classical semantics for classical logic and we use equational semantics, there is also the difference that input output logic puts all the input in one go and detaches with all CTD rules whose output does not contradict it, while we use all possible sequencing of the input, inputting them one at a time. (To understand what "one at a time" means, recall Remark 13.4.1 and example 13.4.3.) There is here a significant difference in point of view. We take into account the obligation progression and given a set of facts as inputs, we match them against the obligation progression. In comparison, Input Output logic lumps all CTD as a set of input output engines and tries to plug the inputs into the engines in different ways and see what you get. The CTD clauses lose their Deontic identity and become just input output engines. See our analysis and comparison in part (c) of example 13.5.3, where this point is clearly illustrated.

Let us do a further comparison. Consider the looping CTD set of Fig. 13.4 which is analysed in example 13.3.6. This has two input output rules

(1) (a, b)
(2) $(b, \neg a)$

Consider the two possible inputs

$$A = \neg a \wedge b$$

and

$$B = \neg a \wedge \neg b$$

A was a solution according to the soft approach option. B was a solution according to the mathematical approach option.

Using the input output approach, we can use $(b, \neg a)$ for A and we cannot use anything for B.

So there is compatibility here with the soft approach.

Let us summarise the comparison of our approach with the input output approach.

Com 1. We use equational semantics, I/O uses classical semantics.

Com 2. We rely on the obligation progression, breaking the input into sequence and modelling it using graphs. I/O does not do that, but uses the input all at once and taking maximal sets of CTDs (x, y) such that the input proves the xs and is consistent with the ys. The question whether it is possible to define violation progression from this is not clear. The I/O approach is a consequence relation/consistency approach. Our graph sequences and input facts sequences can also model action oriented/temporal (real time or imaginary obligation progression 'time'). So for example, we can model something like

$$f \rightarrow O\neg f$$

If you have a fence you should take it down.

Com 3. We remain faithful to the contrary-to-duty spirit, keeping our graphs and equations retain the CTD structure. I/O brought into their system significant AGM revision theory and turn I/O into a technical tool for revision theory and other nonmonotonic systems. See their quoted text 4.

Com 4. The connections are clear enough for us to say we can give equational semantics directly to input output logic, as it is, and never mind its connections with contrary-to-duty. Makinson and Torre defined input output logic, we have our equational approach, so we apply our approach to their logic directly. This is the subject of a separate paper.

13.7 Comparing with Governatori and Rotolo's Logic of Violations

We now compare with Governatori and Rotolo's paper (Governatori and Rotolo 2005). This is an important paper which deserves more attention.

Governatori and Rotolo present a Gentzen system for reasoning with contrary-to-duty obligations. The intuition behind the system is that a contrary to duty is a special kind of normative exception. The logical machinery to formalise this idea is taken from substructural logics and it is based on the definition of a new non-classical connective capturing the notion of reparational obligation.

Given in our notation the following sequence of CTDs

$$\begin{aligned} &A_1, \ldots, A_n \Rightarrow OB_1 \\ &\neg B_1 \Rightarrow OB_2 \\ &\neg B_1 \Rightarrow OB_3 \end{aligned}$$

They introduce a substructural connective and consider it as a substructural consequence relation without the structural rules of contraction, duplication and exchange, and write the above sequence as

$$A_1, \ldots, A_n \Rightarrow B_1, \ldots, B_m.$$

The meaning is: the sequence $A_1, \ldots, A_n$ comports that B_1 is the case; but if B_1 is not satisfied, then B_2 should be the case; if both B_1 and B_2 are not the case, then B_3 should be satisfied, and so on. In a normative context, this means that the content of the obligation determined by the conditions $A_1, \ldots, A_n$. The As are the facts and they are not ordered. So in this respect Governatori and Rotolo approach are like the I/O approach.

Govenatori and Rotolo give proof theoretical rules for manipulating such sequents.

This approach is compatible with our approach in the sense that it relies on the obligation progression. It is also compatible with our proof theory of Sect. 13.5.

For the purpose of comparison, we need not go into details of their specific rules. it is enough to compare one or two cases.

Consider the CTD set represented by Fig. 13.5. Since this set and figure is acyclic, Governatori and Rotolo can represent it by a theory containing several of their sequents. Each sequent will represent a maximal path in the figure. I do not think however that they can represent all possible paths. So the graph representation is a more powerful representation and we could and plan to present proof theory on graphs in a subsequent paper.

From my point of view, Governatori and Rotolo made a breakthrough in 2005 in the sense that they proposed to respect what I call the obligation (or violation) progression and their paper deserves more attention.

They use Gentzen type sequences which are written linearly, and are therefore restricted. We use planar graphs (think of them as planar two dimensional Gentzen sequents) which are more powerful.

I am not sure how Governatori and Rotolo will deal with loops in general. They do find a way to deal with some loops for example, I am sure they can handle the CTD of Fig. 13.9, or of Fig. 13.4, but I am not sure how they would deal with a general CTD set.

By the way, we used ordered sequences with hierarchical consequences in Gabbay and Schlechta (2010).

Governatori and Rotolo do not offer semantics for their system.

We offer equational semantics.

This means that we can offer equational semantics to their Gentzen system and indeed offer equational semantics to substructural logics in general.

This is a matter for another future paper.

Let us quote how they deal with the Chisholm paradox

Chisholm's Paradox The basic scenario depicted in Chisholm's paradox corresponds to the following implicit normative system:

$$\{\vdash_O h, h \vdash_O i, \neg h \vdash_O \neg i\}$$

plus the situation $s = \{\neg h\}$. First of all, note that the system does not determine in itself any normative contradiction. This can be checked by making explicit the normative system. In this perspective, a normative system consisting of the above norms can only allow for the following inference:

$$\frac{\vdash_O h, \quad \neg h \vdash_O \neg i}{\vdash_O (h, \neg i)}$$

Thus, the explicit system is nothing but

$$\{h \vdash_O i, \vdash_O (h, \neg i)]\}.$$

It is easy to see that s is ideal (my words: i.e. no violations) wrt the first norm. On the other hand, while s is not ideal wrt $\vdash_O (h, \neg i)$, we do not know if it is sub-ideal (i.e. there are some violations but they are compensated by obeying the respective CTD) wrt such a norm. Then, we have to consider the two states of affairs $s_1 = \{\neg h, i\}$ and $s_2 = \{\neg h, \neg i\}$. It is immediate to see that s_1 is non-ideal (i.e. all violations throughout, no compensation) in the system, whereas s_2 is sub-ideal.

If so, given s, we can conclude that the normative system says that $\neg i$ ought to be the case.

13.8 Conclusion

We presented the equational approach for classical logic and presented graphs for General CTD sets which gave rise to equations. These equations provided semantics for general CTD sets.

The two aspects are independent of one another, though they are well matched.

We can take the graph representation and manipulate it using syntactical rules and this would proof theoretically model CTD's. Then we can give it semantics, either equational semantics or possible world semantics if we want.

We explained how we relate to Makinson and Torre's input output approach and Governatory and Rotolo's logic of violations approach.

The potential "output" from this comparison are the following possible future papers:

(1) Equational semantics for input output logic
(2) Equational semantics for substructural logics
(3) Development of planar Gentzen systems (that would be a special case of labelled deductive systems)
(4) Planar proof theory for input output logic (again, a special case of labelled deductive systems).
(5) Proof theory and equational semantics for embedded CTD clauses of the form $x \to O(y \to Oz))$.

Chapter 14
Conclusion

Obviously, animals do some reasoning. If an animal sees a lion, it reacts. Seeing the lion is not a problem, but a possible attack by the lion, the animal being killed, is a problem for the animal. In addition, the animal has to abstract from a multitude of visual impressions to identify the lion. Still, we might hesitate to say that all animals are able of reasoning or abstracting in the human sense. There is probably a continuum between hardwired reactions to the flexibility and depth of human reasoning, formalized in logic, and abstraction.

Classical logic might best be described as the theory of its operators. Philosophical intuitionists might disagree, but it seems difficult to imagine situations which do not conform to classical logic. Nonmonotonic logics are more, they are adequate for certain situations and purposes, not for others. Consider general preferential logics. They describe the perfectly normal case. If we change the set, or lose one property of the normal situation, they are not very helpful. In the former case, we cannot move much, in the latter, they will not give any answers any more. But they have nice logical properties, and a clear and intuitive structural semantics, based on case comparison, and a simple and flexible abstract semantics, based on notions of size of different strengths. They are good for reasoning about the text book case. Consider Reiter defaults. They are the logic of the impatient reasoner. They allow to write down utter nonsense like $\{: \phi/\phi, : \neg\phi/\neg\phi\}$, but they allow to treat subideal cases, where we lose some properties of the ideal case, but preserve as many others, as possible. They also allow to change the base set, and are in this sense (almost) as liberal as classical logic. They, or variants of them, are well adapted to reason about a world in which influence changes are rare—otherwise, changing the base set would change the situation too often. This is, of course, relative to the language used. We should not underestimate the importance of the language, see Example 1.2.1. The possible languages are determined by our senses and our brain. This is one reason why knowledge of the brain is important. In spirit, Reiter defaults are suited for 'fast and dirty' reasoning, neglecting exceptions.

D.M. Gabbay and K. Schlechta, *A New Perspective on Nonmonotonic Logics*,
DOI 10.1007/978-3-319-46817-4_14

Thus, various logics correspond to various situations/worlds, languages, reasoning objectives, and reasoning powers. It is desirable, but not done here, to make this formal, resulting in a formal (and quantitative) criterion of adequacy, a kind of meta-semantics, and a classification of different logics.

Note that our logic as described in Chap. 11 is preferential, but we put treatment of conflicts and subideal cases in the construction of the preferential structure. The treatment of subideal cases is coded in the particular order inside the $\omega(A)$. 'Mysterious' cases are put above others, coding the epistemic aspect.

We have various criteria to consider about logics, some of them mentioned already above:

(1) does a logic assure that the set of defaults has some elements where they all hold (the core)? Is this core extended, if possible? Note that we required the first, and had the second by the construction of the (local) relation.
(2) does a logic treat homogeneousness in an adequate way, i.e. offer answers on (most) subsets? Remember that general preferential relations fail here.
(3) is it a logic about the world, or our information about the world? For instance, Nixon is either a hawk, or a pacifist (in a simple world), but we might have contradictory information about this. The natural answer is probably a possible worlds structure.
(4) does the logic have modularity and independence?

    (4.1) our construction is modular, and the local relations are potentially independent, resulting in 'semi-rankedness',
    (4.2) how are sublanguages and products treated (see Chap. 9 for interpolation)?
    (4.3) do we have interpolation?
    (4.4) are there other criteria?

We postulate that language is also a way to make thinking efficient. The manipulation of a word is easier than that of the concept behind it. In addition, the precision is not fixed, but we can make it more or less detailed. It is like manipulating a pointer or the structure to which the pointer points in computer science. (In our construction in Chap. 11, we can 'hide' the surprise exceptions behind a pointer, and follow the pointer only when necessary.)

Living things and man-made objects have a 'plan' according to which they are built, in the former case a genetic plan. Making a plan can go wrong, and the object can be damaged. But we often still see the plan, so a cat born with two legs is still a cat, and not a human being. In addition, critical human objects like aircraft and living beings have a 'desire' to work, and if something goes wrong, they will try do have still limited functionality. If flying for a bird does not work any more, then perhaps walking. If one engine loses power, then the aircraft should still be able to fly with three engines, etc. This often involves groups of predicates with an order: flying is better than walking, walking betten than crawling, etc. To formalize this type of graceful degradation, we have to work with such groups, and not only with a quantitative relation between sets of properties, as is done in our logic.

We think that heuristics give a strong indication about the relation between the world and our thinking about the world. A successful heuristics shows that it is adapted to the world and our thinking. Science and nonmonotonic logic have several points in common. Science is based on a hypothesis of homogeneousness, and this seems best explained by rarity of influence changes. Homogenousness on its own seems difficult to explain, why should the world be homogenous? Explaining it by rarity of influence change seems more natural. In science, there are exceptions to laws, too. But we do not think that the laws are really flawed, we think that they can either be improved, or that the exceptions can be explained by some other effect, like a gust of wind interfering with the effect of gravity on a falling apple. We think that our capacity to explain the world is not a miracle which has to be explained, but is no less surprising than the ability of a cat to catch a mouse. It sometimes fails, but it often succeeds. We are part of the world, after all, and evolution had enough time to develop our brain into a powerful tool of cognition, like the muscles, eyes, etc. of the cat. Our language and its concepts were developed inside the world, they provide an efficient tool to 'cut up the mess' of confusing perceptions—vision, auditory signals ....—into meaningful entities, see, e.g. the 'binding problem', Sect. 8.3.4.4, and, e.g. Malsburg (1994, 1999), the problem to bring different sensory inputs of the same object together, so it is seen as coming from the same object, without the help from a supervising agent in the brain. Again this is no miracle. If it had not happened, then we would not be able to survive; this applies to animals, too, they also have to order the sensory chaos to a meaningful story. This interplay between world, brain, and ideas is essential. Our brain is capable to think about its own activity, this is described by the word 'reentrant' in Edelman's terminology, and to treat its own activity the same way it treats external stimuli. Thus, we can mix thoughts and external stimuli as objects of our thoughts, refine the former, etc.

Homogenousness (or rarity of influence) can be used downward (in NML), sideways (analogical reasoning) or upward (inductive reasoning, science). The common notion is information transfer. When we look at scientific practice, one often choses cases considered representative in the sense that they are supposed to be equivalent to a set of cases, i.e. there is no influence difference between the case and the other elements of the set. For instance, one would not test a physical law especially on Friday the 13th, because the date is supposed to be without influence. One day is as good as the other. But one might test it in a strong magnetic field, and in a weak magnetic field, thinking that the strength of the magnetic field might have an influence. (We are not sure if the 'law of temporal inertia' is the same as rarity of influence, though it is certainly related. Some things change over time, and the rarer influences are, the rarer changes propagate.)

Note that the human enterprise of discovery, finding a limited number of basic constituents of nature, atoms, electrone, quarks ...is of the same spirit. Nature, so the assumption, is ruled by a small number of entities, and the visible world is ruled by the interplay of their effects. The changes of influence are not uniform: for the forces of physics, close objects are often subject to similar forces. For objects like living beings, their specific properties are governed by categories. The grass is right in front of the cow's mouth, but grass and cow are very different. Our categories—

and animal cognition, too—work only because our brain sees the similarities within the categories, and works with the abstraction.

The idea of inheritance is based on homogenousness, too. It expresses that subsets behave similarly to supersets, and if one property fails, the others will probably still hold. A refinement of the idea is that small changes in the base set will result in small changes in the property set. This was expressed (perhaps in too mathematical a way) in Schlechta (1995). Classical logic has a 'discontinuity' in the EFQ rule, where suddenly everything is a consequence.

Occam's Razor This short paragraph is highly speculative. We also think that it might be too early to investigate in depth the astonishing success of the heuristics of Occam's Razor: we just do not know enough so far about the human brain and its relation to the world. (Remark: we do not consider here the complexity problem of reasoning with a simpler theory, but only the far more interesting question, why simple theories often seem to correspond to the world better than more complicated ones.)

It seems evident that our brain (and that of animals) somehow 'fits' the world—otherwise we would not be able to survive. How the world 'really is' might be beyond our understanding, so we should concentrate on the relation brain/world.

It seems reasonable to make the following two assumptions:

(1) Our brain has been made by evolution in a way so we can understand the world. In particular, we—and animals— are able to extract salient features, and are not swamped by an ununderstandable chaos. It is an interplay between nature and our brain, so 'simple', etc. are notions subjective to our brain. This will always be implicit in what follows.
Animals have to do abstractions, too. Every lion looks different, they have to see the common: lion, danger. They have to put order into sensory chaos. They understand the world enough to be able to survive. We do not see objects, say tables, etc. only colours, lines, etc. A table is an interpretation, a construction of our brain.
Evolution has created our brains so that our abstractions will often correspond to entities and influences of nature (forces, etc.), underlying what we see, just as evolution has created the brain of a cat so that it can catch mice. It is a continuation of a development which began for animals, our abstractions are continuations of animal thinking. Our brain (and that of animals) is made to extract the relevant information. On a basic level, this is abovementioned 'binding problem'.
Thus, it is perhaps not surprising that this 'order seeking machine' is able to detect physical laws, and sees them as simple. An essential part of the brain's architecture which allows abstractions and abstract constructions are perhaps the reentrant connections. Human language (and writing) with its abstractions and liberty of manipulation easily allows abstract constructions: Human beings can handle their abstractions like objects 'out there'.

(2) Nature (as we see it) is separated into different levels: physics, chemistry, animals, plants .... It is unlikely that different levels of abstraction 'see' the same or similar laws. Discovery often involves going to deeper levels of abstraction. Thus, it is unlikely that combinations of laws of one level are useful on lower levels.

We think that these two assumptions might offer the basis for a new approach to Occam's Razor as a successful heuristics to find order in the world.

Bibliography

M. Abraham, D. Gabbay, U. Schild, "Analysis of the Talmudic argumentum a fortiori inference rule (Kal Vachomer) using matrix abduction", Studia Logica 92 (3):281–364, 2009
C. Alchourron, P. Gardenfors, D. Makinson, "On the logic of theory change: partial meet contraction and revision functions", Journal of Symbolic Logic, Vol. 50, pp. 510–530, 1985
E. Amir, "Interpolation systems for nonmonotonic reasoning systems", JELIA 2002, pp. 233–244
A. Antonelli, "Defeasible inheritance on cyclic networks", Artificial Intelligence, vol. 92 (1997), pp. 1–23
A. Antonelli, "A directly cautious theory of defeasible consequence for default logic via the notion of general extensions", Artificial Intelligence, vol. 109 (1999), pp. 71–109
A. Antonelli, "Grounded consequence for defeasible reasoning", Cambridge University Press, 2005
S. Ben-David, R. Ben-Eliyahu, "A modal logic for subjective default reasoning", Proceedings LICS-94, July 1994, Paris, France, S. Abramsky, ed. IEEE Computer Science Press, pp. 477–486, 1994.
M. B. A. Bessos, D. Gabbay, "Topological aspects of matrix abduction 2", 2014
R. Booth, S. Chopra, T. Meyer, A. Ghose, "A unifying semantics for belief change", ECAI 2004, 793–797
G. Bossu, P. Siegel, "Saturation, nonmonotonic reasoning and the closed-world assumption", Artificial Intelligence 25 (1985), pp. 13–63
J. Carmo, A. J. I. Jones, "Deontic logic and contrary-to-duties", in: Handbook of Philosophical Logic, Vol. 8, D. Gabbay, F. Guenthner eds., pp. 265–343, Kluwer, 2002
S. Chopra, R. Parikh, "Relevance sensitive belief structures", Annals of Mathematics and Artificial Intelligence, vol. 28, No. 1–4, pp. 259–285, 2000.
R. M. Chisholm, "Contrary-to-duty imperatives and deontic logic", Analysis, 24, 1963.
P. S. Churchland, "Neurophilosophy", MIT Press, 1986
P. M. Churchland, "A Neurocomputational Perspective", MIT Press, 1989
P. M. Churchland, "Neurophilosophy at Work", Cambridge University Press, 2007
W. Craig, "Three uses of the Herbrand-Gentzen theorem in relating model theory and proof theory", Journal of Symbolic Logic 22 (1957), No. 3, pp. 269–285
A. P. Dawid, "Conditional independence in statistical theory", Journal of the Royal Statistical Society, Series B, 41(1) : 1–31, 1979
M. de Boer, D. Gabbay, X. Parent, M. Slavkova, "Two dimensional deontic logic", Synthese, 187(2), 623–660, 2012
M. Denecker, V. Marek, M. Truszczynski, "Reiter's default logic is a logic of autoepistemic reasoning and a good one, too", CoRR, 2011 (arXiv.org/corr)
Gerald M. Edelman, "The remembered present", Basic Books, New York, 1989

D.M. Gabbay and K. Schlechta, *A New Perspective on Nonmonotonic Logics*,
DOI 10.1007/978-3-319-46817-4

Gerald M. Edelman, Giulio Tononi, "A Universe of Consciousness", Basic Books, 2000
Gerald M. Edelman, "Wider than the sky", Yale University Press, New Haven 2004, (German edition "Das Licht des Geistes", Rowohlt, 2007)
N. Friedman, J. Halpern, "Plausibility measures and default reasoning", Journal of the ACM, Vol. 48, pp. 1297–1304, 1996
D. Gabbay, "Compromise update and revision: A position paper", In Dynamic Worlds, B. Fronhoffer, R. Pareschi, eds., pp. 111–148. Applied Logic Series, Volume 12, Springer 1999.
D. Gabbay, "An Equational Approach to Argumentation Networks", Argument and Computation, 2012a, vol 3 issues (2–3), pp 87–142
D. Gabbay, "Temporal deontic logic for the generalised Chisholm set of contrary to duty obligations.", In T. Agotnes, J. Broersen, and D. Elgesem, eds., DEON 2012, LNAI 7393, pp. 91–107. Springer, Heidelberg, 2012b.
D. Gabbay, "Meta-Logical Investigations in Argumentation Networks", Research Monograph, College Publications, 2013, 770 pp
D. Gabbay, "The Equational Aproach to Contrary-to-duty Obligations", in FoIKS 2014, C. Beierle and C. Meghini eds., Springer LNCS 8367, pp 1–61, 2014
D. Gabbay, L. Maksimova, "Interpolation and definability", Clarendon Press, Oxford, 2005
D. Gabbay, K. Schlechta, "Reactive preferential structures and nonmonotonic consequence", Review of Symbolic Logic, Vol. 2, No. 2, pp. 414–450, 2009a
D. Gabbay, K. Schlechta, "Roadmap for preferential logics", Journal of Applied Nonclassical Logic, Hermes, Paris, France, Vol. 19/1, pp. 43–95, 2009b, see also hal-00311941. arXiv:0808.3073.
D. Gabbay, K. Schlechta, "Defeasible inheritance systems and reactive diagrams", Logic Journal of the IGPL, 17:1–54, 2009c
D. Gabbay, K. Schlechta, "Logical tools for handling change in agent-based systems" Springer, Berlin, 2009d
D. Gabbay, K. Schlechta, "Size and logic", Review of Symbolic Logic, Vol. 2, No. 2, pp. 396–413, 2009e
D. Gabbay, K. Schlechta, "Semantic interpolation", Journal of Applied Non-classical Logics, Vol 20/4, 2010a, pp. 345–371
D. Gabbay, K. Schlechta, "A theory of hierarchical consequence and conditionals", Journal of Logic, Language and Information, 19:1, pp. 3–32, Jan. 2010b
D. Gabbay, K. Schlechta, "Conditionals and modularity in general logics", Springer, Heidelberg, August 2011. ISBN: 978-3-642-19067-4
D. Gabbay, K. Schlechta, "A semantics for obligations", hal-00339393. arXiv:0811.2754, 2016
D. Gabbay, C. Strasser, Reactive standard deontic logic, J. Logic and Computation 25(1) 2015 pp. 117–157.
D. Gabbay, L. Gammaitoni, X. Sun, "The paradoxes of permission. An action based solution", Journal of Applied Logic. 12(2) 179–191 (2014)
D. Gabbay, L. Robaldo, L. van der Torre, "A solution to the miner paradox: A Beth semantics approach. In F. Cariani, D. Grossi, J. Meheus, X. Parent (eds.), Deontic Logic and Normative Systems, 12th Intl. Conf., DEON 2014, Springer, 2014 Lecture Notes in Computer Science
D. M. Gabbay, "Theoretical foundations for non-monotonic reasoning in expert systems". K. R. Apt (ed.), "Logics and Models of Concurrent Systems", Springer, Berlin, pp. 439–457, 1989.
D. M. Gabbay, "Labelled Deductive Systems", Oxford University Press, 1996
D. M. Gabbay, "Reactive Kripke semantics and arc accessibility", Proceedings CombLog04, July 28–30, 2004, W. Carnielli, F. M. Dionesio, P. Mateus, Eds., Centre of Logic and Computation, University of Lisbon, 2004, pp. 7–20.
D. M. Gabbay, "Reactive Kripke models and contrary to duty obligations", DEON-2008, Deontic Logic in Computer Science, July 15–18, 2008, Luxembourg, R. van der Meyden, L. van der Torre, eds., LNAI 5076, pp 155–173, Springer, Berlin 2008.
D. M. Gabbay, "Reactive Kripke models and contrary to duty obligations". Part A: Semantics, Journal of applied logic, 11(1): 103–136 (2013)

D.M. Gabbay, O.T. Rodrigues, Equilibrium states on numerical argumentation networks. Logica Universalis 9(4):411–473 (2015); A numerical approach to the merging of argumentation networks. In M. Fisher et al., CLIMA 2012 Proceedings, LNCS 7486, Springer, 2012

H. Geffner, T. Verma, "Inheritance = *Chaining + Defeat*", UCLA Technical Report CSD-890039, R – 129-L, June 1989, also in: Proceed. Fourth International Symposium on Methodologies for Intelligent Systems, North Holland, 1989, pp. 411–418

G. Gottlob, "The power of beliefs or translating default logic into standard autoepistemic logic", IJCAI 93, Vol. 1, 1993

G. Governatori, A. Rotolo, "A Gentzen system for reasoning with contrary-to-duty obligations, a preliminary study. In A. J. I. Jones, J. Horty, eds, Deon 02, pp. 97–116, London, May 2002.

G. Governatori, A. Rotolo, "Logic of violations: a Gentzen system for reasoning with contrary-to-duty obligations", Australasian Journal of Logic, 4, 193–215, 2005.

J. Hawthorne, D. Makinson, "The quantitative/qualitative watershed for rules of uncertain inference", Studia Logica 86(2), pp. 247–297, 2007

B. Hansson, "An analysis of some deontic logics", Nous 3, 373–398. Reprinted in R. Hilpinen, ed. "Deontic Logic: Introductory and Systematic Readings", Reidel, pp. 121–147, Dordrecht 1971

J. F. Horty, R. H. Thomason, D. S. Touretzky, "A Sceptical Theory of Inheritance in Nonmonotonic Semantic Networks", Dept. Comp. Sci., Carnegie Mellon Univ., CMU-CS-87-175, October 1987 also in: Proceedings AAAI-87 (1987), p. 358–363, or: Artificial Intelligence 42 (1990), pp. 311–348

A. J. I. Jones, I. Poern, "Ideality, sub-ideality and deontic logic", Synthese, 65, 1985.

S. Kraus, D. Lehmann, M. Magidor, "Nonmonotonic reasoning, preferential models and cumulative logics", Artificial Intelligence, 44 (1–2), pp. 167–207, July 1990.

G. Kourousias, D. Makinson, "Parallel interpolation, splitting, and relevance in belief change", Journal of Symbolic Logic, 72, pp. 994–1002, Sept. 2007

N. Kolodny, J. MacFarlane, "Ifs and oughts", Journal of Philosophy, 107(3) : 115–43, 2010.

H. Katsuno, A. O. Mendelzon, "On the difference between updating a knowledge base and revising it", Univ. of Toronto Tech. Rept., KRR-TR-90-6, 1990

T. Krishnaprasad, M. Kifer, "An Evidence-based Framework for a Theory of Inheritance", Proceed. IJCAI 89, pp. 1093–1098

T. Krishnaprasad, M. Kifer, D. Warren, "On the Circumscriptive Semantics of Inheritance Networks", Proceed. Fourth International Symposium on Methodologies for Intelligent Systems, North Holland, 1989, pp. 448–456

T. Krishnaprasad, M. Kifer, D. Warren, "On the Declarative Semantics of Inheritance Networks", Proceed. IJCAI 89, pp. 1098–1103

A. Laufer, D. Gabbay, "Topological aspects of matrix abduction 1". In: A. Koslow, A. Buchsbaum, The Road to Universal Logic, Festschrift for the 50th Birthday of Jean-Yves Béziau, Vol. 1, Studies in Universal Logic, Birkhäuser, 2015

D. Lehmann, M. Magidor, "What does a conditional knowledge base entail?", Artificial Intelligence, 55(1), pp. 1–60, May 1992.

D. Lehmann, M. Magidor, K. Schlechta, "Distance semantics for belief revision", Journal of Symbolic Logic, Vol. 66, No. 1, pp. 295–317, March 2001

D. Lewis, "Counterfactuals", Blackwell, Oxford, 1973

J. J. Ch. Meyer, F. P. M. Dignum, R. J. Wieringa, "The paradoxes of deontic logic revisited: a computer science perspective", University of Utrecht, Dept. Comp. Sc., Tech. Rep., 1994

R. Menzel, J. Fischer eds., "Animal thinking", MIT Press, 2011

D. Makinson, "General patterns in nonmonotonic reasoning", in D.M. Gabbay, C.J. Hogger, J.A. Robinson (eds.), "Handbook of logic in artificial intelligence and logic programming", vol. III: "Nonmonotonic and uncertain reasoning", Oxford University Press, 1994, pp. 35–110.

D. Makinson, "Propositional relevance through letter-sharing", Journal of Applied Logic, vol. 195 (2009), pp 377–387

D. Makinson, K. Schlechta, "Floating Conclusions and Zombie Paths", Artificial Intelligence (Research Note) 48 (1991) pp. 199–209

D. Makinson, L. van der Torre, "Input/output logics.", Journal of Philosophical Logic, 29(4): 383–408, 2000.
D. Makinson, L. van der Torre, "Constraints for input/output logics", Journal of Philosophical Logic, 30(2): 155–185, 2001a.
D. Makinson, L. van der Torre, "What is Input/Output Logic?", ESSLLI 2001b.
C.v.d. Malsburg, "The correlation theory of brain function", MPI Biophysical Chemistry, Internal Report 81-2. Reprinted in "Models of Neural Networks II", E. Domany et al. eds., Springer 1994
C.v.d. Malsburg, "The what and why of binding: the modeler's perspective", Neuron 24 (1): 95–104, 111–25, Sept. 1999
Moore, R., "Possible worlds semantics for autoepistemic logic", Technical Note 337, SRI, Menlo Park, CA, 1984
Moore, R., "Semantical considerations on nonmonotonic logic", Artificial Intelligence 25, 1985, pp. 75–94.
J. von Neumann, "The Computer and the Brain", Yale University Press, 1958
X. Parent, D. Gabbay, L. van der Torre, "Intuitionistic Basis for Input/Output Logic", David Makinson on Classical Methods for Non-Classical Problems, Springer, 2014
H. Prakken, M. J. Sergot, "Contrary-to-duty obligations", Studia Logica, 57(1) : 91–115, 1996.
H. Prakken, M. Sergot, "Dyadic deontic logic and contrary to duty obligations", In D. Nute, ed. Defeasible Deontic Logic, pp. 223-262. Synthese Library, Kluwer, 1997.
R. Parikh, "Belief, belief revision, and splitting languages", L.S. Moss, J. Ginzburg and M. de Rijke (eds.) Proceed. Logic, Language and Computation, CSLI 1999, pp. 266–278
J. Pearl, "Probabilistic Reasoning in Intelligent Systems", Morgan Kaufmann, 1988
H. Rott, S. O. Hansson, "Safe Contraction Revisited", in: "David Makinson on Classical Methods for Non-Classical Problems", S. O. Hansson ed., Heidelberg 2014
H. Reichenbach, "Elements of Symbolic Logic", Free Press, New York, 1947.
R. Reiter, "A logic for default reasoning", Artificial Intelligence 13 (1–2), pp. 81–132, 1980.
G. Roth, "Das Gehirn und seine Wirklichkeit", STW 1275 Suhrkamp, Frankfurt 1996
E. Sandewall, "Non-monotonic inference rules for multiple inheritance with exceptions", Proceedings IEEE 74 (1986), pp. 1345–1353
B. Selman, H. Levesque, "The Tractability of Path-Based Inheritance", Proceed. IJCAI 1989, pp. 1140–1145
K. Schlechta, "Semantics for defeasible inheritance", L. G. Aiello (ed.), "Proceedings ECAI 90", Stockholm, Sweden, 1990, Springer, Berlin, pp. 594–597, 1990.
K. Schlechta, "Some results on theory revision", A. Fuhrmann, M. Morreau (eds.), "The Logic of Theory Change", Springer, Berlin, pp. 72–92, 1991
K. Schlechta, "Some results on classical preferential models", Journal of Logic and Computation, Vol.2, No.6, pp. 675–686, 1992.
K. Schlechta, "Directly sceptical inheritance cannot capture the intersection of extensions", Journal of Logic and Computation, Vol. 3, No. 5, pp. 455–467, 1995a
K. Schlechta, "Defaults as generalized quantifiers", Journal of Logic and Computation, Vol. 5, No. 4, pp. 473–494, 1995b.
K. Schlechta, "Logic, topology, and integration", Journal of Automated Reasoning, 14:353–381, 1995c
K. Schlechta, "Some completeness results for stoppered and ranked classical preferential models", Journal of Logic and Computation, Oxford, Vol. 6, No. 4, pp. 599–622, 1996
K. Schlechta, "Nonmonotonic logics - basic concepts, results, and techniques" Springer Lecture Notes in Artificial Intelligence 1187, p.243, Jan. 1997a.
K. Schlechta, "Filters and partial orders", Journal of the Interest Group in Pure and Applied Logics, Vol. 5, No. 5, pp. 753–772, 1997b
K. Schlechta, "Karl's remarks on Kal-Algorithm", Personal Communication to D. Gabbay, 2011
K. Schlechta, "Coherent systems", Elsevier, Amsterdam, 2004.

K. Schlechta, D. Makinson, "Local and global metrics for the semantics of counterfactual conditionals", Journal of Applied Nonclassical Logics, Vol. 4, No. 2, pp. 129–140, Hermes, Paris, France, 1994

Marek Sergot, Dept. of Computing, Imperial College, London, Notes, 2004, 2007, 2010

Yoav Shoham, "A semantical approach to nonmonotonic logics". Proc. Logics in Computer Science, pp. 275–279, Ithaca, NY, 1987, IEEE Computer Society, and Proceedings IJCAI 87, Milan, Italy, August 1987, J. P. McDermott, ed., Morgan Kaufmann,1987, San Francisco, pp. 388–392, 1987

W. Spohn, "Stochastic independence, causal independence, and shieldability", Journal of Philosophical Logic 9 (1980) 73–99

W. Spohn, "Ordinal conditional functions: a dynamic theory of epistemic states". W. L. Harper, B. Skyrms (eds.), "Causation in decision, belief change, and statistics", vol. 2, pp. 105–134, Reidel, Dordrecht, 1988

R. Stalnaker, "A theory of conditionals", N. Rescher (ed.), "Studies in logical theory", Blackwell, Oxford, pp. 98–112, 1968

R. H. Thomason, J. F. Horty, "Logics for inheritance theory", M. Reinfrank, J. de Kleer, M. L. Ginsberg, E. Sandewall (eds.), "Non-monotonic reasoning", Proceedings 2nd Int. Workshop, Grassau, 1988, Springer, Berlin, pp. 220–237, 1989

R. H. Thomason, J. F. Horty, D. S. Touretzky, "A calculus for inheritance in monotonic semantic nets". Tech. Rept. CMU-CS-86-138, Comp. Sci. Dept., Carnegie Mellon Univ., 1986

D. S. Touretzky, "Implicit Ordering of Defaults in Inheritance Systems", Proceed. AAAI 84, pp. 322–325, 1984

D. S. Touretzky, "The mathematics of inheritance systems", Pitman, London/Morgan Kaufmann, Los Altos, 1986

D. S. Touretzky, J. F. Horty, R. H. Thomason, "A clash of intuitions: the current state of nonmonotonic multiple inheritance systems", Proceed. IJCAI 1987, pp. 476–482

D. S. Touretzky, R. H. Thomason, J. F. Horty, "A skeptic's menagerie: conflictors, preemptors, reinstaters, and zombies in nonmonotonic inheritance", Proceed. IJCAI 1991, pp. 478–483

E. Wigner, "The unreasonable effectiveness of mathematics in the natural sciences", Communications on Pure and Applied Mathematics 13: 1–14, 1960

M. Willer, "A remark on iffy oughts", Journal of Philosophy, 109(7) : 449–461, 2012.

Index

D.M. Gabbay and K. Schlechta, *A New Perspective on Nonmonotonic Logics*,
DOI 10.1007/978-3-319-46817-4